FLARE STARS IN STAR CUSTERS, ASSOCIATIONS AND THE SOLAR VICINITY

INTERNATIONAL ASTRONOMICAL UNION

UNION ASTRONOMIQUE INTERNATIONALE

FLARE STARS IN STAR CLUSTERS, ASSOCIATIONS AND THE SOLAR VICINITY

PROCEEDINGS OF THE 137TH SYMPOSIUM OF THE INTERNATIONAL ASTRONOMICAL UNION HELD IN BYURAKAN (ARMENIA), U.S.S.R., OCTOBER 23–27, 1989

EDITED BY

L. V. MIRZOYAN

Byurakan Observatory, Armenia Academy of Sciences, U.S.S.R.

B. R. PETTERSEN

Institute of Theoretical Astrophysics, University of Oslo, Norway

and

M. K. TSVETKOV

Department of Astronomy, Academy of Sciences, Bulgaria

KLUWER ACADEMIC PUBLISHERS

DORDRECHT / BOSTON / LONDON

Library of Congress Cataloging in Publication Data

International Astronomical Union. Symposium (137th : 1989 : Biurakan, Armenian S.S.R.)
Flare stars in star clusters, associations, and the solar vicinity : proceedings of the 137th Symposium of the International Astronomical Union, held in Byurakan (Armenia), U.S.S.R., October 23-27, 1989 / edited by L.V. Mirzoyan, B.R. Pettersen, and M.K. Tsvetkov.
p. cm.
ISBN 0-7923-0770-4. -- ISBN 0-7923-0771-2 (pbk.)
1. Flare stars--Congresses. 2. Stars--Clusters--Congresses. 3. Stellar associations--Congresses. I. Mirzoian, L. V. II. Pettersen, B. R. III. Tsvetkov, M. K. IV. Title.
QB843.F55I58 1989
523.8'5--dc20 90-4437
CIP

ISBN 0-7923-0770-4 (HB)
ISBN 0-7923-0771-2 (PB)

Published on behalf of
the International Astronomical Union
by
Kluwer Academic Publishers, P.O. Box 17, 3300 AA Dordrecht, The Netherlands.

Kluwer Academic Publishers incorporates
the publishing programmes of
D. Reidel, Martinus Nijhoff, Dr W. Junk and MTP Press.

Sold and distributed in the U.S.A. and Canada
by Kluwer Academic Publishers,
101 Philip Drive, Norwell, MA 02061, U.S.A.

In all other countries, sold and distributed
by Kluwer Academic Publishers Group,
P.O. Box 322, 3300 AH Dordrecht, The Netherlands.

Printed on acid-free paper

Printed in The Netherlands

TABLE OF CONTENTS

Preface xi

List of participants xiii

I. FLARE STARS. OPTICAL OBSERVATIONS AND FLARE STATISTICS.

L.V. MIRZOYAN
Flare stars in star clusters, associations and solar vicinity (Invited paper) 1

B.R. PETTERSEN, K.P. PANOV, M.S. IVANOVA, C.W. AMBRUSTER, E. VALTAOJA, L. VALTAOJA, S. AVGOLOUPIS, L.N. MAVRIDIS, J.H. SEIRADAKIS, S.R. SUNDLAND, K. OLAH, O. HAVNES, Ö. OLSEN, J.E. SOLHEIM, T. AANESEN
The flare activity of AD Leo 1972-1988 15

R.E. GERSHBERG, I.V. ILYIN, N.S. NESTEROV, R.G. GETOV, M. IVANOVA, K.P. PANOV, M.K. TSVETKOV, G. LETO
Photometrical, spectral and radio monitoring for EV Lac in 1986 and 1987 19

A.S. MELKONIAN
Electrophotometric observations of the EV Lac flares in U, Hα and Hβ 25

K.P. PANOV, M.S. IVANOVA, A. ANTOV
Rapid spike flares on AD Leo and EV Lac 27

N.D. MELIKIAN, V.S. SHEVCHENKO
Photoelectric observations of flares on UV Ceti 31

V.P. ZALINIAN, A.A. KARAPETIAN, H.M. TOVMASSIAN
Two-colour observations of spikes and flares with high time resolution 33

B.E. ZHILYAEV, Ya.O. ROMANJUK, O.A. SVYATOGOROV
A short-lived flare on EV Lacertae 35

A.V. BERDYUGIN
Diagnostic possibilities of the Walrawen photometric system 37

I.V. ILYIN, N.I. SHAKHOVSKAYA
The flare activity of two peculiar red dwarfs 41

K. ISHIDA
Statistics of flares observed for UV Ceti type stars YZ CMi, AD Leo, and EV Lac at the Okayama Observatory 43

B.R. PETTERSEN
Flare activity levels for fully convective red dwarfs 49

N.I. SHAKHOVSKAYA
Some physical consequences from the flare statistics of the UV Cet type stars in the solar neighbourhood 53

N.I. BONDAR
Search for slow light variations of red-dwarf stars 55

A.T. GARIBJANIAN, V.V. HAMBARIAN, L.V. MIRZOYAN, A.L. MIRZOYAN
Distribution of red dwarfs in general galactic field on the basis of stellar star statistics 59

H.S. CHAVUSHIAN, G.H. BROUTIAN
New observations of flare stars in the Pleiades by the method of stellar tracks 63

J. KELEMEN
Possible new, bright flare star in the Pleiades region 67

G. SZÉCSÉNYI-NAGY
Data filtering in statistical studies of flares recorded in sky fields of stellar aggregates - demonstrated with the example of the Pleiades 71

A.L. MIRZOYAN, M.A. MNATSAKANIAN
On two possible groups of flare stars in Pleiades 77

A.S. HOJAEV
Flare activity of stars in the Taurus region 81

R. ANIOL, H.W. DUERBECK, W.C. SEITTER, M.K. TSVETKOV
An automatic search for flare stars in southern stellar aggregates of different ages 85

L.V. MIRZOYAN, V.V. HAMBARIAN, A.T. GARIBJANIAN
Spectral observations of flare stars 95

H.W. DUERBECK, M.K. TSVETKOV
NTT spectra of the flare stars HM1 and HM2 in the R Coronae Austrinae aggregate 99

R.Sh. NATSVLISHVILI
Catalog of flare stars in Orion nebula region 101

M.K. TSVETKOV, K.P. TSVETKOVA
A catalogue of flare stars in the Cygnus region 105

E.S. PARSAMIAN, G.B. OHANIAN
Slow flares in stellar aggregates and solar vicinity 109

M.A. MNATSAKANIAN, A.L. MIRZOYAN
A prediction of the flare activity of stellar aggregates 113

L. PIGATTO
Flare stars as age indicators in open clusters 117

V.V. HAMBARIAN, A.T. GARIBJANIAN, L.V. MIRZOYAN, A.L. MIRZOYAN
The relative number of flare stars in systems of different ages 121

K.R. LANG
Flare stars at radio wavelengths (Invited paper) 125

A.O. BENZ, M. GÜDEL, T.S. BASTIAN, E. FÜRST, G.M. SIMNETT, L. POINTON
Broadband spectral radio observations of flare stars 139

P.M. HEROUNI, V.S. OSKANIAN
Radio flare on eta Gemini star 145

R. PALLAVICINI, L. STELLA, G. TAGLIAFERRI
X-ray emission from solar neighbourhood flare stars 147

C.J. BUTLER
Balmer and soft X-ray emission from solar and stellar flares 153

J.P. CAILLAULT, S. ZOONEMATKERMANI
X-ray variability in the Orion nebula 159

II. FLARE STARS AND RELATED OBJECTS. T TAURI STARS.

V.A. AMBARTSUMIAN
Some resolutions on T Tauri stars (Invited paper) 163

W. HERBST
The variability of T Tauri stars 169

G.V. ZAJTSEVA
Periodical variations of the brightness of T Tauri 173

R.A. VARDANIAN
The polarization discovery and investigation of T Tau stars in Byurakan 177

P. BASTIEN, F. MÉNARD
Recent results on polarization of T Tauri stars and other young stellar objects 179

N.I. SHAKHOVSKAYA
The UBVRI photometric and polarimetric observations of the T Tau star HDE 283572 185

P.F. CHUGAINOV
Axial rotation of BY Dra-type stars and related objects 189

G.F. GAHM
Flares on T Tauri stars (Invited paper) 193

I. APPENZELLER
T Tauri stars and flare stars: common properties and differences 209

W. GÖTZ
Circumstellar phenomena and the position of T Tauri stars in the colour-magnitude diagram 215

P.P. PETROV
High resolution spectroscopy of RY Tauri: variability of the Na D line profiles 219

R.D. SCHWARTZ
Herbig-Haro phenomena associated with T Tauri: evidence for a precessing jet? 221

K. ISHIDA
Survey observations of emission line stars and Herbig-Haro objects at the Kiso Observatory 225

III. FLARE STARS AND RELATED OBJECTS.
FUORS AND OTHER NON-STABLE OBJECTS.

B. REIPURTH
FU Orionis eruptions and early stellar evolution (Invited paper) 229

L.G. GASPARIAN, A.S. MELKONIAN, G.B. OHANIAN, E.S. PARSAMIAN
Subfuors in Orion association 253

M.A. IBRAHIMOV, V.S. SHEVCHENKO
Slow flares and eruptive phenomena in early stages of stellar evolution 257

H.M. TOVMASSIAN, R.Kh. HOVHANNESSIAN, R.A. EPREMIAN
A new fuor? 261

V.S. SHEVCHENKO, S.D. YAKUBOV
Correlation between flare stars and other populations in young clusters and star forming regions 263

T.Yu. MAGAKIAN, T.A. MOVSESSIAN
The results of spectral observations of collimated outflows 267

L.I. MATVEYENKO
H_2O megamaser in Orion KL 271

I.V. GOSACHINSKIJ, N.A. YUDAEVA, R.A. KANDALIAN,
F.S. NAZARETIAN, V.A. SANAMIAN
Flares of radio line emission H_2O in Ori A and W49N 275

R.L. GYULBUDAGHIAN, R.D. SCHWARTZ, L.F. RODRIGUEZ
On the connection between the IRAS point sources and galactic nonstable objects 279

T.A. MOVSESSIAN
The statistical analysis of the optical and infrared luminosities of young stars 283

V. TSIKOUDI
Far infrared emission from post-T Tauri stars 287

J. KRELOWSKI, J. PAPAJ, W. WEGNER
Residual extinction effects in spectra of newly formed stars 293

IV. THEORETICAL PROBLEMS AND INTERPRETATION OF OBSERVATIONS.

V.P. GRININ
Review of the theoretical models of flares of the UV Ceti-type stars (Invited review) 299

E.R. HOUDEBINE, C.J. BUTLER
Quiescent chromospheric response to the (E)UV/optical flare radiation field on dMe stars 313

E.A. BRUEVICH, M.A. LIVSHITS
The hydrogen atom kinetics in flare star chromospheres 317

M.M. KATSOVA
The Balmer decrement in red dwarfs spectra during the flares and quiescent state 321

J.G. DOYLE, G.H.J. VAN DEN OORD, C.J. BUTLER, T. KIANG
A model for the observed periodicity in the flaring rate on YY Gem 325

V.S. HAYRAPETYAN, A.G. NIKOGHOSSIAN
Pinch-model of flares and its observational consequences 329

H.A. HARUTYUNIAN, V.S. HAYRAPETYAN
On the gamma-activity of stellar flares 333

C.H. PAPAS
On a differential equation for electromagnetic wave transmission in flare stars and the possible existance of cohesive wave solutions 337

V.P. GRININ, A.S. MITSKEVICH
The stochastical approach to the modelling of the line profiles in T Tauri stellar winds 343

C.J. LADA
On the origin of dwarf stars (Invited review) 347

T. MONTMERLE
Magnetic activity and evolution of low-mass young stars 363

V. PROSPECTS FOR STELLAR FLARE RESEARCH AND CONCLUDING REMARKS.

M. RODONO
Prospects for studies of UV Cet-type flare stars (Invited review) 371

C.J. LADA
Closing remarks 393

Author index 395

Subject index 399

Object index 405

PREFACE

Stellar flares represent one of the most challenging problems of contemporary astrophysics. Both solar and stellar observations have shown the flare phenomenon to be very complex, and in recent years important progress has been made from simultaneous observations over wide wavelength ranges. Some similarities exist between solar and stellar flares, but important differences have also been established. Such topics, as well as theoretical aspects, were discussed in detail at the Palo Alto IAU Colloquium No. 104, *Solar and Stellar Flares*, in 1988.

Another approach to the study of stellar flares is through observations of flare stars in physical systems. The possibility of detecting flare stars in star clusters and associations with wide angle telescopes have allowed observations of systems with quite different ages. The classical works of G. Haro and V. A. Ambartsumian demonstrated the evolutionary nature of the flare phenomenon. Flares occur at the earliest stages of dwarf star evolution. The photographic observations of flare stars in systems of different ages turned out to be significant not only for the evolutionary study of flare stars, but also for the study of their physical nature. This observational fact was conditioned by very large diversity of flare star luminosities, i.e. of scales of flares produced by them and by peculiarities of stellar flares observed in star clusters and associations.

The IAU Symposium No. 137, *Flare Stars in Star Clusters, Associations and Solar Vicinity* in Armenia was the first IAU conference to bring together the scientists working in both fields: on the UV Ceti stars in the solar vicinity, and on the flare stars in the star clusters and associations.

In fact, it was the fourth Symposium held at Byurakan devoted to flare stars. The first three were held in: 1976 – *Flare Stars*, 1979 – *Flare Stars, Fuors and Herbig-Haro Objects* and 1984 – *Flare Stars and Related Objects*.

The decision of the IAU Executive Committee to hold this Symposium in Byurakan can by considered as a sign of recognition for the significant contribution by the Byurakan Observatory to the study of stellar physics and evolution, particularly, the study of flare stars.

This IAU Symposium was sponsored by the Armenian Academy of Sciences and by the Byurakan Observatory, who paid all expenses for the foreign participants inside the Soviet Union. Additional travel grants were provided by the IAU.

The present Symposium was planned for May, 1989 but because of the tragic earthquake in Armenia it was postponed to October, 1989. It was held on October, 23-27, 1989 at the Byurakan Astrophysical Observatory. There were 93 participants from 19 countries. Unfortunately, some of the USA scientists who were invited could not come to Byurakan because of the earthquake in San-Francisco.

The official opening of the symposium took place in Conference-Hall of the Byurakan Observatory, where all sessions were held. The participants were welcomed by academician V.A. Ambartsumian, the President of Armenian Academy of Sciences, Miss G.V. Danielian, the Representative of the Local Government and Dr. E.E. Khatchikian, the

Director of the Observatory. Dr. B. R. Pettersen presented the congratulating telegram on behalf of IAU General Secretary Dr. D. McNally.

The Scientific Organizing Committee for the Symposium was chaired by V. A. Ambartsumian. The members of the SOC were P. F. Chugainov (USSR). D.S. Evans (USA), G. Haro (Mexico), I. Jankovics (Hungary), K. Kodaira (Japan), L. V. Mirzoyan (USSR), B. R. Pettersen (Norway), L. Rosino (Italy) and M. K. Tsvetkov (Bulgaria). During 6 sections of the Symposium 8 invited and 42 oral papers were presented including 20 poster presentations.

In addition to different aspects of flare stars several papers were devoted to the study of related objects (T Tauri stars, Fuors (objects like FU Ori), Herbig-Haro objects etc.). They suggest a similarity in physical nature among several non-stable phenomena – flares, fuor-like changes, irregular variations etc. It is most likely that the different manifestations of stellar activity are the results of release of some unknown kind of energy in outer layers of young stars. Further study of flare stars and related objects will contribute to a better understanding of stellar activity, in particular to stellar flares.

The main achievement of the Symposium, perhaps, was the presentation of the flare star problem from both a physical and an evolutionary aspect. The discussion showed that the interaction between investigators working on flare stars in star clusters and associations, and on the UV Ceti stars in solar vicinity, can be significant for the understanding of the physics and evolution of red dwarf stars.

The weather, as is usual for this time of the year in Armenia, was sunny and warm. The participants of the Symposium enjoyed many social events and received lasting impressions of more than 3000 years of Armenian history and contemporary Armenia. They could visit many interesting museums, ancient citadels and temples (Garni I - III Cent. B.C.), Matenadaran-Institute of ancient manuscripts, cathedral church (IV cent. A.D.) in Echmiadzin where they were received by Vasgen I - the Katolicos of all Armenians as well as the the memorial of the world-wide known genocide of 1915. The concert of the Armenian State Dance assembly directed by V. Khanamirian in the Aram Khachaturian Concert-Hall in Yerevan was impressive.

At the time of the Symposium conditions in Armenia were very difficult: a blockade of all railway and auto roads connecting Armenia with the world was in effect. In despite of such unfavorable environment, excellent conditions of work were provided to all participants, helped by the warm hospitality of the Armenian astronomers.

To all of them we would like to express our heartiest thanks.

LUDVIG V. MIRZOYAN, BJORN R. PETTERSEN
and MILCHO K. TSVETKOV

February 1990

LIST OF PARTICIPANTS

AMBARTSUMIAN V.A.	Byurakan Astrophysical Observatory, Armenian Academy of Sciences,378433, USSR
ANIOL R.R.	Astronomisches Institut, Uiversitat Munster, Wilhelm-Klemm-Str.10, D-4400 Munster, FRG
ANTOV A.	Department of Astronomy and National Astronomical Observatory, Academy of Sciences, Blvd. Lenin 72, 1184 Sofia, Bulgaria
APPENZELLER I.J.	Landessternwarte, Heidelberg-Konigstuhl, D-6900 Heidelberg 1, FRG
BASTIEN P.	Departement de Physique, Universite de Montreeal, B.P. 6128, Succ. A,Montreal, Quebec, Canada
BENZ A.O.	Institut fur Astronomie, ETH-Zentrum, CH-8092 Zurich, Switzerland
BERDYUGIN A.V.	Crimean Astrophysical Observatory, Academy of Sciences, 334413, USSR
BLAAUW A.	Kapteyn Astronomical Laboratory, Landleven-12 Postbus 800-9700 AV Groningen, Netherlands
BONDAR N.I.	Crimean Astrophysical Observatory, Academy of Sciences, 334413, USSR
BROMAGE G.E.	Science and Enginereering Research Council, Rutherford Appelton Laboratory, Chilton, Didcot, Oxfordshire OX 11 OQX, England
BROUTIAN G.H.	Byurakan Astrophysical Observatory, Armenian Academy of Sciences, 378433, USSR
BUTLER Ch.J.	Armagh Observatory, Armagh BT 61 9 DG, N. Ireland, UK
CHAVUSHIAN H.S.	Byurakan Astrophysical Observatory, Armenian Academy of Sciences,378433, USSR
CHUGAINOV P.F.	Crimean Astrophysical Observatory, Academy of Sciences, 334413, USSR
DOYLE J.G.	Armagh Observatory, Armagh BT 61 9 DG, N. Ireland, UK
DUERBECK H.W.	Astronomisches Institut, Universitat Munster, Wilhelm-Klemm-Str.10, D-4400 Munster, FRG
EPREMIAN R.A.	Byurakan Astrophysical Observatory, Armenian Academy of Sciences, 378433, USSR
GAHM G.F.	Stockholm Observatory, S-13300 Saltsjobaden, Sweden

GARIBJANIAN A.T. Byurakan Astrophysical Observatory, Armenian Academy of Sciences, 378433, USSR

GASPARIAN L.G. Byurakan Astrophysical Observatory, Armenian Academy of Sciences, 378433, USSR

GERSHBERG R.E. Crimean Astrophysical Observatory, Academy of Sciences, 334413, USSR

GIAMPAPA M.S. National Solar Observatory, P.O. Box 26732, Tucson, AZ 85726-6732, USA

GOTZ W. Akademie der Wissenschaften der DDR, Sternwarte Sonneberg, PSF 55 27-28 Sonneberg 6400, DDR

GRININ V.P. Crimean Astrophysical Observatory, Academy of Sciences, 334413, USSR

GYULBUDAGHIAN A.L. Byurakan Astrophysical Observatory, Armenian Academy of Sciences, 378433, USSR

HAMBARIAN V.V. Byurakan Astrophysical Observatory, Armenian Academy of Sciences, 378433, USSR

HARUTYUNIAN H.A. Byurakan Astrophysical Observatory, Armenian Academy of Sciences, 378433, USSR

HAYRAPETIAN V.S. Byurakan Astrophysical Observatory, Armenian Academy of Sciences, 378433, USSR

HERBST W. Van Vleck Observatory, Wesleyan University, Midletown, Connecticut CT 06457, USA

HEROUNI P.M. Radiophysical Measurement Institute, Armenian Academy of Sciences, 375014, USSR

HOJAEV A. Astronomical Institute, Uzbek Academy of Sciences, 700052, USSR

HOVHANNESIAN R.Kh. Byurakan Astrophysical Observatory, Armenian Academy of Sciences, 378433, USSR

IBRAHIMOV M.A. Astronomical Institute, Uzbek Academy of Sciences, 700052, USSR

ILJIN I.V. Crimean Astrophysical Observatory, Academy of Sciences, 334413, USSR

ISHIDA K. Kiso Observatory, University of Tokyo, Mitakamura, Kiso-gun, Nagano-gen, 397-01, Japan

JANKOVICS I. Konkoly Observatory, 1525 Budapest, Box 67, Hungary

KALLOGHLIAN A.T. Byurakan Astrophysical Observatory, Armenian Academy of Sciences, 378433, USSR

KANDALIAN R.A. Byurakan Astrophysical Observatory, Armenian Academy of Sciences, 378433, USSR

KARAPETIAN A.A.	Byurakan Astrophysical Observatory, Armenian Academy of Sciences, 378433, USSR
KATSOVA M.M.	Sternberg State Astronomical Institute, Moscow University, 119899, USSR
KELEMEN J.	Konkoly Observatory, 1525 Budapest, Box 67, Hungary
KHACHIKIAN E. Ye.	Byurakan Astrophysical Observatory, Armenian Academy of Sciences, 378433, USSR
KONSTANTINOVA-ANTOVA R.	Department of Astronomy and National Astronomical Observatory Academy of Sciences, Blvd. Lenin 72, 1184 Sofia, Bulgaria
KRELOWSKI J.	Institute of Astronomy, N.Copernicus University, Chopina 12/18 PL-87-100 Torun, Poland
KRIKORIAN R.A.	LAT Institut d'Astrophysique, 98 bis, Bd Arago 75014 Paris, France
LADA Ch.J.	Steward Observatory, University of Arizona, Tucson, AZ 85721, USA
LANG K.R.	Department of Physics and Astronomy, Tufts University, Robinson Hall, Medford, MA 02155, USA
LIVSHITS M.A.	Institute of Terrestrial Magnetism, Ionosphere and Radio Propagation, Academy of Sciences, USSR
LOVKAJA M.N.	Crimean Astrophysical Observatory, Academy of Sciences, 334413, USSR
LUTHARDT R.	Sternwarte Sonneberg, PSF 55-27/28, Sonneberg 6400, DDR
MAGAKIAN T.Yu.	Byurakan Astrophysical Observatory, Armenian Academy of Sciences, 378433, USSR
MATVIENKO L.I.	Space Research Institute, Academy of Sciences, Moscow, 117810, USSR
MELIKIAN N.D.	Byurakan Astrophysical Observatory, Armenian Academy of Sciences, 378433, USSR
MELKONIAN A.S.	Byurakan Astrophysical Observatory, Armenian Academy of Sciences, 378433, USSR
MENCHENKOVA E.	Astronomical Observatory, Odessa University, Odessa, 270014, USSR
MIRZOYAN A.L.	Byurakan Astrophysical Observatory, Armenian Academy of Sciences, 378433, USSR
MIRZOYAN L.V.	Byurakan Astrophysical Observatory, Armenian Academy of Sciences, 378433, USSR
MITSKEVICH A.S.	Crimean Astrophysical Observatory, Academy of Sciences, 334413, USSR

MNATSAKANIAN M.A.	Byurakan Astrophysical Observatory, Armenian Academy of Sciences, 378433, USSR
MONTMERLE T.	Section d'Astrophysique, Centre d'Etudes Nucleaires, Saclay B.P. No.2 F-91190 Gif-sur-Yvette, France
MOURADIAN Z.M.	Observatoire de Paris-Meudon, F-92190 Meudon, France
MOVSESSIAN T.A.	Byurakan Astrophysical Observatory, Armenian Academy of Sciences, 378433, USSR
MURADIAN R.M.	Byurakan Astrophysical Observatory, Armenian Academy of Sciences, 378433, USSR
NATSVLISHVILI R.Sh.	Abastumani Astrophysical Observatory, Georgian Academy of Sciences,383762, USSR
NAZARETIAN F.S.	Department of Astronomy, University of Yerevan, 375070, Yerevan, USSR
NIKOGHOSSIAN A.G.	Byurakan Astrophysical Observatory, Armenian Academy of Sciences, 378433, USSR
OGANIAN G.B.	Byurakan Astrophysical Observatory, Armenian Academy of Sciences, 378433, USSR
PALLAVICINI R.T.	Osservatorio Astrofisico di Arcetri, Largo Enrico Fermi 5, I-50125 Firenze, Italy
PANOV K.P.	Department of Astronomy and National Astronomical Observatory, Academy of Sciences, Blvd. Lenin 72, 1184 Sofia, Bulgaria
PARSAMIAN E.S.	Byurakan Astrophysical Observatory, Armenian Academy of Sciences, 378433, USSR
PETROV P.P.	Crimean Astrophysical Observatory, Academy of Sciences, 334413, USSR
PETTERSEN B.R.	Institute of Theoretical Astrophysics, University of Oslo, P.O. Box 1029 Blindern, N-0315 Oslo 3, Norway
PIGATTO L.	Osservatorio Astronomico di Padova, Vicol. dell'Osservatorio 5-35122 Padova, Italy
REIPURTH B.J.	ESO, Karl Schwarzschild Strasse 2, D-8046 Garching bei Munchen, FRG
RODONO M.	Instituto di Astronomia, Citta Universitaria, I-95125 Catania, Italy
SANAMIAN V.A.	Institute of Radiophysics and Electronics, Armenian Academy of Sciences, Ashtarak, USSR

SCHWARTZ R.D.	Department of Physics, University of Missouri-St.Louis, 8001 Natural Bridge Road, St.Louis, MO 63121, USA
SEITTER W.	Astronomisches Institut, Univrsitat Munster, Wilhelm-Klemm-Str. 10 D-4400, Munster, FRG
SHAKHOVSKAJA N.I.	Crimean Astrophysical Observatory, Academy of Sciences, 334413, USSR
SHEVCHENKO V.S.	Astronomical Institute, Uzbek Academy of Sciences. 700052, USSR
SIKORSKI J.	Institute of Theoretical Physics and Astrophysics, University of Gdansk, Ul. Wita Stwosza 57, 80-952 Gdansk, Poland
SZECSENYI-NAGY G.	Department of Astronomy, Eotvos University,Kun Bela Ter 2. H-1083 Budapest, Hungary
TERZAN A.	Observatorie de Lyon, F-69230. Saint Genis Laval, France
TOVMASSIAN H.M.	Byurakan Astrophysical Observatory, Armenian Academy of Sciences, 378433, USSR
TSIKOUDY V.	University of Ioannina, Department of Physics, Division of Astro-Geophysics, 45332 Ioannina, Greece
TSVETKOV M.K.	Department of Astronomy and National Astronomical Observatory, Academy of Sciences, Lenin Blvd. 72. 1184 Sofia, Bulgaria
TSVETKOVA A.	Department of Astronomy and National Astronomical Observatory, Academy of Sciences, Lenin Blvd. 72, 1184 Sofia, Bulgaria
TSVETKOVA K.	Department of Astronomy and National Astronomical Observatory, Academy of Sciences, Lenin Blvd. 72, 1184 Sofia, Bulgaria
VARDANIAN R.A.	Byurakan Astrophysical Observatory, Armenian Academy of Sciences, 378433, USSR
YAKUBOV S.D.	Astronomical Institute, Uzbek Academy of Sciences, 700052, USSR
ZAJTSEVA G.V.	Sternberg State Astronomical Institute, Moscow University, 119899, USSR
ZALINIAN V.P.	Byurakan Astrophysical Observatory, Armenian Academy of Sciences, 378433, USSR
ZARATSIAN S.V.	Byurakan Astrophysical Observatory, Armenian Academy of Sciences, 378433, USSR

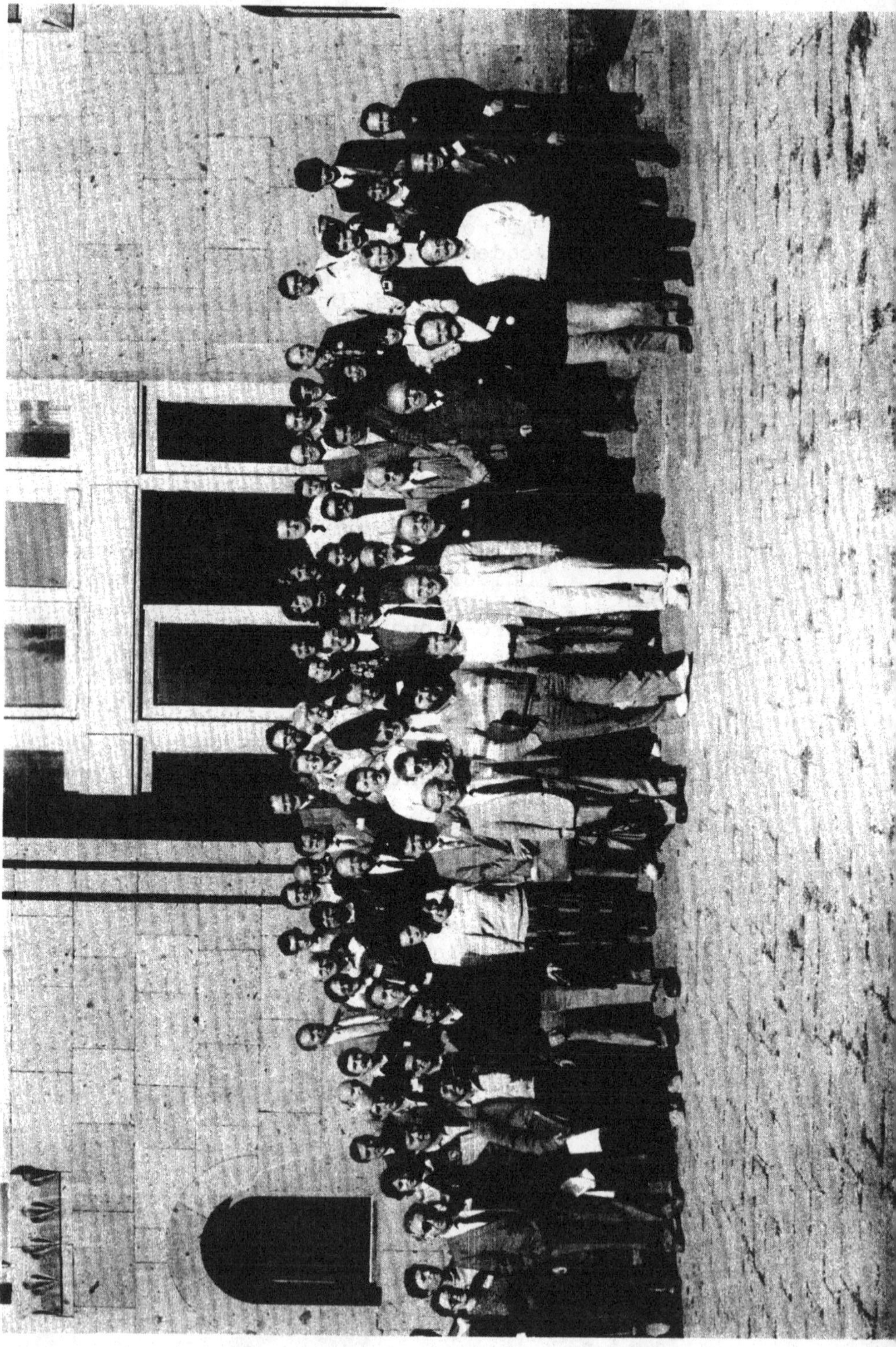

KEY TO PHOTOGRAPH

1. Katsova M. M.
2. Grinin V. P.
3. Terzan A.
4. Andreassian N. K.
5. Mouradian Z.
6. Krikorian A.
7. Shevchenko V. S.
8. Hambarian V. V.
9. Nikoghossian A. G.
10. Götz W.
11. Luthardt R.
12. Hayrapetian V. S.
13. Herbst W.
14. Panov K. P.
15. Vardanyan R. A.
16. Bastien P.
17. Gahm G.
18. Zaytseva G. V.
19. Sikorski I.
20. Bondar' N. I.
21. Lada Ch.
22. Jankovics I.
23. Magakian T. Yu.
24. Bromage G. E.
25. Hojaev A. S.
26. Ishida K.
27. Petrov P. P.
28. Krelowski J.
29. Pigatto L.
30. Doyle J. D.
31. Aniol R.
32. Mitskevich A. S.
33. Tsvetkov M. K.
34. Szecsenyi-Nagy G.
35. Shakhovskaya N. I.
36. Duerbeck H.
37. Ivanova N. L.
38. Giampapa M.
39. Lovkaya M. H.
40. Lang K. R.
41. Seitter W. C.
42. Appenzeller I.
43. Tsvetkova A. G.
44. Antov A. P.
45. Antova R. K.
46. Rodono M.
47. Mirzoyan L. V.
48. Pallavicini R.
49. Gasparian L. G.
50. Blaauw A.
51. Chugainov P. F.
52. Gershberg R. E.
53. Kandalian R. A.
54. Mikaelian A.
55. Ambartsumian V. A.
56. Ibrahimov M. A.
57. Tsikoudi V.
58. Natsvlishvili R. Sh.
59. Montmerle T.
60. Tsvetkova K. P.
61. Ilyin I. V.
62. Yakubov S. D.
63. Pettersen B. R.
64. Benz A. O.
65. Israelian G.
66. ----
67. Mirzoyan A. L.
68. Butler J.

FLARE STARS IN STAR CLUSTERS, ASSOCIATIONS AND SOLAR VICINITY

L.V.MIRZOYAN
Byurakan Astrophisical Observatory
Armenian Academy of Sciences, USSR

ABSTRACT. The observational data on flare stars observed in star clusters and associations as well as in the solar vicinity (the UV Ceti type stars) are discussed. The analysis of these data show that they constitute one common class of objects possessing flare activity and the differences between them are conditioned by the age differences. The stage of flare activity is an evolutionary stage, one of the earliest stages of evolution passed by all red dwarf stars. It comes before the end of their T Tau stage of evolution. The UV Ceti type flare stars in the solar vicinity seem to be the population of the general galactic field, which were formed in the systems, already desintegrated. Most probably the stellar flares are the result of the release of the the surplus energy having intra-stellar origin.

1. INTRODUCTION

The discovery of flare stars in associations and star clusters of different ages turned out to be significant for the problem of stellar evolution. Namely, this discovery allowed to establish that the flare star stage is an evolutionary stage for all red dwarf stars. At the same time it has been shown that this stage of flare activity is one of the early stages of evolution of dwarf stars. Thus, owing to the study of flare stars a possibility has appeared to receive a definite idea on the evolution of dwarf stars based on the observational approach (see, for example, [1-5]).

On the other hand, the study of flare stars, on the whole, is of great importance for the physics of stars, as the discovery of flare phenomenon in young stars was compeletely unexpected for the contemporary theories of stellar evolution and has not yet its generally assepted

L. V. Mirzoyan et al. (eds.), Flare Stars in Star Clusters, Associations and the Solar Vicinity, 1–13.

explanation. Therefore, because of the wide range of flare star luminosities observed in star clusters and associations of different ages indicating on large scale differences in conditions of formations of stellar flares their study can bring to a better understanding of the nature of this unusual phenomenon (see, for example, [4]).

Owing to these two circumstances the further investigations of flare stars became one of the actual problems of modern astrophysics.

In this paper we discuss some results of the flare star study in the light of the investigations carried out at the Byurakan Astrophysical Observatory.

2. EVOLUTIONARY STATUS OF FLARE STARS

An estimate of the total number of flare stars in the comparatively young (age $7x10^7$ years [6]) star cluster Pleiades, obtained by Ambartsumian [7] showed that all (or almost all) stars of low luminosities in this system are flare ones. This unexpected result based on an analysis of observational data had extraordinary importance. It brought to natural conclusion that all dwarf stars at a definite stage of their evolution possess an ability to show flares of radiation i.e. possess flare activity.

The fundamental conclusion showing that the flare activity characterizes an evolutionary stage of red dwarf stars has been confirmed by further investigations of flare stars in the Pleiades and other stellar systems of different ages (see, [4,8].

Owing to the fact, that young dwarf stars, in one of the early stages of evolution possess high flare activity, an unusual rich abundance of flare stars is observed in associations and in comparatively young star clusters. The following Table 1, which contains the estimates of the numbers of them - N in some nearest clusters and associations, is a good illustration of this.

The estimations of total numbers of flare stars presented in Table 1 have been received by the simple formulae [7] giving the number of unknown flare stars in system

$$n_o = \frac{n_1^2}{2n_2}$$

where n_1 and n_2 are numbers of flare stars observed in one and two flares respectively. (This formulae gives the lower limit of the number n_o [9]).

The data of Table 1 show rich abundance of flare stars in young and comparatively young stellar systems, which is a strong confirmation of the above mentioned evolutionary status of flare stars. They testify, that the stage of flare activity is indeed one of the early stages of evolution for red dwarf stars.

TABLE 1. The numbers of flare stars in some nearest star clusters and associations [10]

System	n	n_o	N
Pleiades	546	448	994
Orion I	482	989	1471
TDC*	102	430	532
Cygni(NGC 7000)	67	336	403
Praesepe	54	161	215
Monocerotis I (NGC 2264)	42	400	442
	1293	2764	4057

* TDC - Dark Clouds in Taurus

3. DURATION OF THE FLARE ACTIVITY STAGE

According to generally accepted idea the rates of evolution depend on the mass of a star, in the sense that the duration of any evolutionary stage decreases with the stellar mass (luminosity). In particular the duration of the flare activity stage should decreas with the increase of luminosities of flare stars. There are several observational manifestations of this phenomenon.

1. The luminosity of the brightest flare star in a given system depends on its age. The older the system the higher this luminosity (the earlier the spectral class of the brightest flare star). This regularity has been detected by Haro and Chavira [8].

2. The relative number of flare stars among all red dwarf stars in each system is increasing to the lower luminosities. This number for a definite luminosity seems to be decreasing with the increasing of the age of the system [11].

3. The older the system containing flare stars the lower their luminosities [12].

These regularities can also be considered as evidences in favour of the evolutionary status of flare stars.

The correlation between the mean luminosity of flare stars and the age of the corresponding system, based on the photographic observations of flare stars in the nearest star clusters and associations is presented in Fig.1.

This correlation allows to estimate the age of a system by the mean luminosity of flare stars in it, that is the mean luminosity of flare stars is a definite indicator of its age.

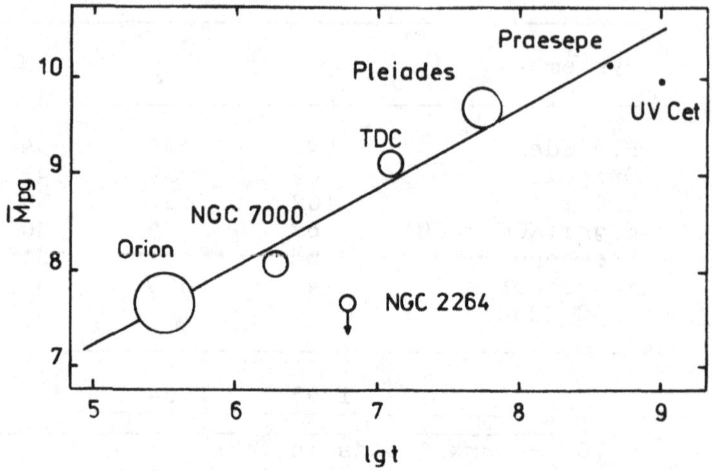

Figure 1. Correlation between the mean absolute photographic magnitude ($\overline{M}_{pg}$) of flare stars and age of the system (t, in years) containing them. The circle areas are proportional to the numbers of stars used. After the paper [14].

Fig.1 indicates, that the age of flare stars can reach hundreds of millions and more years. Indeed, calculations made by Kunkel [13], for example, show that for stars having luminosity equal to $M=15^m$ the duration of flare activity stage approaches 10^{10} years that is the age of our Galaxy.

Thus, in spite of the fact that the evolutionary stage of flare activity is one of the early stages in the evolution of red dwarf stars a star of low enough luminosity can stay in this stage hundreds of millions and more years.

For our further discussion it should be noted that Fig.1 shows in particular that the UV Ceti type flare stars of the solar vinicity have very low luminosities, which correspond to their old age.

This is a very important conclusion. In the light of this conclusion the existence of a large number of flare stars possessing very low luminosity in the general star field of the Galaxy should be explained namely by the large duration of the flare activity stage for very low luminosity stars (mainly of M-type).

4. THE UV CETI TYPE STARS- THE POPULATION OF GENERAL GALACTIC STAR FIELD

Let us assume that the space distribution of the UV Ceti stars in the Galaxy is uniform. In this case the observational data on these stars in the vicinity of the Sun allow us to estimate the number of flare stars which can be discovered (during the definite time) by photographic observations made with the wide-angle telescopes in the regions of the sky free of star clusters and associations. A comparison of such estimations with the results of the flare star observations showed [15] that they are in agreement. It means that our assumption is right: the space distribution of the UV Ceti stars at least in the observed part of the Galaxy is practically uniform. Speaking more exactly they are distributed approximately in the same way as all faint stars. Taking into account that the UV Ceti stars are similar to the flare stars in star clusters and associations in their physical properties [14] one can conclude that they constitute a population of general galactic field [15]. Moreover, it is very probable to assume that the density of number of the UV Ceti type flare stars does not differ in other parts of our Galaxy. They were formed in associations and clusters but having very low luminosities the UV Ceti type stars survived their maternal systems which were desintegrated already.

The mean density of the number of flare stars in stellar aggregates exceeds their mean density in the general star field in some orders of magnitude. For example, the mean density of flare stars in the Pleiades cluster is larger than that in the general galactic field at least by two orders of magnitude. Calculations showed [16] that the relative number of the background flare stars among flare stars observed in the regions of star clusters and associations is less than 20% . As their mean flare frequencies are lower than the mean frequencies for the members of aggregates this relative number can not be larger than 10% of known flare stars.

Assuming that the UV Ceti stars present a population of the general galactic field and their space distribution is uniform in the sense that their percentage among all stars is the same everywhere and equal to that near the Sun one can estimate the total number of them in the Galaxy. It turns out to be equal to 4.2×10^{9}. According to the luminosity functions of the corresponding groups the total number of non-flaring red dwarf stars must be essentially larger. Assuming further that the ratio of flaring and non-flaring red dwarf stars is the same for the whole Galaxy then for the total number of non-flaring red dwarfs one can obtain 2.1×10^{10}.

Therefore, the total number of red dwarf stars in the

Galaxy will be $2.5x10^{10}$, that is about one forth of the total number of all stars in our stellar system. Taking into account that the masses of red dwarf stars in the most cases are not larger than $0.1\mathfrak{M}_\odot$ for the total mass of all red dwarf stars in the Galaxy we obtain a magnitude of the order of $10^9\mathfrak{M}_\odot$. This does not contradict to Oort's estimation [17].

The observational data on the UV Ceti stars of the solar vicinity are in agreement with the idea that they are coming out from the desintegrated systems - star clusters and associations. Apparently, this conclusion is correct also for non-flaring red dwarfs of the general galactic field. In favour of this conclusion one can consider in particular the calculations of the numbers of stars which could be formed in the OB- and T - associations, correspondingly, during the life of our Galaxy, made by Ambartsumian [18] as far back as in 1950. They testify that all stars of the flat and intermediate subsystems in the Galaxy could be formed in stellar associations.

5. FLARE STARS - POST T TAURI STARS

Due to the fact that the duration of each evolutionary stage depends on the mass (luminosity) of stars in associations we observe stars which are in different stagesof evolution. The existance of the stars being in different evolutionary stages in the same system allowed to determine the sequence of the early stages which were passed through by stars on the base of their observations in stellar associations only.

For example.in stellar associations the flare stars coexist with the T Tauri stars. which represent the typical (characteristic) population of the T-associations. At the same time they are numerous in the comparatively old star clusters like the Pleiades and Praesepe, where the T Tauri stars are absent.

This means, that the T Tauri stars are younger than flare stars and naturally are in the earlier stage of evolution. At the same time this observational fact indicates, that the duration of the T Tauri stage in average is shorter than that of the flare activity stage.

These important circumstances gave grounds to Haro [19] to conclude for the first time that the evolutionary stage of flare activity of red dwarf stars follows immediately the stage of the T Tauri stars. This conclusion became firmly established as a regularity of stellar evolution after Ambartsumian's [7] first estimation of the total number of flare stars in the Pleiades cluster and its qualitative confirmation by observations .

Indeed, as one can see from Table 1 the number of the known flare stars in the Pleiades region is at present 546.

Out of them not more than 10% can be the general galactic field stars. Therefore, the number of known flare stars in this cluster already exceeds 500 i.e. larger than we had from the initial estimate of it.

The conclusion on the connection of the stage of flare activity with the T Tauri stage brings to two serious questions:

1.Whether all T Tauri stars transform into flare stars;

2.Whether it is the possible transition from the T Tauri stage to the stage of flare activity. if we consider the masses (luminosities) of the corresponding stars.

The first question have been discussed in Ambartsumian's paper [20]. It has been shown that in the Orion association about one forth of all T Tauri stars were at the same time flare stars (known or having ability to show photographic flares). This estimate of the relative number of the T Tauri stars having flare activity is an approximation. The new estimations show that it can reach 50% (see, for example. [21]).But it is more important that there is no doubt at present that not all T Tauri stars are flaring-up. However, the analisis has shown that the answer to the first question is apparently positive: it is difficult to assume that the red dwarf stars have two different ways of evolution. As to the second question, the transition between the T Tauri and UV Ceti stages seems to be quite possible from the point of view of the corresponding obsereved luminosities.

6. PHYSICAL PROPERTIES OF THE FLARE STARS

Thus, at present there are enough grounds to consider that the flare stars observed in star clusters and associations and the UV Ceti stars of the solar vicinity have common physical nature and the observed differences betweenthem are due to the defferences in their ages.

Indeed, the flare stars in star clusters and associations being formed comparatively recently, are connected with a more or less dense diffuse matter, they have higher luminosities and are more active than the UV Ceti stars of the solar vicinity. The latter, on the contrary, left already their maternal systems, where they were formed, the diffuse matter co-existing with them in their youth has dispersed long ego, now they have lower luminosities and are less active. According to the space distribution the UV Ceti stars constitute a population of the general galactic field.

Due to their low luminosities the UV Ceti type stars can be observed only in the nearest vicinity of the Sun. That is the reason that the physical study of stellar flares' phenomenon practically is completely based on the observations of the UV Ceti stars of the solar vicinity.

Almost all information on physical properties of stellar flare radiation at present was obtained from the UV Ceti flare observations. These observations include the wide region of spectrum, beginning from radio and finishing with far ultraviolet and X-ray wavelengths.

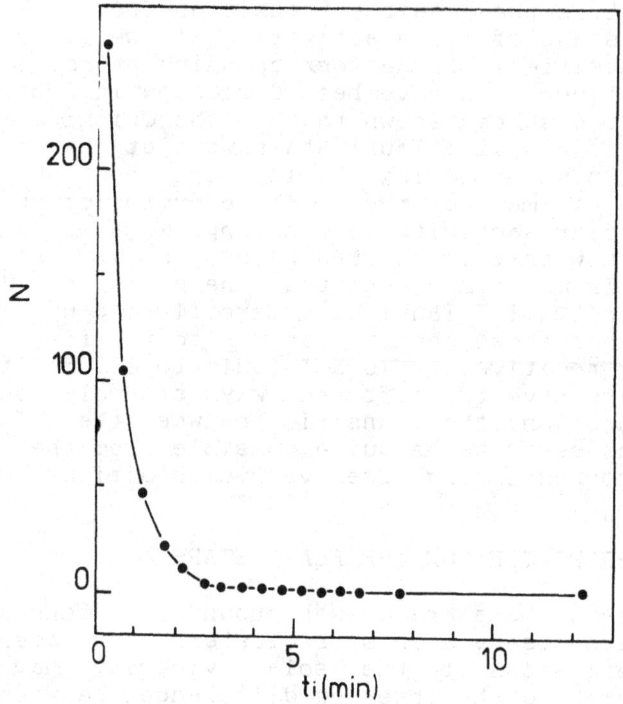

Figure 2. The distribution of the increasing times of flares - t_i for the UV Ceti stars according to Moffett's photoelectric observations [23]. N - number of flares. After the paper [24].

On the contrary, the photographic observations of stellar flares in star clusters and associations were very important for the evolutionary study of flare stars. At the same time they helped us for better understanding of the nature of flare phenomenon.

In this connection, it should be particularly mentioned that namely owing to photographic observations of stellar flares in systems Haro [22] has divided all flares into two classes, according to their increasing times: "fast" flares with this time shorter than 20-30 min, and "slow" flares for which it is longer.

The flares of these two classes possess some, quite different pecularities. Though as it has been shown recently the distribution of flares by the increasing time (Fig.2) is a continuous one. i.e. quite different increasing times are observed Haro's classification is of great significance for understanding of the flare phenomenon.

The matter is that, the "fast" and "slow" nature of flares does not depend on the physical state of a flare star having a flare on the whole , but due to physical conditions in the active zone of a star, where the energy of flare is liberated. In favour of this conclusion one can consider the eloquent obsevational fact, that the most flare stars, observed in"slow" flares have shown also "fast" flares.

As an illustration of this observational fact on Fig.3 light curves of two complex flares are presented,which are combinations of two flares, happened on the same star one after another. The first of them is a combination of light curves of "fast" and "slow" flares.

Without considering the physical pecularities of stellar flares belonging to two mentioned classes we would like to note that their differences (velocity of light increasing, mean frequency, colour of flare emission, total energy of optical radiation) can be explained satisfactorily by Ambartsumian's hypothesis [26,27] on the flare activity.

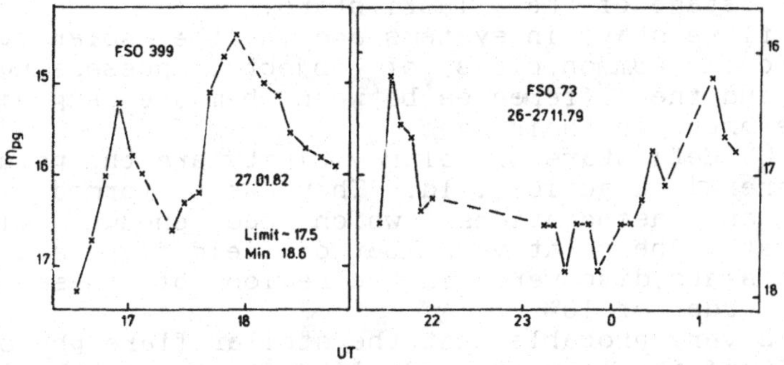

Figure 3. Photographic light curves of two complex flares observed on the Orion flare stars No.73 and No.399, which are combinations of light curves of "fast"-"slow" and "slow"-"slow" flares, respectively. After the paper [25].

According to this hypothesis (see, in detail, [4]) the differences between "fast" and "slow" flares are due to the differences existing in the star atmospheric layers, where a flare is held (above or below the photosphere).

It should be noted that the flare phenomenon has a physical similarity with the fuor phenomenon. Namely, the differences between radiations in prefuor and postfuor phases are like the differences between those of "fast" and "slow" flares [27].

In favour of the common physical nature of the flares and fluor-like variations, as well as the T Tauri type irregular changes, one can consider the observational fact that sometimes these phenomena happen on the same stars. This is known for T Tauri stars and fuors since 1970. But for flare stars it became known recently when Natsvlishvili [21] discovered fuor-like variations on the flare star FSO 435 in Orion.

It is important to add that there are some grounds to assume [27] (see, also [4,28]) that the energy responsible for both flare and fuor phenomena comes from internal layers of the star that is it has an intra-stellar nature.

7. CONCLUSION

The study of flare stars having different ages in star clusters, associations and in the general galactic field (the UV Ceti type stars) in last decades brought to the following results.

1. The stage of flare stars is an evolutionary stage, one of the early stages of evolution for all red dwarf stars. This stage is characterized by flare activity, which appears before the end of the earliest stage of evolution of these stars, the stage of the T Tauri stars.
2. The flare stars in systems and in the solar vicinity constitute one common class of objects, possessing flare activity and the differences between them are explained by the differences in their ages.
3. The UV Ceti stars of solar vicinity are the population of the general galactic field. They were formed in star clusters and associations, which had enough time to desintegrate. The relative number of field flare stars among the flare stars discovered in the regions of these systems is of the order of 10% .
4. It is very probable that the stellar flare phenomenon is the result of the leaving and liberation of the surplus energy having intra-stellar origin.

It should be added that in spite of the serious achievments in the study of flare stars there are many unsettled problems in this field. Many of them have special importance for the physics and evolution of stars.

Concluding the present paper, let us note some of these problems.

a). Are there flare stars of very low luminosities ($M_{pg} > 15^m$) like the faint stars of the UV Ceti type of the solar vicinity in star clusters and associations?

b). What kind of correlation does exist between the appearences of the stellar flare phenomenon in different spectral regions from radio to X-ray?

c). What is the nature of stellar instability in the early tages of stellar evolution (T Tauri stars and related objects)?

EFERENCES

. Ambartsumian,V.A. and Mirzoyan,L.V.(1971),'Flare stars', in: New Directions and New Frontiers in Variable Star Research, Veroff, Bamberg,Bd. 9, Nr.100, 98-108.
. Ambartsumian,V.A. and Mirzoyan,L.V.(1975)'Flare stars in star clusters and associations', in W.Sherwood and L.Plaut (eds.), Variable Stars and Stellar Evolution, Reidel Publishing Company, Dordrecht, pp.3-14.
. Mirzoyan, L.V.(1981) Stellar Instability and Evolution, Acad. Sci.Armenian SSR, Yerevan.
Ambartsumian,V.A. and Mirzoyan,L.V.(1982)'An observational approach to the early stages of stellar evolution', Astrophys.Sp.Sci.,84, 317-330.
. Mirzoyan,L.V.(1984)'Flare stars', Vistas in Astronomy, 27, 77-109.
. Allen,C.W.(1973), Astrophysical Quantities, University of London, The Athlon Press.
. Ambartsumian,V.A.(1969)'On the statistics of flaring objects',in V.V.Sobolev, Stars, Nebulae, Galaxies. Acad. Sci.Armenian SSR,Yerevan, pp.283-292.
. Haro,G. and Chavira,E.(1966)'Flare stars in stellar aggregates of different ages',Vistas in Astronomy,8, 87-107.
. Ambartsumian,V.A., Mirzoyam,L.V., Parsamian,E.S., Chavushian H.S. and Erastova,E.L.(1970)'Flare stars in the Pleiades', Astrofizika,6,3-30.
0.Mirzoyan,L.V. and Ohanian,G.B.(1986)'Flare stars in star clusters and associations', in L.V.Mirzoyan (ed.),Flare Stars and Related Objects, Acad. Sci. Armenian SSR, Yerevan,68-78.
1.Mirzoyan,L.V.,Hambarian,V.V.,Garibjanian,A.T.and Mirzoyan, A.L. (1989)'Statistical study of flare stars IV. Relative number of flare stars in the Orion association, Pleiades cluster and in the solar vicinity', Astrofizika, 31,pp.258-269.
2.Mirzoyan,L.V. and Brutian,G.A.(1980)'On the statistics of flare stars', Astrofizika, 16, 97-106.
3.Kunkel,W.(1975)'Solar neighborhood flare stars - a review', in V.Sherwood and L.Plaut (eds.), Variable Stars and Stellar Evolution, Reidel Publishing Company, Dordrecht, pp.15-38
4.Mirzoyan,L.V. and Hambarian,V.V.(1988)'Statistical study of flare stars I.UV Ceti type stars of solar vicinity and flare stars in star clusters and associations', Astrofizika, 28, 375-389.

15.Mirzoyan,L.V., Hambarian,V.V., Garibjanian,A.T. and Mirzoyan,A.L.(1988)'Statistical study of flare stars II On the origin of the UV Ceti type stars',Astrofizika, 29, 44-57.
16.Mirzoyan,L.V., Hambariam,V.V., Garibjanian,A.T. and Mirzoyan,A.L.(1988) 'Statistical study of flare stars III. Flare stars in the general galactic star field', Astrofizika, 29, 531-540.
17.Oort,J.H.(1957)'Dynamics and Evolution of the Galaxy, in so far as relevant to the problem of the papulations', in D.J.K.O'Connell, S.J.(ed.), Stellar Populations, Vatican Obs., pp.418-433.
18.Ambartsumian,V.A.(1950)'Stellar associations and the origin of stars',Izw.Acad.Sci.USSR,Phys.Series, 14, No.1.15-24.
19.Haro,G.(1975)'The possible connection between T Tauri stars and UV Ceti stars', in G.H.Herbig (ed.), Non-Stable Stars, University Press, Cambridge, pp.26-30.
20.Ambartsumian,V.A.(1970)'On the percentage of flare stars among variables of the RW Aur type in the Orion association', Astrofizika, 6, 31-38.
21.Natsvlishvili,R.Sh.(1987), Flare Stars in Orion and Pleiades, Byurakan Astrophys.Obs.
22.Haro,G.(1964)'Flare stars in stellar aggregates', in F.J.Kerr and A.W.Rogers (eds.), The Galaxy and the Magellanic Clouds, Australian Acad.Sci.. Canberra. pp.30-37.
23.Moffett,T.J.(1974) 'UV Ceti flare stars: observational data', Astrophys.J.Suppl .Series, 29, 1-42.
24.Mirzoyan,L.V. and Melikian,N.D.(1986)'Some characteristics of stellar flares', in L.V.Mirzoyan(ed.), Flare Stars and Related Objects, Acad.Sci.Armenian SSR,Yerevan, pp.153-161.
25.Mirzoyan,L.V. and Natsvlishvili,R.Sh.(1987) 'Remarkable combinations of stellar flares', Astrofizika, 27,605-608.
26.Ambartsumian,V.A.(1954)'Phenomenon of continuous emission and sources of stellar energy',Comm.Byurakan Obs., 13,3-35.
27.Ambartsumian,V.A.(1971)'Fuors',Astrofizika, 7, 557-572.
28.Mirzoyan,L.V.(1990), Early Stages of Stellar Evolution, Acad.Sci.Armenian SSR, Yerevan, in press.

LANG: Fast (minutes) and slow (hours) flares are also detected at radio wavelengths from the same flare star, but we also detect low-level emission lasting for days. Is there evidence for this longer, slower variation in the optical region of the spectrum?

MIRZOYAN: Yes, this has been discussed by Ambartsumian. But maybe there is some confusion since the optical distinction between fast and slow flares as introduced by Haro has been made from the rise time parameter, not the total duration. Concerning the latter, I may note that some flares observed in the optical had durations longer than that of radio flares.

LANG: What is the relation between rise times and total duration?

MIRZOYAN: The rise times for individual flares do not depend on their total duration.

APPENZELLER: Do you have any information on the mass distribution (from the location of the objects in the HR-diagram) of the large number of flare stars which you have detected?

MIRZOYAN: The masses of flare stars are not known as we have no direct determination of them. We can only estimate masses by using the mass-luminosity relation for flare stars.

THE FLARE ACTIVITY OF AD LEO 1972-1988

B.R. PETTERSEN[1], K.P. PANOV[2], M.S. IVANOVA[2], C.W. AMBRUSTER[3], E. VALTAOJA[4], L. VALTAOJA[5], S. AVGOLOUPIS[6], L.N. MAVRIDIS[7], J.H. SEIRADAKIS[6], S.R. SUNDLAND[1], K. OLAH[8], O. HAVNES[9], Ø. OLSEN[9], J.-E. SOLHEIM[9], T. AANESEN[9].

[1] Institute of Theoretical Astrophysics, University of Oslo, P.O. Box 1029 Blindern, N-0315 Oslo 3, Norway.
[2] Department of Astronomy and National Astronomical Observatory, Academy of Sciences, 72 Lenin Blvd., Sofia 1184, Bulgaria.
[3] Department of Astronomy and Astrophysics, Villanova University, Villanova, PA 19085, USA.
[4] Metsähovi Radio Research Station, Helsinki University of Technology, SF-02150 Espoo, Finland.
[5] Department of Physical Sciences, Turku University, SF-20500 Turku 50, Finland.
[6] Department of Astronomy, University of Thessaloniki, Greece.
[7] Department of Geodetic Astronomy, University of Thessaloniki, Greece.
[8] Konkoly Observatory, P.O. Box 67, H-1525 Budapest XII, Hungary.
[9] Institute of Mathematical and Physical Sciences, University of Tromsø, P.O. Box 953, N-9001 Tromsø, Norway.

ABSTRACT. Time resolved U-filter lightcurves of stellar flares on the dM4e star AD Leo have been measured to determine the flare frequency for individual observing seasons between 1972 and 1988. This report adds new data for 1984-1988. We use the number of flares with energy larger than 10**30 erg, and the number of flare maxima with amplitudes larger than 0.3 as separate parameters to estimate flare frequencies. Chi-square tests do not reveal statistically significant variations of the flare frequency with time for 1972-1988.

1. INTRODUCTION

AD Leo is one of the most extensively observed flare stars in the solar neighbourhood. In U-filter photometry with S/N>10 (referring to the quiet star) one detects an outburst on the average every 1 - 1.5 hours. In time the flares appear to occur at random. Flare importance, measured by the size of the flare amplitude or by the amount of energy emitted,

L. V. Mirzoyan et al. (eds.), Flare Stars in Star Clusters, Associations and the Solar Vicinity, 15–18.

span 4-5 orders of magnitude. Small flares are much more common than large flares, and in about 1/3 of the cases the event is composed of several individual flare-ups. Pettersen et al.(1984, 1986) have analyzed previous observations of AD Leo to investigate if the flare activity varies on a timescale of years, in analogy with the solar activity cycle. In the present paper we update the analysis, using new observations between 1984 and 1988 from efforts at several observatories.

2. OBSERVATIONS

U-filter photometry with a time resolution of 1-9 seconds was done with 0.5-1.0 m telescopes at National Astronomical Observatory, Rozhen, Bulgaria; Stephanion Observatory, Greece; Piszkéstetö, Hungary; Skibotn, Norway; Kitt Peak National Observatory, Arizona, USA; and McDonald Observatory, Texas, USA. Each observing season lasted from November till May, and the effort reported here took place between November 1984 and May 1988. The results are summarized in Table 1. We detected a total of 149 flares in 190 hours of observations. We rejected all observations with $S/N<10$, and usually had $20 < S/N < 50$.

3. ANALYSIS AND RESULTS

We have extracted the following quantities from our observations: Flare coverage, i.e. the time actually spent monitoring the star for flares; the number of flares detected with a peak flux exceeding 3.8 10**28 erg/s (corresponding to a flare amplitude of $I_f/Io > 0.3$); and the number of flares (some with several maxima) where the emitted energy during the flare lifetime exceeded 10**30 erg. The limits of peak flux and flare energy were chosen to ensure complete statistical detection in our sample.

Since flares on AD Leo are randomly distributed in time we can estimate the flare frequency by $N/T \pm \sqrt{N}/T$, where N is the number of flares in the sample and T is the flare coverage in hours. We do this analysis on two separate datasets, one using the flare energy as a key parameter and the other using the flare amplitude. The two panels of Fig. 1 give the results, where data before 1985 are from Pettersen et al (1984, 1986). The 1973 point is slightly modified by including recently published data by Herr and Opie (1987).

The two flare parameters mimic each other extremely well. Although certain structures can be noticed in each diagram, no excursions are more than one or two error bars away from the average value (except for the 1973 point). Chi-square tests do not show conclusively that variability has been detected over the 1972-1988 baseline. The flare activity of AD Leo varies by less than a factor of two on a timescale of years and decades.

TABLE 1. Summary of flare monitoring

Year	Coverage (hours)	Number of flares observed Total	log $E_U \geq 30.0$	$\Delta U \geq 0.3$
1985.2	59.73	39	27	23
1986.2	55.85	47	34	39
1987.2	24.98	18	15	14
1988.2	49.06	45	36	41

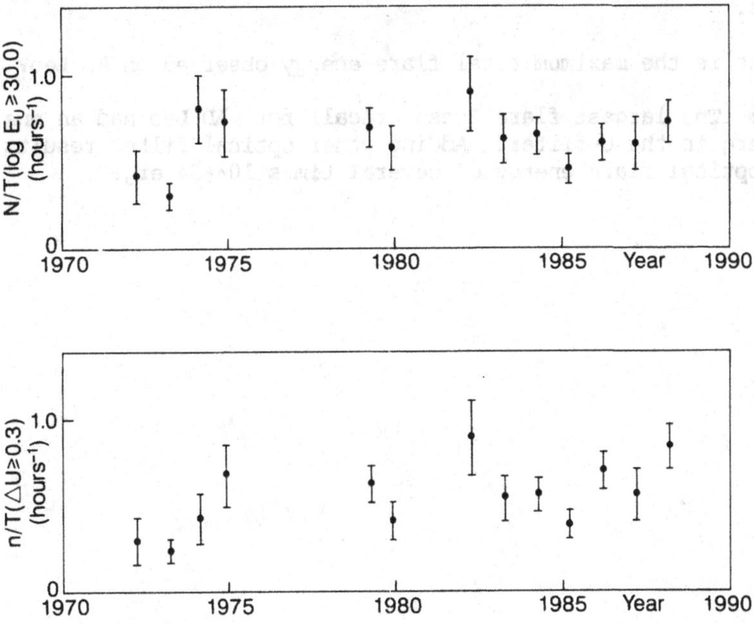

Figure 1. The flare activity of AD Leo as a function of time. The upper panel shows the frequency of flares with energy larger than 10^{30} erg. The lower panel shows the frequency of flare-ups with amplitudes exceeding I_f/I_o=0.3.

References

Herr, R.B., Opie, D.B., 1987, Inf. Bull. Var. Stars # 3069.

Pettersen, B.R., Coleman, L.A., Evans, D.S., 1984, Ap.J. suppl. 54, 375.

Pettersen, B.R., Panov, K.P., Sandmann, W.H., Ivanova, M.S., 1986, Astr. Ap. suppl. 66, 235.

GIAMPAPA: Do you observe pre-flare "dips" in the flare events on AD Leo?

PETTERSEN: In our data we have no significant detections of pre-flare dips for AD Leo. Most of our data are in the U-filter, but some observers also include BVRI-filters. Even in those we do not see any dips, and certainly not anything like your own beautiful result for EQ Peg (Astrophys. J. 252, L39).

PARSAMIAN: Would you like to say the magnitude increase of the largest flare amplitude of AD Leo?

PETTERSEN: As far as I recall the largest U-amplitude seen is 4.5 - 4.7 magnitudes.

GAHM: What is the maximum total flare energy observed on AD Leo?

PETTERSEN: The largest flare I can recall for AD Leo had an energy of 5 10**33 erg in the U-filter. Adding other optical filter results would imply an optical flare energy of several times 10**34 erg.

PHOTOMETRICAL, SPECTRAL AND RADIO MONITORING
FOR EV LAC IN 1986 AND 1987

R.E.Gershberg, I.V.Ilyin, N.S.Nesterov,
N.I.Shakhovskaya Crimean Astrophysical Observatory,
Nauchny-Simeiz, USSR

R.G.Getov, M.Ivanova, K.P.Panov, M.K.Tsvetkov,
A.G.Tsvetkova National Astronomical Observatory,
Rozhen, Bulgaria

G.Leto Osservatorio Astrofisico, Catania, Italy

ABSTRACT. During 9 nights in 1986-1987 we have carried out the patrol observations of red dwarf EV Lac within the wavelength range from 3500 Å to 8 mm and registered more than 50 optical flares. Some features of stellar flares, spots and active regions for the periods of cooperative observations are discussed.

1. Introduction

In 1987 and 1987 during 9 September nights we carried out photometrical, spectral and radio monitoring for the red flaring dwarf EV Lac. The stellar photometry has been fulfilled at the 125cm telescope in the Crimea, at the 60cm telescope in Rozhen and at the 91cm telescope in Catania. Spectral and radio monitorings have been carried out in the Crimea at the Shajn 2.6m reflector in Nauchny and at the 22m radio telescope in Simeiz.

2. Flare photometry

The most complete photometrical data have been obtained at the 125cm telescope with Piirola's (1984) UBVRI photometer-polarimeter and Fig 1 presents the EV Lac light curves in the U band registered with this equipment; these curves yeild a rather certain image of a general level of flare activity in periods of our observations. The plot shows a significant variety of light curves of the star: there are single fast brightness flashes as well as prolonged - up to 2-3 hours - intervals of an enhanced brightness with a few overlapped flares. Differences in flare activity levels in consecutive nights are well noticeable. Let us remind that the axial rotation period of the star is near to 4 days, therefore in both seasons we observed EV Lac from the all sides, and differences noted mean an essential heterogeneity of different parts of the stellar surface.On the other hand, Fig 1 shows that in 1986 stellar

L. V. Mirzoyan et al. (eds.), Flare Stars in Star Clusters, Associations and the Solar Vicinity, 19–23.

flares were in general more frequent and shorter than in 1987.

3. Spot photometry

During observations at the 125cm telescope, we alternated 30-40 minute monitoring of the EV Lac brightness with 1-2 minute measurements of the brightness of the comparison star SAO 52337 (K2). Having omitted the EV Lac brightness estimations affected by close flares, we found gradual variations of the stellar brightness with small amplitudes that are caused by a spottedness of EV Lac. In Fig.2 brightness differences ΔV= V(EV Lac) - V(SAO) are confronted to the phase of the periods Min=$2444435^{d}.447$ + $4^{d}.379$E, that was obtained by Rojzman (1984) and coincides practically with one obtained earlier by Pettersen et al (1983). The plot shows that

i) in 1986 the periodic brightness variations with the amplitude $\Delta V = 0^{m}.15$ took place that is in agreement with the results by Kleinman et al.(1987), but in 1987 an amplitude of such oscillations decreased to $0^{m}.07$;

ii) in 1986 and 1987 phases of the EV Lac periodic brightness variations differed by about half a period and none of them coincided with the phase in 1980-81;

iii) in 1986 the EV Lac maximum brightness was at least by $0^{m}.055$ higher as compared with one in 1987, and the 1986 minimum brightness was about $0^{m}.025$ lower as compared with one in 1987; it means that the decrease of the brightness oscillation amplitude in 1987 is due mainly to more uniform distribution of spots over the stellar surface but not to a decrease of a total area of spots.

4. Spectral study

A spectral monitoring for EV Lac was carried out with the CCD matrix of the Helsinki University (Poutanen,1987): in the coude focus of the Shajn reflector we registered the stellar spectrum in the Hα region.During 9 nights we have obtained 143 spectrograms with expositions from 12 to 30 min. In Fig 3 the mean spectra of quiescent EV Lac for each night are given, they were obtained by summarizing all the spectra registered out of flares. The plot shows significant variations of mean Hα profiles from night to night, and given phases correspond to the stellar spottedness degree according to Fig 2. We have found no correlation between nightly mean equivalent widths of the Hα emission line and sums of equivalent durations of flares for each night. It means that measured values W(Hα) out of flares give a total intensity of the stellar active regions but are not due to an afterglow effect of strong flares in the line. However, it is seen in Fig 3, that 13 and 14.9.86 the Hα intensity was noticeably lower as compared to previous nights, and it is naturally to suggest that in noted two nights EV Lac faced the Earth with the least active side possessing a low brightness chromosphere. The stellar spottedness phases in Fig 3 are in agreement with such a suggestion. But in 1987 such simple correlation was not seen.

Spectral monitoring for EV Lac was carried out at the 2m reflector in Rozhen too.

5. Radio data

Radio monitoring for EV Lac was carried out at 37 GHz. More than a hundred measurements with a 2.5min integration time were fulfilled, but no significant differences are found between the powerful flare on 11.9.86 19^h18^mUT and rather quiet states of the star.

6. Conclusion

Thus, our photometric and spectral observations suggest that in two consecutive seasons the general picture of the EV Lac surface activity was noticeably different. Such qualitative difference may be due to different characteristic distances between stellar spots, since according to Gershberg et al.(1987), with respect to such distances one may expect the appearance in a star the MHD disturbances of different kinds which correspond to stellar flares of different energy deposits and durations.

7. References

1. Gershberg R.E., Mogilevskij E.I., Obridko V.N. (1987), Kinematika i fizika nebesnykh tel V.3, N 5, P.3.
2. Kleinman S.J., Sandmann W.H., Ambruster C.W. (1987), IBVS N 3031.
3. Pettersen B.R., Kern G.A., Evans D.S. (1983), Astron.Astrophys.V.123, P. 184.
4. Piirola V. (1984), Observatory and Lab.Astrophys.Univ.Helsinki Rept. N 6, P. 151.
5. Poutanen M. (1987) 'The charge-couple device in stellar spectroscopy'. Ph.D.Thesis. Helsinki University Observatory.
6. Rojzman G.Sh. (1984), Letters to Sov.Astron. V.10, P. 279.

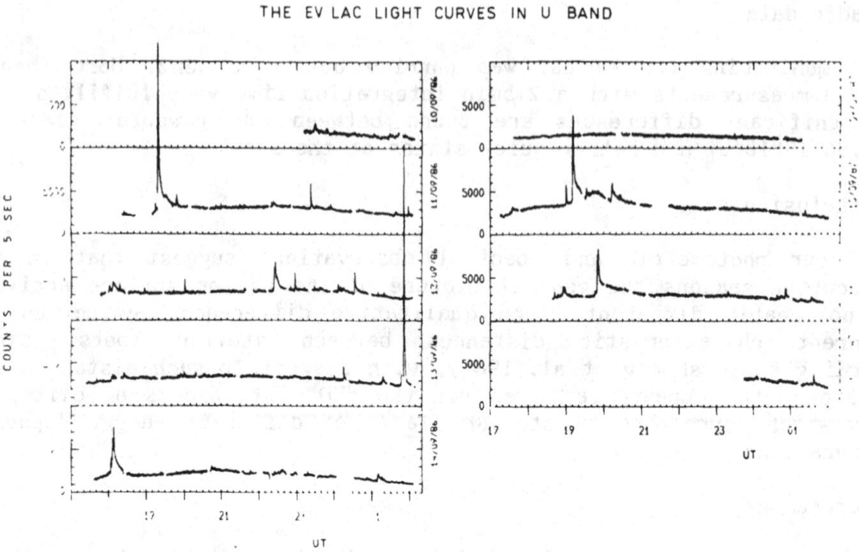

Fig. 1. The EV Lac light curves in the U band.

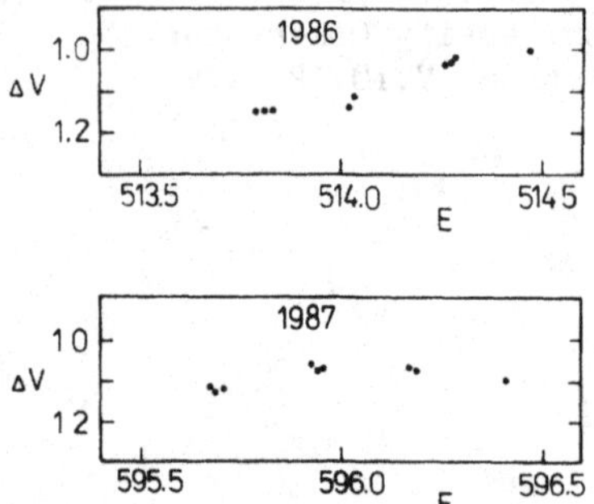

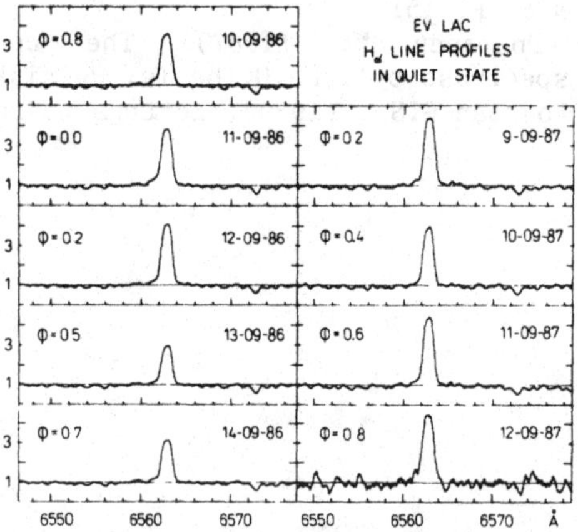

Fig. 2. Small amplitude variations of the EV Lac brightness due to stellar spots.

Fig.3. The nigthly averaged Hα line profiles in a quiet state of the star.

PETTERSEN: The flare time scale of the H-alpha emission line is much longer than e.g. in the U-filter (continuum). Identifying quiet intervals in the H-alpha studies from U-filter monitoring may therefore be risky. Sometimes H-alpha never reaches "quiet level" during nights of active flaring.

GERSHBERG: We had long duration exposures - up to 30 minutes, therefore this effect was weakened on the selected CCD-spectra.

MIRZOYAN: What was the correlation between the behavior of EV Lac in optical and radio wavelengths during your synchronuous observations?

GERSHBERG: We have no detection of radio flares during very strong optical flare.

BENZ: Have you observed a radio flare at 8 mm? How long was your observing time and what was the sensitivity? I have observed UV Ceti for 6 hours with the Very Large Array at 1.3 cm but not seen a flare.

GERSHBERG: We did not detect radio flares during the powerful optical one. The radio coverage was about 1 hour and the noise was several tens of milliJansky.

LANG: The radiation mechanism at radio and optical wavelengths may be very different, so you might not expect to see flares simultaneously at these two regions of the electromagnetic spectrum.

GERSHBERG: I completely agree.

ELECTROPHOTOMETRIC OBSERVATIONS OF THE EV LAC FLARES IN U, Hα AND Hβ

A.S.MELKONIAN
Byurakan Astrophysical Observatory
Armenian Academy of Sciences, USSR

ABSTRACT. Results of the patrol observations of the EV Lac flares in U, Hα and Hβ bands are given. The observations are carried out simultaneously in U and Hα or in U and Hβ. During (U,Hα) observations 21 U -flares are registered of which only comparatively strong one is accompanied by a real Hα - flare. And during (U,Hβ) observations among 7 U-flares the strongest and a "slow" one is accompanied with an increase of Hβ -signal. Small amplitude periodic light variations of EV Lac are detected with a period of $3^d.3$ and amplitude $0^m.02$ in B and V bands.

1. INTRODUCTION

It is difficult to interpret the correlations between the photometric characteristics of the flares observed in the continuum and in the spectral lines. In 1979-1980 and 1985 during 145.5 hours the patrol U,Hα and Hβ observations of EV Lac was organized. Hα and Hβ interference filters with 12A and 6A of FWHM correspondigly were used. The observations were made by the 48-cm cassegrain telescope of the Byurakan Astrophysical Observatory in the analogue system. Using the results of the simultaneous spectral and photoelectric observations of the flares of AD Leo, made by Gershberg and Chugainov [1], the expected amplitudes of the Hα and Hβ flares for our instrumental photoelectric system were estimated. The values of $0^m.4$ and $0^m.8$ were calculated for the Hα and Hβ flares correspondingly.

2. OBSERVATIONS

21 U flares during the 72.2 hours of (U,Hα) observations were registrated. The increas of Hα-signal was registered in 4 cases, but only one of them was reliable, i.e. only one of 21 U-flares was accomponied by reliable Hα flare, with

L. V. Mirzoyan et al. (eds.), Flare Stars in Star Clusters, Associations and the Solar Vicinity, 25–26.

amplitude $0^m.3$. Probably only the most powerful U-flares were accompanied with the increase of Hα-signal. 7 U-flares during the 40 hours of (U,Hβ) observations are registered. The increase of Hβ-signal in only one case is registered.The duration of increased Hβ-signal was as long as 45 min.

To search the small-amplitude variability of EY Lac observed from 10.08.85 to 14.10.85 during 24 nights 64 UBV measurments were made. It is probable the variability of EV Lac with the period of $3^d.3$ and amplitude of $0^m.02$.

3. DISCUSSION

28 U-flares during U, Hα and Hβ observations were registrated. The shapes of the light-curves of the flares confirm the idea about the complexity and multiple behavour of flare phenomenon. The increase of the Hα and Hβ signals was registrated during only most powerfull U-flares. Especially we would like to mention the flare illustrated in Fig.1, which was accompanied with the increase of Hβ signal. This flare is the most powerfull and has a small velocity of increase. It seems, that this event confirms Ambartsumian's idea [2], that the energy of a slow flare is liberated in the dipper layers of the star atmosphere.

Our observational evidence estabiishes, that 12A of FWHM is much wide for the Hα filter to discover Hα flares on the flare stars. The discussion of Hβ observations becomes impossible because of low sensitivity of our equipment.

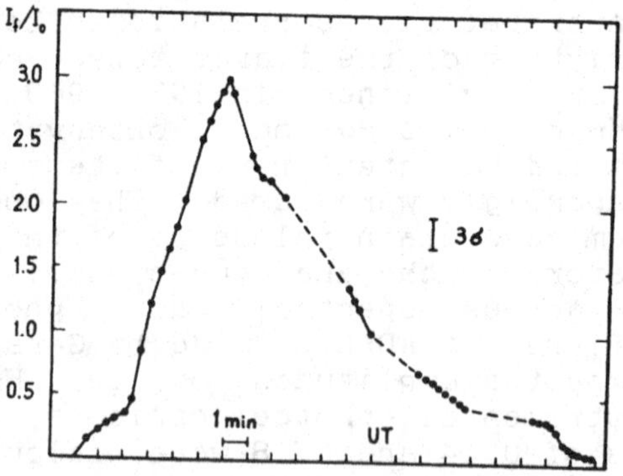

Figure 1. Flare light curve of EV Lac 27-Aug-1985.

REFERENCES

1. Gershberg,R.E. and Chugainov,P.F.(1966) 'Photoelectric and spectrographic observations of flares on AD Leo in 1965',Astron.J.,43, 1168-1178.
2. Ambartsumian,V.A.(1971)'Fuors', Astrofizika,7, 557-572.

RAPID SPIKE FLARES ON AD Leo AND EV Lac

K. P. PANOV, M. S. IVANOVA, A. ANTOV
Department of Astronomy and National Astronomical
Observatory, Bulgarian Academy of Sciences,
Lenin Blvd 72, 1784 Sofia, Bulgaria

ABSTRACT. Photoelectric U - band observations of the flare stars A Leo and EV Lac during the last 9 years obtained at the Bulgarian National Astronomical Observatory revealed 8 rapid spike flares on AD Leo and 9 rapid spike flares on EV Lac which duration is less than 6 seconds. The corresponding total monitoring time is 173.6 hours for AD Leo and 173.3 hours for EV Lac.

1. Introduction

During the last years special attention has been focused on the problem of rapid spike flares. Their time-scales could provide the clues to discriminate between thermal and non-thermal flare theories. Thermal flare models require rise times of flares longer than 0.2 - 0.6 s, and for the star EV Lac this value is 0.3 - 0.4 s (Shvartsman et al. /1988/. Beskin et al. /1988/ and Shvartsman et al. (1988/ carried out high time resolution studies of stellar flares and found no significant flare activity on time-scales 3.10^{-7} s - 10^{-1} s. Their results seem to confirm the thermal origin of stellar flares. However, Tovmassian and Zalinian /1988/ reported a number of rapid spike flares on EV Lac and BY Dra with rise time of about 0.2 s and amplitudes in the U filter up to 5 mag. Gershberg and Petrov /1986/ reported a 3.1 mag spike flare on EV Lac with rise time of less than 0.6 s. Millisecond radio spikes on AD Leo were detected by Lang and Willson /1986/. These results clearly show the need for further flare studies with high time resolution.

2. Observations and discussion

During the last 9 years regular monitoring U - filter observations were carried out at the Rozhen National Astronomical Observatory using the 60 cm telescope and the one-channel photon counting photometer. Observations were taken with an integration time of 1 s. This time resolution can provide information for the occurrence of rapid flares with a duration of several seconds in a longer time period.

In Table 1, spike flares of AD Leo are listed, which have a duration of less than 6 s. The total monitoring time was 173.63 hours.

L. V. Mirzoyan et al. (eds.), Flare Stars in Star Clusters, Associations and the Solar Vicinity, 27–30.

In Table 2, spike flares of EV Lac with duration of less than 6 s are listed. The total monitoring time for EV Las was 173.34 hours. Figs 1 and 2 show some of the light curves. During the 9 years of operation the photometric equipment proved to be rather stable. Therefore we believe these events are not due to instrumental problems. From Table 1, the frequency of rapid flares on AD Leo is 0.05 flares per hour. The cumulative flare frequency for this star included all flare energies greater than log E_U = 30 is 0.66 /Pettersen et al., 1986/. From Table 2, the frequency of rapid flares on EV Lac is 0.05 flares per hour. The cumulative flare frequency included all energies greater than log E_U = 31 for this star is 0.25 /Lacy et al., 1976/. Tipical energies of rapid spike flaves on both stars are log $E_U \approx 29.5$ ergs.

References

Beskin, G.M., Gershberg, R.E., Zhuravkov, A.V., Mitronova, S.N., Neizvestny, S.I., Plakhotnichenko, V.L., Pustilnik, L.A., Chekh, S.A., and Shvartsman, V.F., /1988/, Pisma v A.J. /Soviet/ 14, 156.
Gershberg, R.E., Petrov, P.P., /1986/, Vspihivajustie zvezdi i rodstvenie obekti, Erevan, ed. Acad. Sci. of the Arminian S S R, p. 38.
Lacy, C.H., Moffett, T.J., and Evans, D.S., /1976/, Astrophys. Jour. Suppl. Ser. 30, 85.
Lang, K.R. and Willson, R.F., /1986/, Astrophys. Jour. 305, 363.
Pettersen, B.R., Panov, K.P., Sandmann, W.H., Ivanova, M.S., /1986/, Astron. Astrophys. Suppl. Ser. 66, 235.
Shvartsman, V.F., Beskin, G.M., Gershberg, R.E., Plakhotnichenko, V.L., and Pustilnik, L.A., /1988/, Pisma v A.J. /Soviet/ 14, 233.
Tovmassian, H.M., Zalinian, V.P., /1988/, Astrofizika 28, 131.

TABLE 1. Spike flares on AD Leo

Date	UT	Flare rise time /s/	Amplitude $/I_{o+f} - I_o/I_o$	Noise σ/I_o
1982 Mar 25	$22^h51^m52^s$	1	0.47	0.03
1983 Mar 15	21 27 06	2	0.67	0.06
1983 May 14	20 16 28	1	2.93	0.10
1983 May 14	20 39 54	1	4.07	0.10
1984 May 17	20 38 31	2	1.65	0.06
1986 Jan 14	22 28 53	1	0.79	0.05
1986 Mar 6	21 58 50	4	2.45	0.04
1989 Feb 3	21 42 16	1	0.31	0.04

TABLE 2. Spike flares on EV Lac

Date	UT	Flare rise time /s/	Amplitude $/I_{o+f} - I_o/I_o$	Noise σ/I_o
1983 Oct 6	$21^h05^m08^s$	2	1.08	0.11
1985 Aug 17	00 17 48	1	0.39	0.10
1985 Aug 17	00 17 51	1	0.39	0.10
1985 Aug 17	01 23 31	3	1.00	0.10
	01 23 34	1	1.00	0.10
	01 23 36	1	1.00	0.10
	01 23 38	1	1.00	0.10
1986 Dec 3	17 40 51	1	2.61	0.06
1987 Sep 13	00 12 58	2	4.39	0.12

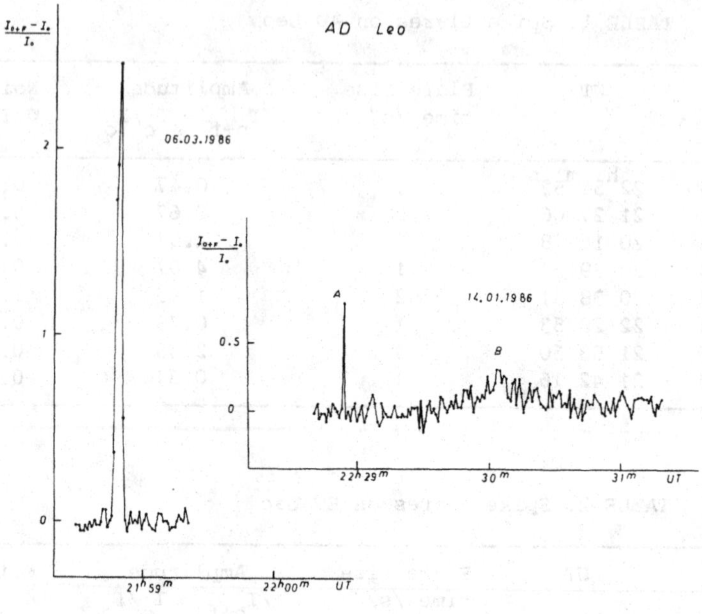

Figure 1

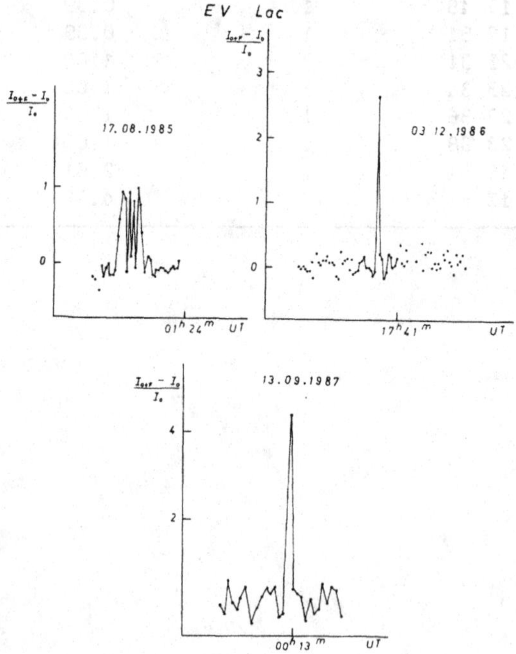

Figure 2

PHOTOELECTRIC OBSERVATIONS OF FLARES ON UV CETI

N.D.MELIKIAN
Byurakan Astrophysical Observatory Armenian Academy of Sciences, USSR
V.S.SHEVCHENKO
Tashkent Astronomical Institute, Uzbek Academy of Sciences, USSR

ABSTRACT. Synchronous observations of the UV Cet flares were carried out in 1987. Two 60-cm telescopes in U-band were used. Accuracy of syncronous registration on two telescopes is 0.001s. 15 flares were registered simultaneously during observations. Single short-time (up to 2s) light increases were observed in quiescent state and flares, which are the result of observation errors. No light increase with duration shorter than 10s was registered simultaneously on two telescopes.

For synchronous photoelectric observations of UV Ceti. Two 60-cm telescopes of Maydanak high-moutnain station of Tashkent Astronomical Institute were used, both in U - passband. The precission of their synchronization was 0.001s. The measurements were made by the photon counting method. The duration of each measurement was 2s,and the time interval between them - 0.4s (see [1]).

15 flares were detected on both telescopes during 12 hours, in October - November, 1987.

Analysis of the observed flare light curves showed that the secondary increase of star brightness on both telescopes was observed only in three cases (NN 6,14,15), duration of which in all cases exceeds 10 seconds. The light curve one of these flares are presented in Fig.1, where I_o is the stellar intensity in normal state, and I_f is the additional intensity.

In four cases "spike-shaped" increase of star brightness were observed, with a duration less or equal to two seconds (Fig.2), but only with one telescope.

Such "spike-shaped" increases of light were detected earlier during the EV Lac flare observations (see.for example, [2,3]). Analysing the AD Leo flare observations Pettersen et al [4] have concluded that no instrumental effects could explain the observed spikes, but cosmic ray

L. V. Mirzoyan et al. (eds.), Flare Stars in Star Clusters, Associations and the Solar Vicinity, 31–32.

impacts at the photomultiplier cathode could also produce such short-lived phenomenon.

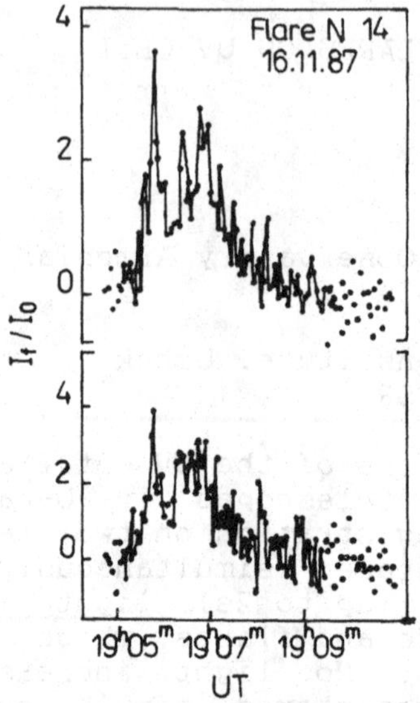

Figure 1.

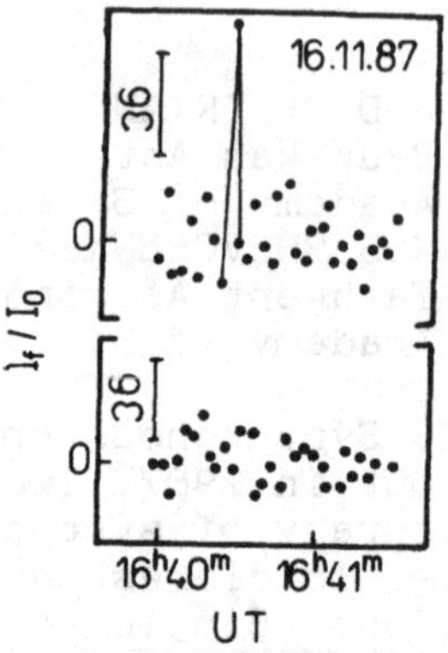

Figure 2.

We think, that such increases of brightness can be explained by the Poissonian distribution of the observational errors. The number of the points on our registrograms is about 20000. In this case the mathematical expectation for the number of "spike-shaped" increases of the brightness with the amplitudes larger or equal to 4σ according to Poissonian distribution must be

$$N_{(\Delta m \geq 4\sigma)} \simeq 3 .$$

This is in good agreement with our results.

REFERENCES

1. Kiljachkov,N.N., Melikian, N.D., Mirzoyan, L.V. and Shevchenko,V.S.(1979) 'Synchronous UBV-Observations of the UV Ceti Flares. I.' Astrofizika, 15, 423-430.
2. Tsvetkov,M.K., Tsvetkova,K.P. and Melikian,N.D. (1986) 'Flare Observations of EV Lac', IBVS, 2954.
3. Zalinian,V.P. and Tovmassian,H.M.(1986) 'Observations of Spiky Flares of BY Dra and EV Lac', IBVS, 2992.
4. Pettersen,B.R., Panov,K.P., Sandman,W.H. and M.S.Ivanova (1986) 'Analysis of the Flare Activity of AD Leo', Astron.Astrophys.Suppl.Ser., 66, 235-253.

TWO-COLOUR OBSERVATIONS OF SPIKES AND FLARES WITH HIGH TIME RESOLUTION

V.P.ZALINIAN, A.A.KARAPETIAN, H.M.TOVMASSIAN
Byurakan Astrophysical Observatory
Armenian Academy of Sciences, USSR

Observations of flare stars with high time resolution permit to register spiky type flares and also to study in detail bright curves of flares. In Byurakan observatory observations of EV Lac have been made with 0.1 s time resolution simultaneouslyin two filters "U" and "B" [1-3]. These observations permited to detect flares of a burst type (Fig.1) and also to reveil multipeak structure of long duration flares. It is necessary to stress that simultaneous observations in two colours increas appresiably the thruustworthiness of the registered spiky flares.

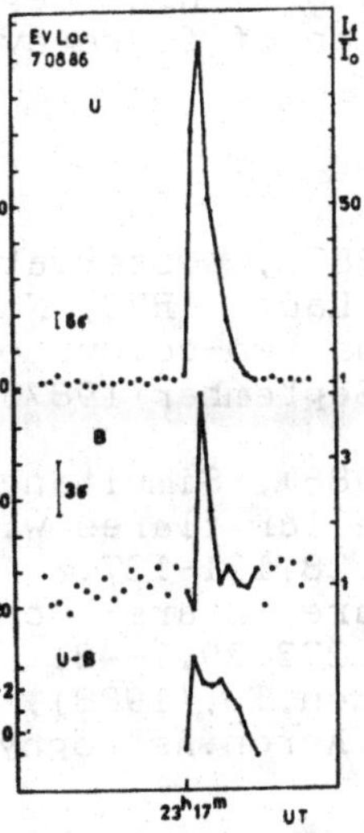

Figure 1.

L. V. Mirzoyan et al. (eds.), Flare Stars in Star Clusters, Associations and the Solar Vicinity, 33–34.

On Fig.2 the dependance of the colour U-B of the star at flare maximum from the amplitude ΔU of the flare is presented. The data on separate defenite peaks on the light curves of a long duration flares agree with the data on single spikes, which permits to suggest that long duration flares may be a superprosition of short flares.

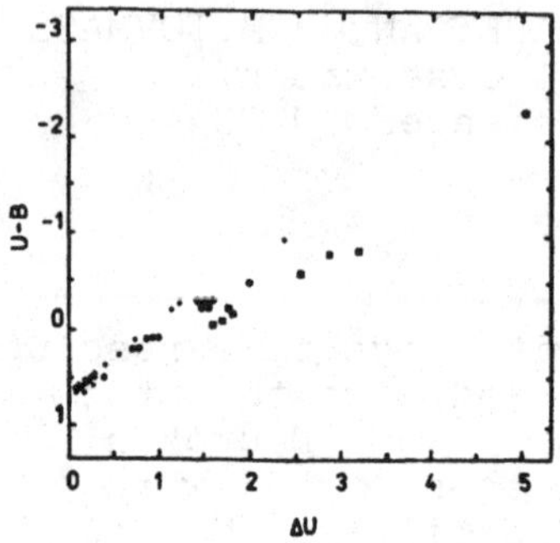

Figure 2. • - [5], ∘ -[4], ▪ - our observations.

The data obtained by observations with 1 s [4] and 4 s [5] time resolutions are also presented in Fig.2. These data show that the slope of the colour - flare amplitude dependance decreases with the decreas of time resolution. The determination of the true slope of this dependance is very important for the interpretation of flare events.

REFERENCES

1. Zalinian,V.P.,Tovmassian,H.M.(1987), 'Observations of spiky flares of BY Dra and EV Lac', IBVS, No.2992.
2. Zalinian,V.P.(1988),'Simultaneous two-colour observations of the flare stars EV Lac in September 1987, IBVS,No.3142.
3. Tovmassian,H.M.,Zalinian,V.P.(1988),'Simultaneous two-colour observations of stellar flares with a high time resolution', Astrofizika,28,131-137.
4. Moffett,T.J.(1974),'UV Ceti flare stars: observational data', Astrophis.J.Suppl.Ser.,273,29,1-42.
5. Panov,K.P.,Piirola,V. and Korhonen,T.,(1988),'Five-colour photometry of EV Lac flares', Asron.Astrophys. Suppl.Ser.,75,53,53-65.

A SHORT-LIVED FLARE OF EV LACERTAE

Zhilyaev B. E., Romanjuk Ya. O., Svyatogorov O. A.
Main Astronomical Observatory of the Ukrainian Academie of Sciencies, Kiev, USSR

Flare events on a time scale of the order of one second were observed on EV Lac by Gershberg and Petrov [1], Zalinian and Tovmassian [2],Tsvetkov, Antov and Tsvetkova [3]. The nature of such phenomena is very mysterious. We hope that monitoring of EV Lac with high time resolution will yield the information about growing and decay times as well as fine structure of the light curve and color one. A short-lived flare of EV Lac was recorded in U and V bands in 1989, September 3 2h 14m 30s UT. The duration of the event was about 150 milliseconds.

The observations were made with the 60-cm telescope at the high altitude Observatory Peak Terskol (3100 m) in the North Caucasus. The high-speed photometer used has two independent synchronous channels with photon counting system. Two operating modes can be used. In the first mode we'll be able to observe two different stars. In the second one a single star will be observed in the two channels by using the beam-splitter. The latter operating mode was used for the photometry of EV Lac in U and V bands simultaneously.

Both search and record of the flare are carried out on-line. Using special routine it is possible to analyse the photon count rate in real time. If some critical level is achived the control programme begins the recording of the light curve. 8192 samples of photon counts may be stored in the computer memory.

Monitoring of EV Lac was carried out with time resolution of 50 milliseconds. The light curve of the event is shown in Fig. 1 in magnitude scale. The background is substracted. The bars show rms caused by Poisson statistics of the star and background intensities together. The mean intensities are equal to 0.4 and 33 counts per 50 ms in U and V bands. In the peaks they are equal to 6.2 and 184 respectivelly.

Some features of the flare are seen:
a) the growing time of event is shorter than time resolution (50 ms); b) the U beginning precedes the V one; c) the duration of flare is about 150 ms;
d) the amplitudes of the flare in U and V bands are equal to $3^{m}.0 \pm 0^{m}.6$ and $1^{m}.9 \pm 0^{m}.2$ respectively.
The full time of our monitoring was 4.5 hours.

L. V. Mirzoyan et al. (eds.), Flare Stars in Star Clusters, Associations and the Solar Vicinity, 35–36.

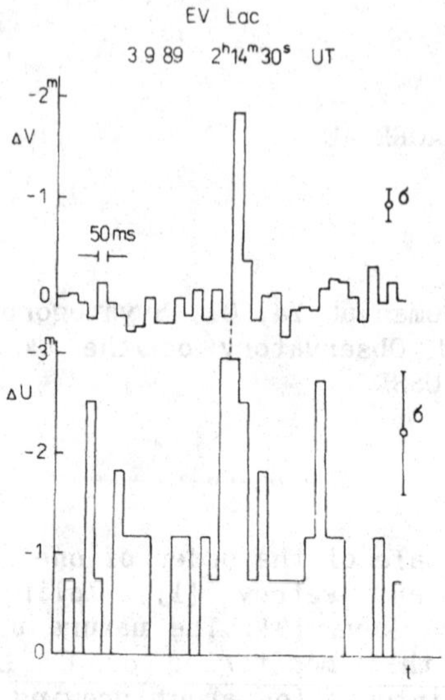

Fig 1

REFERENCES

1. Gershberg R. E. ,Petrov P. P., in 'Flare stars and related objects' Ed. by L. V. Mirzoyan, Yerevan 1986, p. 38-42.
2. Zalinian V. P. ,Tovmassian H. M. , Commun. Konkoly Obs. Hung. Acad. Sci.,1986, No 86, p. 435-436.
3. Tsvetkov M. K. ,Antov A. P.,Tsvetkova A. G. ,ibidem, p. 423-424.

DIAGNOSTIC POSSIBILITIES OF THE WALRAWEN PHOTOMETRIC SYSTEM

A. V. BERDYUGIN
Crimean Astrophysical Observatory
Nauchny
Crimea 334413
USSR

ABSTRACT. Diagnostic possibilities of the Walraven photometric system to analyze the UV Cet type flare stars have been studied. It was shown that the Walrawen system marks out the hydrogen spectrum features better than the standard UBV system. Observations of the UV Cet flare of 1985 December 23, carried out in the Walraven system were analyzed with computed two-colour diagrams, temperature and gaseous density at the maximum of brightness were estimated.

1.1 Introduction

Resently some statements on necessity to use non-traditional photometric systems for the observations of flare stars have been published. Some authores suggested the Walrawen system. Thus, diagnostic possibilities of the Walrawen photometry are to be considered.

1.2 Walrawen middle-band photometric system

The main information about Walrawen system see in [1]. The positions of the passbands have been specially chosen to measure the hydrogen spectrum features. This circumstance is important for observations of flare stars since strong hydrogen emission is observed during flares.

1.3 Two-colour diagrams computations

For two-colour diagram computations two kinds of the hydrogen emission spectra were considered. These spectra correspond to two different models of the flare radiative region.

The first model was computed by using the computer program taken from [2]. The program permits us to calculate the continuum and line emission formed in the non-LTE moving gas. The computations were made for gaseous temperatures T_e=10000-20000K, gaseous densities $n_e = 10^{12} - 10^{14} cm^{-3}$, the star's temperature T_*=3000K, dilution factors W=0.1-0.5. The Lα quantum escape probability β_{12} was varied from 10^{-7} to 1.

The second model of a flare radiative region advanced by Grinin and Sobolev [3] for the explanation of the flare continuum emission at the

L. V. Mirzoyan et al. (eds.), Flare Stars in Star Clusters, Associations and the Solar Vicinity, 37–39.

maximum of brightness was also considered .The gas was adopted to be denser - the atomic concentrations $n_H=10^{15}-10^{17}cm^{-3}$. Under the temperatures 10000 - 20000K such gas becomes partly non-transparent for radiation, and when the thickness of the region increases,the gaseous radiation approaches to the black body one.

For this two emission models colour indexes in the Walrawen system were computed. For determination the scale coefficients which are need to reduce calculated indexes to observational ones, we used the results of the spectrophotometry of α Lyr [4].

1.4 Two-colour diagrams' features

Among different possible kinds of two-colour diagrams, two of them - (V-U,U-W) and (V-L,L-W) are the most interesting. The passbands V,U and W are more sensible for the continuum emission while the passbands B and L are more sensible for the emission of lines. Therefore, the (V-U,U-W) diagram must mark out better the continuum behaviour, but the (V-L,L-W) mark out better the behaviour of lines. These diagrams are given on figure 1 where black body radiation are given by thick line. The curves for first kind of spectra were drawn by narrow solid lines and marked by numbers 1 and 2.The curves for second kind were drawn by dash lines and marked by numbers 3 and 4. The radiation of a hydrogen gas is seen to be separated from the black body radiation on both diagrams.

To illustrate the possibilities of Walraven system to mark out the flare behaviour , position of the real UV Ceti flare observed at 1985, 23 December by De Jager et al [5] was marked on both diagrams. The colour indexes of this flare were corrected for the contribution of the star's radiation. On the figure 1 the full circle designates the first moment of flare, open circle - the moment of the maximum of brightness. The next five minutes of flare development was marked also by crosses.

The flare colours at the maximum of brightness are seen close to coloures of the black body radiation with a temperature near 15000K. In the second model this position corresponds to the gaseous density which can be appreciated as $10^{16}cm^{-3}$.

The displacement of the flare position during its development on the diagrams is seen to be real and this displacement really marks out the changes occurring in the flare spectrum. Such changes in the broad-band UBV system can be smoothed and not displayed,see for example [6].

The continuum of this flare was short-lived and at once after maximum of brightness the line emission given the main contribution to the flare radiation. As it may be seen from behaviour of V-L, during the flare,the strong line emission in the L-band was present.

There is no conformity beetwen computed and observed coloures on both diagrams. Therefore, more detailed computations of the emission spectrum is needed (taking into account non-homogenious structure of the radiative region.) It should be noted that L-band can be influenced by H and K emission of Ca II. If we take this circumstance into account,the curves 1 and 2 on the (V-L,L-W) diagram move leftward and the disagreement between computed and observed coloures increases.

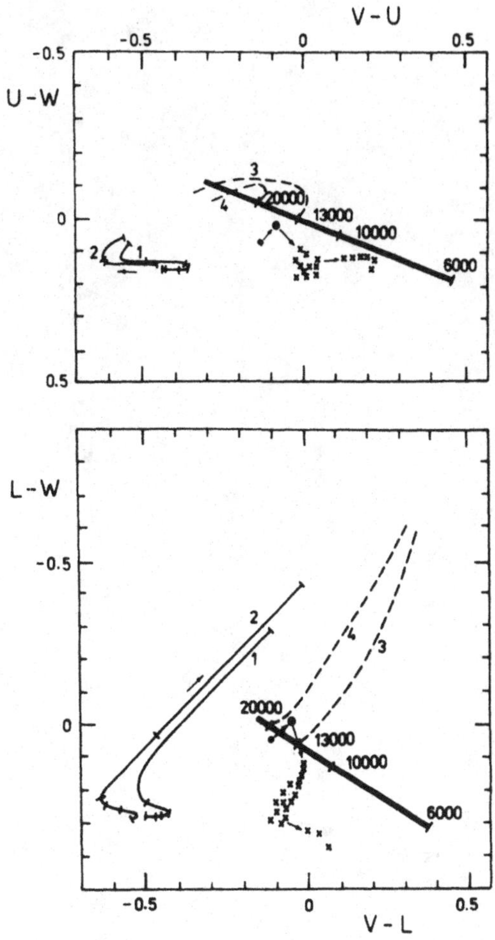

Figure 1. The (V-U,U-W) and (V-L,L-W) diagrams. The thick line is a black body radiation. The curves for the first kind of hydrogen spectrum are marked by 1 and 2. 1 - for $n_e=10^{12}cm^{-3}$, 2 - for $n_e=10^{14}cm^{-3}$. Both curves were calculated for temperature T_e=10000K. The decreasing of β_{12} was shown along both curves. The curves for the second kind of spectrum are marked by 3 and 4. 3 - for T_e=13000K, 4 - for T_e=20000K. Both curves were calculated for $n_H=10^{16}cm^{-3}$. The position of the UV Cet flare of 1985, 23 December was marked on diagrams, too.

1.5 Conclusions

In general, the Walrawen system is more preferable for separation different radiative sources and for marking features of the flares than the traditional UBV. The (V-U,U-W) and (V-L,L-W) diagrams are more useful for diagnostics.

I wish to thank Drs. Gershberg and Grinin for their interest to my work and valuable recommendations.

References

1. Lub, J. and Pel, J. W. (1977) 'Properties of the Walraven VBLUW photometric system' Astron. & Astrophys. 54,137-158
2. Grinin, V. P. and Katysheva, N. A. (1980) 'Relative intensities of the hydrogen lines in the moving mediumes' Bull.Crimean Astrophys. Obs. 62, 66-78
3. Grinin, V. P. and Sobolev, V. V. (1977) 'On the theory of the flare stars' Astrophysics 13, 587- 603
4. Alekseev, N. L., Alekseeva, G. A. et al (1978) 'Results of the astrophysical expedition of the USSR Academy of Sciences in Chile (1971 - 1973)' Izv. Glav. Astron. Obs. Pulkovo 83, 4-147
5. C. de Jager, Heise, J. et al (1989) 'Coordinated observations of a large impulsive flare on UV Ceti' Astron. & Astrophys. 211, 157-172
6. Cristaldi, S. and Longitano, M. (1979) 'The colour behaviour of nine flares of BY Dra, CR Dra, EV Lac and AD Leo variable stars' Astron. & Astrophys. Suppl. 38, 175-180

THE FLARE ACTIVITY OF TWO PECULIAR RED DWARFS

I.V. ILYIN, N.I. SHAKHOVSKAYA
Crimean Astrophysical Observatory
Nauchny, Crimea 334413, USSR

ABSTRACT. Photoelectric observations of flares on the two red dwarfs Gliese 171.2A and CM Dra are reported.

We have carried out photoelectric observations of Gliese 171.2A in U-band on 4 and 5 December 1983 with the 1.25 m telescope of Crimean Astrophysical Observatory using a one-channel photon-counting photometer. During the first night five flares were observed, but the next night the star showed no activity. The photoelectric recordings are shown in Fig. 1.

The short period, low mass, eclipsing binary CM Dra was observed concurrently with IUE on July 5, 1986. Two partially overlapping flares (Fig. 2) were detected in the U-band.

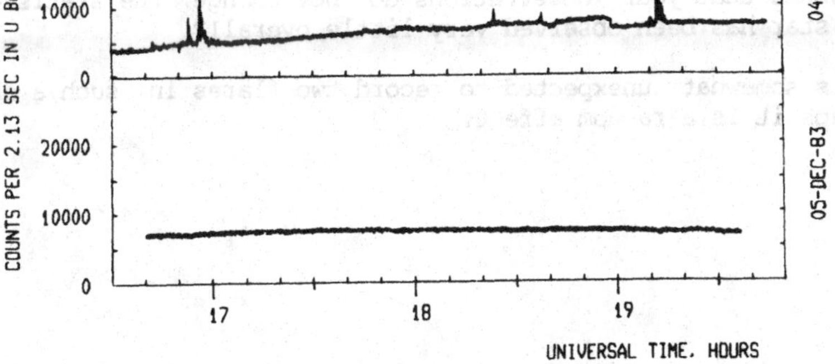

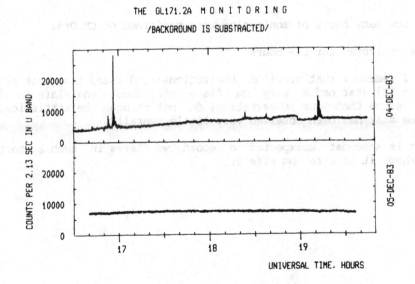

Figure 1. Flare monitoring of Gliese 171.2A.

L. V. Mirzoyan et al. (eds.), Flare Stars in Star Clusters, Associations and the Solar Vicinity, 41–42.

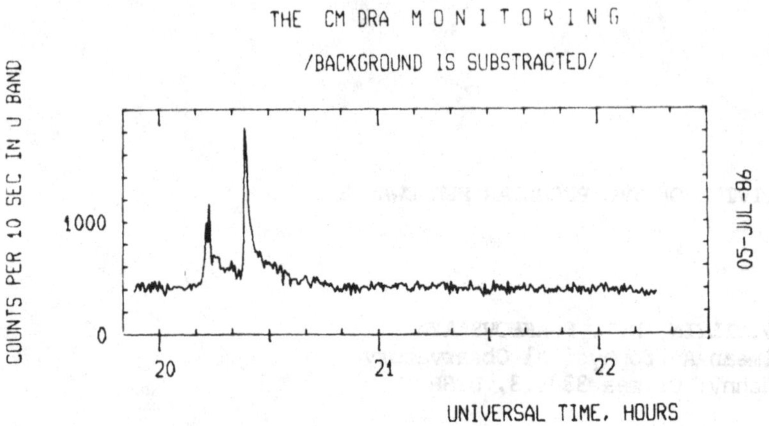

Figure 2. Flare monitoring of CM Dra.

PETTERSEN: From my own (unpublished) observations at McDonald Observatory I can confirm the flare activity of Gliese 171.2A=V833 Tau. My flares are longer than yours and I do not have a feature in my results like the sharp drop in your data during the first night. What is the cause of this unusual feature?

ILYIN: The cause of this feature is not clear. It may be an equipment effect. The photometer system does not allow simultaneous observations of a comparison star so further discussion of this feature cannot be made.

BROMAGE: How many hours of monitoring were performed on CM Dra?

ILYIN: We monitored for 2.5 hours.

BROMAGE: I remember that previous observations indicated that this old (population II) star had a very low flare rate, about one flare in 40 hours. I assume that your observations do not change the statistics, because the star has been observed very little overall?

ILYIN: It is somewhat unexpected to record two flares in such a short time. Perhaps it is a random effect.

STATISTICS OF FLARES OBSERVED FOR UV CETI TYPE STARS YZ CMI, AD LEO, AND EV LAC AT THE OKAYAMA OBSERVATORY

K. ISHIDA
Institute of Astronomy, University of Tokyo
2-21-1 Osawa, Mitaka, Tokyo 181 Japan

ABSTRACT. Photoelectric monitoring of flare stars YZ CMi, AD Leo, and EV Lac has been done at the Okayama Observatory since early 1970s. This is a simultaneous UBV observations with a high time resolution. Some statistics of the flares of the UV Ceti type stars are presented.

1. Introduction

A photoelectric patrol observation of several flare stars was started at the Okayama Observatory in 1968 by the request of the "Working group of UV Ceti type stars" of the IAU Commission No. 27 (Osawa et al. 1968). Dr. Chugainov of the Crimean Astrophysical Observatory, the chairman of the working group, kindly supplied necessary information concerning the international cooperative programme.

At Okayama a photometer was designed to observe rapid phenomena like flare-ups of UV Ceti type stars (Shimizu and Norimoto 1972). A three color filter set on a wheel is rotated continuously with a rate of 20 Hz. The light beam collected by the 91 cm reflector is received through the three color filter set by an EMI 6256B photomultiplier and a synchronous detection is made. The time constant of the system is 0.2 second of time. A three-channel strip chart recorder has been used for the observations with the rolling speed of 20 mm per minute on a paper.

This is a simultaneous UBV photoelectric observation of stellar flares with a time resolution higher than one second. The detection threshold of the flare-ups is $\Delta U = 0.2$ mag and depends on the brightness of the quiescent intensity of the stars and sky conditions. It seems to me that it is statistically complete only for those brighter than $\Delta U = 1.0$ mag. Several UV Ceti type stars have been monitored with the photometer since early 1970s by Ichimura et al. (1974, 1978, 1981, 1986), and still on-going at present in 1989. In this brief report, some statistics of the flares observed during 1974 to 1988 in YZ CMi, AD Leo, and EV Lac are presented.

Name	U-B	B-V	V	Mv	Sp.	Period	Remarks
YZ CMi	1.01	1.61	11.20	12.3	M4.5Ve	2.781d	0.12mag var
AD Leo	1.08	1.54	9.43	11.0	M4.5Ve	-	UC? 26.5y
EV Lac AB	0.75	1.37	10.05	11.5	M4.5Ve	4.378d	UC 45y 2au

L. V. Mirzoyan et al. (eds.), Flare Stars in Star Clusters, Associations and the Solar Vicinity, 43–47.

2. Number frequency of flares in YZ CMi, AD Leo, and EV Lac

The total watch time is 147 hours for YZ CMi, 299 hours for AD Leo, and 229 hours for EV Lac, distributing over 15 years from 1974 to 1988. The number of flares recorded is 68 in YZ CMi, 69 in AD Leo, and 46 in EV Lac, which are large enough number to see average number frequency in unit time. The average number frequency of flares in an hour is 0.46 for YZ CMi, 0.23 for AD Leo, and 0.20 for EV Lac. It is interesting to see if there is any time variation of the average number frequency year to year. Dividing the whole period of 15 years into four terms, the average frequencies per an hour appear 0.29, 0.22, 0.25, 0.15 for AD Leo, and 0.17, 0.25, 0.19, 0.20 for EV Lac. For YZ CMi, only the last three terms have large enough length of watching time to see the average number frequencies, which are 0.70, 0.28, and 0.43. Difference of the number frequencies between term to term is noticed for YZ CMi, and barely noticed for AD Leo and EV Lac.

Name	year	t(h)	n	n/t(h)	(U-B)+ rms	(B-V)+ rms	n'
YZ CMi	77-80	49	34	0.70	-1.09±0.22	0.18±0.22	23
	82-84	51	14	0.28	-1.23±0.27	0.22±0.21	10
	85-88	47	20	0.43	-1.16±0.15	0.20±0.20	13
	(77-88	147	68	0.46)			
AD Leo	74-76	65	19	0.29	-1.11±0.19	0.17±0.32	14
	77-80	51	11	0.22	-1.02±0.11	0.16±0.13	4
	81-84	116	29	0.25	-1.19±0.21	0.24±0.35	19
	85-88	67	10	0.15	-0.59±0.58	0.47±0.28	8
	(74-88	299	69	0.23)			
EV Lac	74	29	5	0.17	-1.05±0.23	0.16±0.15	4
	79	32	8	0.25	-1.07±0.12	0.00±0.11	5
	80-84	119	23	0.19	-0.64±0.20	0.38±0.39	17
	85-88	49	10	0.20	-0.78±0.35	0.34±0.21	9
	(74-88	229	46	0.20)			

3. Average color of flares in YZ CMi, AD Leo, and EV Lac

Two colors of (U-B)+ and (B-V)+ of the flares at peak brightness within a few seconds of time from the maximum appeared on the paper chart were read out. The colors of the 126 flares at the peak brightness apparently show similar values in most of the terms, that are (U-B)+ = -0.99±0.21 and (B-V)+ = 0.23±0.23 for the average of three stars in the four terms and standard deviation from the average. The similarity is even more clear if three discrepant values are ignored, that are (U-B)+ = -1.12±0.07 and (B-V)+ = 0.17±0.07 for 92 flares. The present values can be compared with (U-B)+ = -0.88±0.31 (n=153) and (B-V)+ = 0.34±0.44 (n=77) of Moffett (1974) and (U-B)+ = -0.64±0.61 (n=45) and (B-V)+ = -0.02±0.72 (n=12) of Mirzoyan et al. (1983).

4. Energy release by flare events

The intensity at flare maximum is expressed in terms of the quiescent star's brightness in the U band for stars on which flares were detected, because the U band shows the most prominent flares. Average brightening of 183 flares is ΔU = 1.36$\pm$0.28 mag at flare maximum for the present three UV Ceti type stars in the four terms which covers 1974-1988. The intensity range is large from 0.46 mag to 3.30 mag in average.

The more reliable flare parameter to show the energy released in flare event is the equivalent duration of the flare. The equivalent duration is the energy of flare event expressed in time of the relative intensity of a flare event normalized in unit of the quiescent star intensity in the U band and the relative intensity is a function of time as observed during flare event. The total equivalent durations over the length of the watching time is a rough relative measure of the energy released by flare events. It looks far from a constant value and changes in nearly two orders of magnitude from one term to the other. Cumulative number count of the equivalent durations with frequency of occurrence is expressed by a power law function with an exponent around -0.4 over 3 to 4 orders of energy range of equivalent duration, when summed up for 1974 to 1988. It shows that a few of the biggest flare events are so efficient that one flare event could release more than half or even 95% of the whole amount of energy released by many flare events in a term of a few years.

Acknowledgments. I thank Profs. Y. Yamashita, Director of the Okayama Observatory, and K. Ichimura for showing me material in advance of publication and discussion to prepare this manuscript.

References

Ichimura, K. and Shimizu, Y. (1978) Tokyo Astron. Bull., Ser. 2, No. 255.
Ichimura, K. and Shimizu, Y. (1981) Tokyo Astron. Bull., Ser. 2, No. 264.
Ichimura, K. and Shimizu, Y. (1986) Tokyo Astron. Bull., Ser. 2, No. 276.
Ichimura, K., Shimizu, Y. and Okida, K. (1974) Tokyo Astron. Bull., Ser. 2, No. 230.
Mirzoyan, L. V., Chavushyan, O. S., Melikyan, N. D., Natsvlishvili, R. Sh., Ambaryan, V. V., and Brutyan, G. A. (1983) Astrofizika, **19**, 725.
Moffett, T. J. (1974) Astrophys. J. Suppl. Ser., **29**, 1.
Osawa, K., Ichimura, K., Noguchi, T. and Watanabe, E. (1968) Tokyo Astron. Bull., Ser. 2, No. 180.
Shimizu, Y. and Norimoto, Y. (1972) Tokyo Astron. Obs. Report, 15, 820 (in Japanese)

Name	year	ΔU	rms	n	minΔU	maxΔU
YZ CMi	77-80	1.55±0.85		34	0.43	3.50
	82-84	1.44±0.54		14	0.55	2.86
	86-88	1.77±0.69		20	0.53	2.97
				(68)		
AD Leo	74-76	1.54±1.12		19	0.34	5.01
	77-80	1.02±0.69		11	0.37	2.30
	81-84	1.17±0.91		29	0.23	>5.17
	85-88	1.12±0.71		10	0.53	>3.14
				(69)		
EV Lac	74	1.45±0.55		5	0.81	2.24
	79	1.80±0.90		8	0.62	3.50
	80-84	1.08±0.60		23	0.24	2.67
	85-86	1.00±0.70		10	0.46	2.90
				(46)		
		1.36±0.28		183	0.46	3.30

Name	year	ΣP/t(h)	No. of Event	P of 1st biggest event	(% in ΣP)	P of 2nd biggest event	(% in ΣP)
YZ CMi	77-80	1.13	31	>3120min	(>94)	86.8min	(2)
	82-84	0.020	10	>13	(>22)	12.5	(20)
	86-87	0.14	18	259	(65)	>27.7	(>7)
			(59)				
AD Leo	74-76	0.35	19	>1318	(>95)	39.4	(1)
	77-80	0.072	11	190	(87)	17.8	(8)
	81-84	0.064	27	>350	(>79)	29.9	(6)
	85-88	0.020	10	>65	(>82)	5.2	(6)
			(67)				
EV Lac	74	0.0032	5	2.7	(49)	0.8	(15)
	79	0.038	8	37	(50)	14.0	(19)
	80-84	0.013	19	51	(55)	25.7	(28)
	85-86	0.018	9	46	(86)	>2.6	(>3)
			(41)				

Color band	Filter	λ(A)	Δλ(A)
U	UV-DIC 1mm	3700	430
B	VV-42 + UV39	4300	890
V	VO-51	5500	950

with an EMI 6256B photomultiplier

PANOV: Did you find any changes of the flare colours during the development of a flare?

ISHIDA: I did not study that problem.

FLARE ACTIVITY LEVELS FOR FULLY CONVECTIVE RED DWARFS

B.R. PETTERSEN
Institute of Theoretical Astrophysics, University of Oslo,
P.O. Box 1029 Blindern, N-0315 Oslo 3, Norway.

ABSTRACT. Current stellar models predict fully convective structures for masses less than about 0.3 Mo. We have monitored low luminosity dMe and dM stars for flare activity with the 2.1 m telescope at McDonald Observatory. The dMe stars produced 2-5 detectable U-filter flares per hour while none were seen in the dM stars. Typically, 15-30% of the U-filter flux received from the dMe stars is due to flares. The flare activity level of dMe stars (on an absolute scale) decreases as one approaches the lower end of the main sequence. Non-linear frequency distributions of flare energy are seen in some binary flare stars, suggesting that both stars contribute to the ensemble of flares observed from the binary. Upper limits for the dM stars suggest activity levels about one thousand times smaller than in the dMe stars.

1. INTRODUCTION

Flare stars on the lower main sequence may be separated into at least two types of structures. The brightest stars (0.3<M/Mo<1) have radiative cores and convective envelopes, while those fainter than log L/Lo=-2.2 (corresponding to M/Mo<0.3) are fully convective from the center to the photosphere. It is unclear what consequences, if any, this structural difference may have on generation of magnetic fields and the phenomena of stellar activity. We have monitored some low luminosity dMe and dM stars for flare activity, and report here some results of that observing program.

2. OBSERVATIONS

Photometry with a time resolution of 1-5 seconds was done through a U-filter, using the 2.1 m telescope at McDonald Observatory. Data were taken on 18 nights distributed over 6 runs between 19 November 1979 and 10 December 1983. Table 1 contains a summary of the results for each star. For the dMe stars we detected 2-5 flares per hour, while none were seen in the dM stars.

L. V. Mirzoyan et al. (eds.), Flare Stars in Star Clusters, Associations and the Solar Vicinity, 49–52.

Table 1. Flare summary for some fully convective red dwarfs.

Star	Giclas	M_V(A)	M_V(B)	U(AB)	σ/I_o	T(h)	N	L_f(U)
FL Vir	G12-43	15.05	15.05	15.52	0.08	20.31	57	8.0 10^{26}
V780 Tau	G100-28	15.12	16.12	17.9	0.3	13.11	34	6.0 10^{26}
UV Cet	G272-61	15.45	15.95	14.90	0.07	38.55	154	3.1 10^{26}
EI Cnc	G9-38	15.47	16.33	16.7	0.08	4.47	24	3.5 10^{26}
DX Cnc	G51-15	17.00		19.0	0.3	6.70	17	3.6 10^{25}
HH And	G171-10	14.80		15.7	0.06	1.33	0	$< 4\ 10^{23}$
	G157-77	15.40		18.0	0.4	0.83	0	$< 6\ 10^{23}$

Notes: N is the number of flares detected during the monitoring time T. L_f(U) is the time averaged flare luminosity, in units of erg/s.

3. ANALYSIS AND DISCUSSION

We have integrated the flare lightcurves and calibrated the results to determine the amount of energy emitted over the U-filter bandpass for each flare. The total energy for all the observed flares on a star was then divided by the monitoring time to determine the time averaged flare luminosity in erg/s, given in the last column of Table 1, where the upper part contains data for dMe stars and the lower part for dM. The discriminating feature is the hydrogen Balmer lines in emission. The numbers correspond to 15-30% of the U-filter flux being due to flares. The largest proportions are seen at the lower end of the main sequence. On an absolute scale the flare luminosity of the dMe stars in Table 1 decreases as one moves towards fainter stars. This trend is confirmed if a wider range of stellar luminosity is considered (Fig. 1). It is also evident from Table 1 and Fig. 1 that dM stars are about 1000 times less active than their dMe counterparts, but we caution that this number is based on very limited sets of observations.

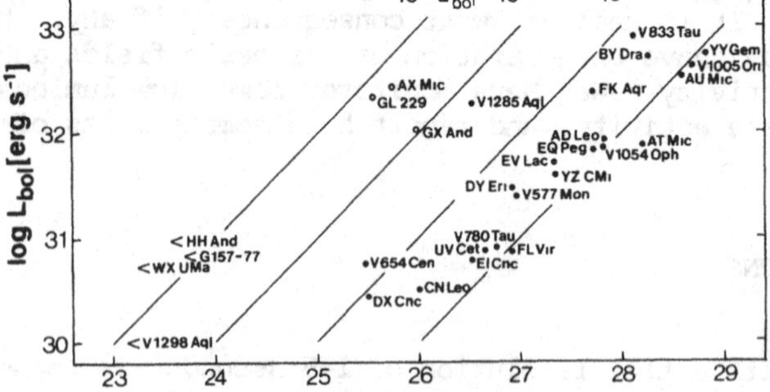

Figure 1. The flare activity for dM (open circles) and dMe (filled circles) stars. Upper limits are shown as <.

In Fig. 2 we have plotted the cumulative frequency distributions of flare energy for the dMe stars in Table 1. Each is the result of two or more observing runs and therefore represents some average in time. We note that the binaries have non-linear distributions, perhaps reflecting that both stars have contributed to the flare sample in accordance with their intrinsic luminosity. The slopes above 10**30 erg are close to unity, while for smaller energies they are 2-3 times steeper. This causes considerable uncertainties when trying to determine if all flare stars are characterized by the same slope in this diagram.

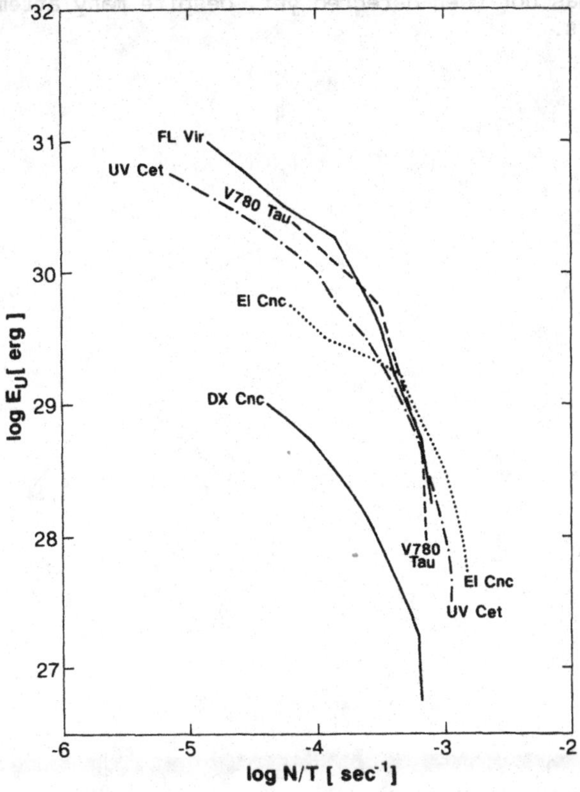

Figure 2. Cumulative flare energy distributions for four late type flaring binaries and one single flare star (DX Cnc).

MONTMERLE: Is flare activity linked with binarity? In other words, is anything known about the possible connection between rotation and activity in dMe and dM stars?

PETTERSEN: Many flare stars are members of binary systems. Indeed, all but one in the sample discussed here are binaries. G 51-15 is believed to be single, but should be investigated in more detail. For the fully convective flare stars we do not have rotation information. Very high spectral resolution is difficult for these very faint stars (to determine v sin i), and photometric periodic variability due to spots on a rotating star has not been detected yet, despite many attempts on fully convective stars.

SOME PHYSICAL CONSEQUENCES FROM THE FLARE STATISTICS OF THE UV CET TYPE STARS IN THE SOLAR NEIGHBOURHOOD

N.I. SHAKHOVSKAYA
Crimean Astrophysical Observatory
Nauchny, Crimea 334413, USSR.

ABSTRACT. The observational data permit us to establish statistical correlations between different parameters of stellar flare activity and the characteristics of quiet stars. We discusss physical consequences following from these relationships.

This paper has been published in Solar Physics (1989) 121, 375-386.

L. V. Mirzoyan et al. (eds.), Flare Stars in Star Clusters, Associations and the Solar Vicinity, 53.

SEARCH FOR SLOW LIGHT VARIATIONS OF RED-DWARF STARS

N.I.BONDAR'
Crimean Astrophysical Observatory
Nauchny Crimea, 334413, USSR

ABSTRACT. The results of a search for slow light variations of 13 K2e-M4.5e dwarf stars using the Sternberg Institute plate collection are presented.These variations are discovered for PZ Mon and V577 Mon,confirmed for BY Dra and BD+26^{0}730 and suspected for V639 Her and V654 Her.

1. Introduction

There are many evidences from many years of spectral and photometric observations as well as photographic measurements of brightness that for late-type active dwarfs that there exist cycles in periodic light variations. The plate collection of the Sternberg Institute was founded in 1895 and there is a possibility to search for cycles of long-term brightness variations about 20% of the known flare stars.For several of them these variations would sugnificantly exceed the measurements' accuracy. In this paper the results of measurements for 13 stars selected from [1,2] are given. For three stars-BY Dra,BD+26^{0}730 and V577 Mon-the long-term variations are already known [2-5].

2. Measurements and results

The brightness value for studied stars was measured relatively to the comparison stars and to estimate the measurement's accuracy the control late-type stars were used.When the magnitude for these stars was not known,they were measured relatively to photometric standards at the plates obtained on the 40-cm astrograph whose photographic magnitudes are close to B band with the accuracy of about 0.05-0.2 mag. The list of investigated stars is exhibited in Table 1.,and in Figures 1,2 the black circles indicate their light curves, and plotted standard deviations.The light variability for all control stars is random and its amplitude did not exceed 0.1 mag.It is not more than standard deviation equal to 0.05-0.08 mag. Figure 1. shows the yearly mean magnitudes for

L. V. Mirzoyan et al. (eds.), Flare Stars in Star Clusters, Associations and the Solar Vicinity, 55–58.

stars with variability exceeding 0.3 mag.For BY Dra and BD+26^{0}730 these data were supplemented from [2,3] - crosses and dashes, and from [4] - the pluses.The magnitudes for V577 Mon in [5] were measured relatively to comparison stars, which were observed photoelectrically. These magnitudes are brighter than those from the Sternberg Institute plates.By two coinsided dates was found the magnitude's discrepancy about of 0.25 mag. Thus, in Figure 1. for V577 Mon open circles show the yearly mean data from the table in [5],and the triangles -data from the Sternberg Institute plates with the reduction at the indicated correction.

Table 1.

Dwarf stars	Spectral type	Time scale (years)	Plate's number	Amplitude for mean magnitude (Δ B)	Observed cycle (years)
BD+26^{0}730	K5e	1905-1988	370	0.6	60
V577 Mon	M4.5-7e	1909-1988	91	0.4-0.5	40
PZ Mon	K2e	1896-1988	130	$\gtrsim$1	50
BF CVn	M1.5e	1899-1989	104	0.2	-
DT Vir	M1.5-2e	1911-1967	66	0.2	-
EQ Vir	K5e	1914-1988	53	0.1	-
V654 Her	K4V	1907-1988	295	0.3	?
V639 Her	M4e	1910-1983	191	0.5	?
BY Dra	M0e	1904-1989	131	0.3-0.4	60
V1216 Sgr	M4.5e	1960-1988	237	0.13	-
V1285 Aql	M2-3e	1899-1987	108	0.2-0.3	-
V1396 Cyg	M3e	1896-1982	184	0.15	-
DO Cep	M4.5e	1899-1989	152	0.25	-

3. Discussion

The sufficiently good agreement for all known data for BY Dra and BD+26°730 confirmed the cyclical light variations for them with time scale of about 50-60 years and amplitude of 0.3-0.6 mag.A possibility,that BY Dra has the shorter activity cycle-about 10-15 years still exists. V577 Mon shows the light variability with an amplitude of 0.4-0.5 mag that occures approximately once in 40 years. PZ Mon has the light amplitude of about 1 mag.There are two minima for the average magnitudes with possible interval of 50 years. The time scale for the data of V639 Her and V654 Her is exhibited in several series of mean light and one may only notice its changes by 0.3-0.5 mag. Figure 2. shows the light curves for stars whose brightness amplitude does not exceed 0.2 mag.Probably,for several of them it can be explained by the absence of data for individual years.

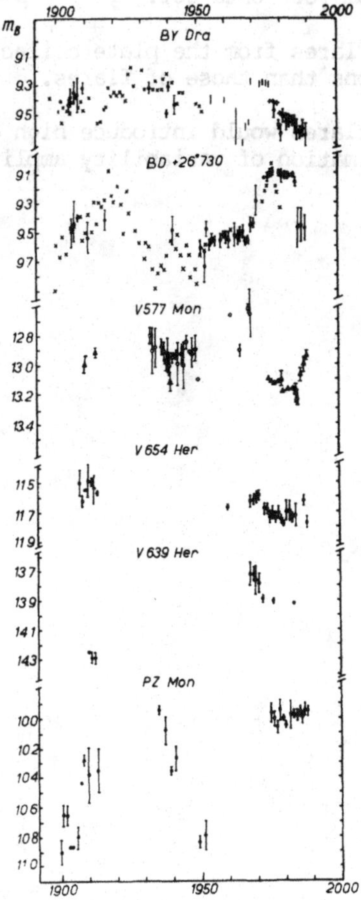

Fig.1.-The cyclical light variations

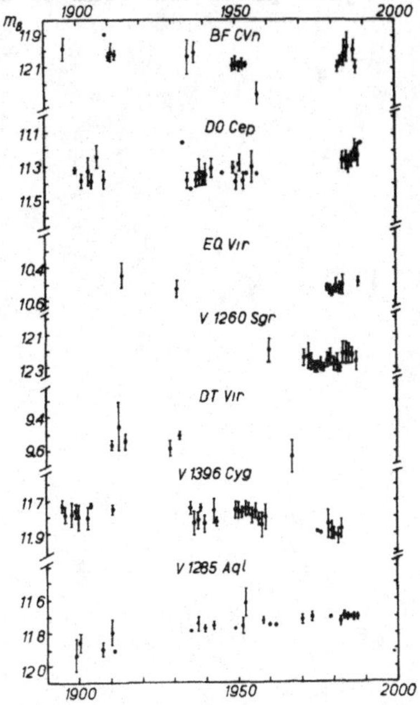

Fig.2.-The long-term light variations with small amplitude.

5. References

1. Gershberg, R.E. (1978) Low Mass Flare Stars, Nauka, Moscow.
2. Phillips,M.J.and Hartmann,L.(1978) 'Long-term variability of dMe stars', Asrophys.Journal 224,162-189.
3. Hartmann,L.,Boop,B.W.,Dussault,M.,Noah,P.V.,and Klimke,A.(1981) 'Evidence for a starspot cycle on BD+26^{0}730 ',Astrophys.Journal 249,662-665.
4. Mavridis,L.N.,Asteriadis, G.,Mahmoud,F.M.(1982) 'Long-term changes of the flare stars EV Las and BY Dra', Contributions Department Geodetic Astronomy University Thessaloniki No.41,253-276.
5. Corben,F.M.,Harding,G.A.,and Thomas,Y.Z.R.(1970) 'Search for new flare stars', Monthly Notes Astronomical Society Southern Africa 29, 57-77.

PETTERSEN: How did you remove flares from your data set?

BONDAR: It was not possible to exclude flares from the plate collection data. We have looked for longer variations than those of flares.

PETTERSEN: My point is that individual flares would introduce high data points in your plots and lead to overestimation of variability amplitudes

DISTRIBUTION OF RED DWARFS IN GENERAL GALACTIC FIELD ON THE BASIS OF STELLAR STAR STATISTICS

A.T.GARIBJANIAN, V.V.HAMBARIAN, L.V.MIRZOYAN, A.L.MIRZOYAN
Byurakan Astrophysical Observatory
Armenian Academy of Sciences, USSR

ABSTRACT. The mathematical expectation for detection of stellar flare on UV Ceti type stars in the solar vicinity during photographic patrol observations with 40" Shcmidt camera of the Byurakan Astrophysical Observatory is estimated. We use the luminosity function of the flaring red dwarfs the assume a uniform distribution in the general galactic field. Comparison with the results of photographic patrol supports this assumption. The numbers and total mass of the flare and non-flare red dwarf stars in the Galaxy for the uniform distribution are determined. They are not in contradiction with Oort's estimate of total mass of red dwarfs.

1. INTRODUCTION

Existing observational data demonstrate beyond doubt the physical similarity of the UV Ceti type flare stars in solar vicinity and flare stars in star clusters and associations [1]. On the other hand the flare stars are in one of the earliest stages of red dwarf evolution [2-4].

Proceeding from these facts two possible hypotheses on the origin of the UV Ceti stars in solar vicinity were suggested, one by Ambartsumian [5] (they were formed together and now are members of a system) and the other by Herbig [6] (they are remnants of already disintegrated systems and constitute a population of general star field in Galaxy).

In the present report an evidence in favour of the last hypothesis is presented.

L. V. Mirzoyan et al. (eds.), Flare Stars in Star Clusters, Associations and the Solar Vicinity, 59–62.

2.MATHEMATICAL EXPECTATION OF FLARES CAUSED BY THE FLARE ACTIVITY OF THE UV CETI STARS

The mathematical expectation for detecting N flares caused by the flare activity of the UV Ceti stars during photographic observations assuming the uniform distribution of these stars in Galaxy is equal to [7]

$$N=\frac{1}{3}\omega t\sum_{M} D^{*}(M)\sum_{\Delta m} R^{3}(M,\Delta m)\nu(M,\Delta m) \qquad (1),$$

where ω is the field of the used telescope, t- total time of observations, $D^{*}(M)$- space density of the UV Ceti stars having absolute magnitude M,R(M,Δm)- distance up to which a stellar flare with an amplitude Δm can be detected on these stars, and $\nu(M,\Delta m)$- mean frequency of such flares. In the formula (1) summing up is taken for all values of M and Δm.

As the number of known UV Ceti stars in solar vicinity is not enough to determine directly the luminosity function $D^{*}(M)$ it was calculated using the luminosity function D(M) of emission line red dwarfs which have a common nature with the UV Ceti stars [1,2]. The function D(M) itself was determined on the basis of Gliese's catalogue of nearby stars [8,9]. The observation selection in this catalogue was taken into account by using the limiting distances for which its data are complete [10,11].

The function $\nu(M,\Delta m)$ was determined from Moffett's photoelectric observations of the UV Ceti type flare stars [12].This function is essentially different from that for photoelectric observations became of large differences in integrations times. Not all flares registered by Moffett [12] could be detected by photographic method.

3. RESULTS OF CALCULATIONS

The estimate of the time T during which one can expect a single photographic flare caused by flare activity of the UV Ceti stars with 40" Schmidt camera of the Byurakan Astrophysical Observatory are presented in Table 1.

TABLE 1. Estimates of the time T

Passband	Exposure(min)	T(hours)
B	5	94
"	10	161
U	5	11
"	10	27

In agreement with these estimates during about 190 hours of photographic observations carried out with the Byurakan 40" Schmidt camera only one flare was detected in general galactic field [13].

4. CONCLUSION

The comparison of mathematical expectations of stellar flare detections on UV Ceti stars with results of photographic observations made by the Byurakan 40" Schmidt camera in general galactic star field allows to conclude that the UV Ceti stars have uniform distribution in Galaxy.Most probably they are remnants of already desintegrated star clusters and associations and constitute now population of galactic star field.

For uniform distribution the numbers of flare and non-flare red dwarf stars in Galaxy are 4.2x10**9 and 2.1x10**10 respectively [14]. Their total mass ~ 10**9 solar mass is not in contradiction with Oort's estimate of the total mass of red dwarf stars in Galaxy [15].

REFERENCES

1. Mirzoyan,L.V. and Hambarian,V.V.(1988) 'Statistical study of flare stars I. UV Ceti type stars of solar vicinity and flare stars in star clusters and associations', Astrofizika, 28, 375-389.
2. Haro,G. and Chavira,E. (1966) 'Flare stars in stellar aggregates of different ages', Vistas in Astronomy, 8, 89-107.
3. Ambartsumian,V.A. and Mirzoyan,L.V.(1971) 'Flare stars', in: New Directions and New Frontiers in Variable Star Research, Veroff. Bamberg, 9,Nr.100, 98-108.
4. Mirzoyan,L.V.(1984) 'Flare stars', Vistas in Astronomy, 27, 77-107.
5. Ambartsumian,V.A.(1957) 'Introductionary remarks',in M.A.Arakelian (ed.),Non-Stable Stars, Armenian Acad. Sci., Yerevan, pp. 9-16.
6. Herbig,G.H.(1962), Symposium of Stellar Evolution, J.Sahade (ed.), Astron. Obs. Nat. Univ. La Plata, p. 45.
7. Mirzoyan,L.V.,Hambarian,V.V.,Garibjanian,A.T. and Mirzoyan,A.L. (1988) 'Statistical study of flare stars II.On the origin of the UV Ceti type stars',Astrofizika, 29, 44-57.
8. Gliese,W.(1969) 'Catalogue of nearby stars',Veroff.Astron. Rechen Inst.Heidelberg,No.22,1-116.
9. Gliese,W. and Jahreiss,H.(1979) 'Nearby star data published 1969-1978', Astron. Astrophys. Suppl. Series, 38,423-448.

10. Arakelian,M.A.(1970) 'On the statistics of flare stars in the solar vicinity', Comm. Byurakan Obs., 41, 56-67.
11. Wielen.R., Jahreiss,H.and Kruger,R.(1983) 'The determination of the luminosity function of nearby stars',in A.G.D.Philip and A.R.Upgren (eds.) The Nearby Stars and the Stellar Luminosities Function,Davis Press New York, pp.163-169.
12. Moffett,T.J.(1974) 'UV Ceti flare stars: observational data', Astrophys. J. Suppl. Series, 29, 1-42.
13. Chavushian,H.S.(1979) A Study of Flare in the Pleiades Agregate Region, Byurakan Astrophys. Obs..
14. Mirzoyan,L.V.,Hambarian,V.V.,Garibjanian,A.T.,and Mirzoyan,A.L. (1988) 'Statistical study of flare stars III. Flare stars in the general galactic star field', Astrofizika, 29, 531-540.
15. Oort,J.H.(1957) 'Dynamics and evolution of the Galaxy in so far as relevant to the problem of the populations', in D.J.K.O'Connell, S.J.(ed.),Stellar Populations, Vatican Obs.,pp.418-433.

NEW OBSERVATIONS OF FLARE STARS IN THE PLEIADES BY THE METHOD OF STELLAR TRACKS

H.S.CHAVUSHIAN, G.H.BROUTIAN
Byurakan Astrophysical Observatory
Armenian Academy of Sciences, USSR

ABSTRACT. In 1984-1986 new observations of flare stars in the Pleiades region were carried out by the method of stellar tracks with 40" Schmidt telescope of the Byurakan observatory. The effective observational time is 100 hours in U-rays (ORWO ZU-21 plates combined with UG-1 filter were used). The stellar tracks have 1-1.5 cm length, the limiting stellar magnitude on the plates is 13^m-14^m, time resolution is 5^s-20^s.17 flares of 13 stars were detected. Six of these stars were detected as flare stars for the first time. According to the Hertzsprung catalogue all these stars are the Pleiades cluster members. The flare amplitudes reach up to 1.5-3.0 magnitudes, the duration vary from 2.5 to 10.5 seconds.

The observations of flare stars in stellar aggregates up to now were carried out generally by multiexposure method. As a result of this method no information about rapid and faint flares of relatively bright stars was possible to get. Meanwhile, the problem of bright flare stars and rapid flares in aggregates is actual. The method of star tracks, proposed and tested by one of the authors [1], allows to get relatively large information on rapid and faint flares of bright stars by photographic observations on wide-angle telescopes. This method is something betweenthe electro-photometric and photographic multiexposure methods. It makes possible to combine the advantages of the method of survey photographic research of flare stars in aggregates by wide-angle cameras, with those of the detailed research of each star, which is typical for the electrophotometric method.

The observations were carried out in 1984-1986 with 40" Schmidt camera of Byurakan astrophysical observatory. In order to get stellar tracks, the correction mechanism of this telescope was used. It was intended for comets'

L. V. Mirzoyan et al. (eds.), Flare Stars in Star Clusters, Associations and the Solar Vicinity, 63–66.

observations and allows to move the telescope in any direction during the exposure (with a wide range of velocities). The exposure time and the velocity of the telescope were choosen so, that it was possible to detect the flares of relatively bright stars of 13-14 magnitudes. For our telescope the optimum was 1-1.5 cm length of tracks got during 30 min. The photometry of flare tracks is done as in the case of ordinary photographic observations.

During 100 hours 17 flares of 13 stars were detected. Six of them are new flare stars. The data about these flares are presented in Table 1, where in corresponding colomns are given: Haro's [2] catalogue number, Hertzsprung [3] number, U-magnitude at minimum [4], ΔU amplitude of the flare, n- number of all known flares before our observations, t- total duration of the flare, τ- duration of the flash phase, and the date. As an illustration some of the light curves of the detected flares are represented in Fig.1.

For a comparison with the usual multiexposure method it can be noted that according to Haro's catalogue [2] 144 flares of bright stars ($U_{min} < 15^m$) were detected during 3600 hours of multi exposure observations, whereas we have detected 17 such flares during 100 hours of observations by tracks method. It means that for bright stars this method is much more effective (more than 4 times) than the usual one.

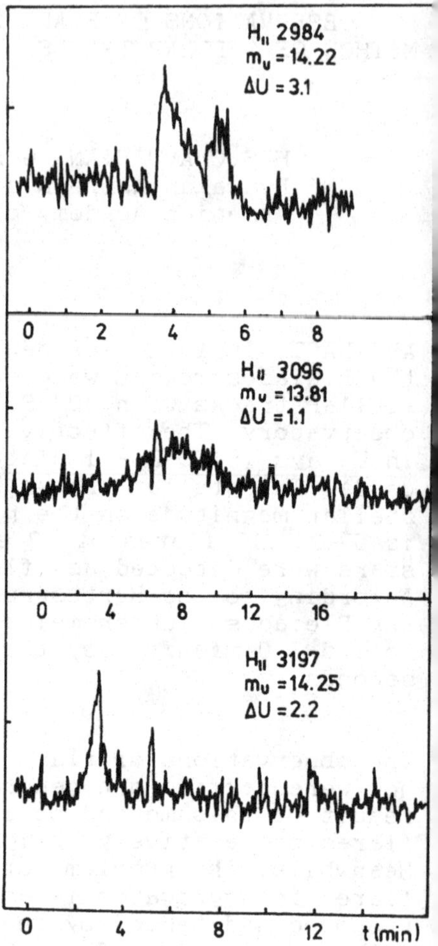

Figure 1. Light curves of some flares obtained by the method of stellar tracks.

Table 1. The flare stars observed by tracks method.

N	HII	U_{min}	ΔU	n	t(min)	τ(min)	Date
85	5	13.46	1.5	2	9	0.9	27.11,84
	738	14.14	1.7	0	7.3	1.0	24,10,84
	1332	14.30	1.4	0	7.7	<0.1	28,10,86
278	1454	14.88	2.3	1	4.4	0.4	23.10,84
290	1553	13.25	1.8	2	7.2	1.5	28,10,86*
"	"	"	1.3	"	3.5	<0.2	"
"	"	"	2.0	"	5.9	0.1	"
331	1883	14.37	1.9	3	8.9	<2.0	03,09,84
"	"	"	1.6	"	3.3	0.2	25,10,84
348	2034	14.20	1.8	1	10.6	2.0	27,08,84
369	2244	14.33	1.2	1	6.8	1.6	23,10,84
	2381	14.50	1.7	0	3.1	1.1	24,10.84
	2870	14.41	1.5	0	2.4	0.3	15,12.84
	2984	14.22	1.4	0	5.1	0.1	15,12,84
	"	"	3.1	"	>1.8	0.1	27,10,84
	3096	13.81	1.1	0	5.7	1.3	24,10,84
448	3197	14.25	2.2	4	5.9	0.3	26,11,84

*) The three mentioned flares of this star have taken place during the same night, the second flare-up was two hours after the first one, and the third was 0.5 hour after the second.

REFERENCES

1. Chavushian.H.S.,(1986) 'Observations of flare stars in the Pleiades by the stellar tracks method', in L.V.Mirzoyan (ed.) Flare stars and related objects, Academy of Sciences of Armenia, Yerevan, pp. 125-129.
2. Haro, G., Chavira E., and Gonzales, G.(1982) 'A catalogue and identification charts of the Pleiades flare stars', Bol. Inst. Tonantzintla 3, No.1, 3-68.
3. Hertzsprung, E. et al (1947) 'Catalogue de 3259 etoiles dans les Pleiades', Ann. Leiden Obs., 19, No. IA.
4. Johnson, H.L. and Mitchell, R.I. (1958) 'The color-magnitude diagram of the Pleiades cluster. II', Astrophys. J. 128, 31-40.

BLAAUW: In your search did you investigate all stars occurring on the plate or did you limit it to the known members of the Pleiades?

BROUTIAN: We have investigated all the stars that occur on our plates, but may be a little more attention was payed to the stars of the Hertz-sprung catalogue.

BLAAUW: The fact that all flare stars found are cluster members has impostant implications for the study of the dimensions of the Pleiades. Since for the outer regions there are very few early epoch plates, p.m. cannot be very accurately determined here; so flare stars may be the best approach to this problem.

POSSIBLE NEW, BRIGHT FLARE STAR IN THE PLEIADES REGION

JANOS KELEMEN
Konkoly Observatory, Budapest
Box.67
1525 Budapest
Hungary

ABSTRACT. With the aid of the star track method we discovered a possible new, relatively bright, flare star in the Pleiades region.

During the last few years in the mountain station of the Konkoly Observatory we introduced the star track method (which originally was developed in Bjurakan, Chavushian (1986)) for photographic flare star observations.

To get the star image to move at a constant velocity on the photographic plate, we changed the original clock frequency of the clockwork of our 60/90 cm Schmidt to a lower one. Technically it means that we replaced the original 1 kHz frequency standard with a new quartz controlled device in which we could change the output frequency in 0.1 Hz steps. Corresponding to the speed of the movement the limiting magnitude on our U plates (with KODAK 103aO emulsions + UG2 filter) is between 14 - 14.5 magnitude.

For the evaluation of the observed material we used a modified Zeiss Microdensitometer (Kelemen, (1989)). Using this instrument we were able to control the scanning procedure and the data handling with the aid of a microcomputer.

Unfortunately we had to preselect the plate material, because this method is very sensitive to the short timescale weather instabilities, which was smoothed out in the previous multiexposure technique.

As a preliminary result we found a previously unknown i.e. not catalogued (Haro et al. (1982)) star, which showed an approximately 30 minute long brightening (see Chart 1). The brightening had a flare like light curve.

The date of the event:	02.10.1986
time of the maximum :	01:46 a.m. (UT)
m(U) quiescent :	13.1 mg(U)
amplitude :	0.8 - 0.9 mg(U)

The light curve resembles and is characteristic of the so

L. V. Mirzoyan et al. (eds.), Flare Stars in Star Clusters, Associations and the Solar Vicinity, 67–70.

called slow flares (Fig. 1.). The rising time was about 10 minutes long and after the maximum the fading went on a moderate rate.

Because the exposure was ended before the star reached its quiescent brightness the decay time must be at least 30 min. The estimated quiescent U magnitude of the star is 13.9 mg(U) and the amplitude of the brightening was 0.8 - 0.9 magnitude. For comparison Fig. 2. shows a parallel record of a 13.9 mg(U) star. (Star "T" in the photoelectric sequence (based on Johnson et al. (1958)) from Henden et al. (1982))

REFERENCES

H. S. Chavushian (1986)
'Observations of Flare Stars in the Pleiades by the Stellar Tracks Method'
'Flare stars and realated objects', symposium
ed. L. V. Mirzoyan
Publ. House of the Acad. of Sci., Yerevan

G. Haro, E. Chavira, and G. Gonzalez (1982)
'A catalog and identification charts of the Pleiades flare stars'
Boletin del Inst. de Tonantzinta, No.1, Vol.3, july 1982

A. A. Henden and R. H. Kaitchuk (1982)
'Astronomical Photometry'
Van Nostrand

Johnson, H. L., and Mitchell, R. I. (1958)
'The color-magnitude diagram of the Pleiades cluster II.'
Ap.J. 128, 31., 1958

J. Kelemen (1989)
'Photographic photometry with a modified Zeiss Microdensitometer'
contr. paper, 1989 Sonneberg

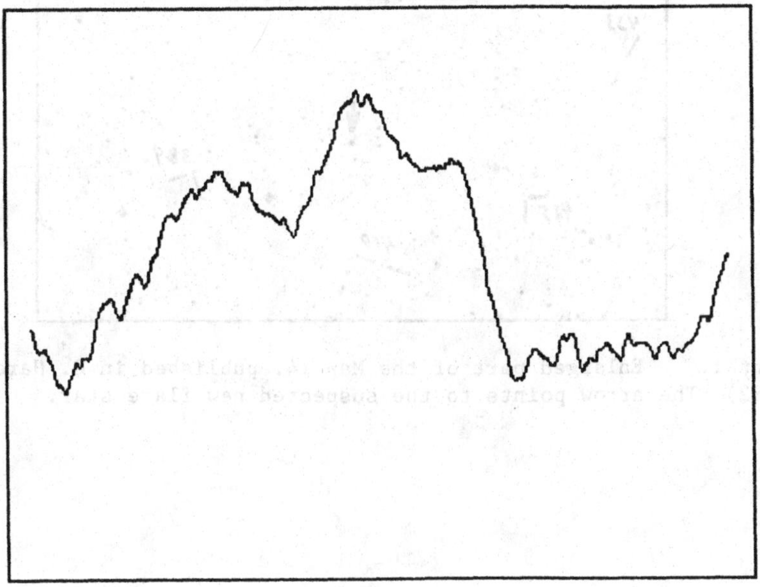

Figure 1. Moving average plot of the flare up.

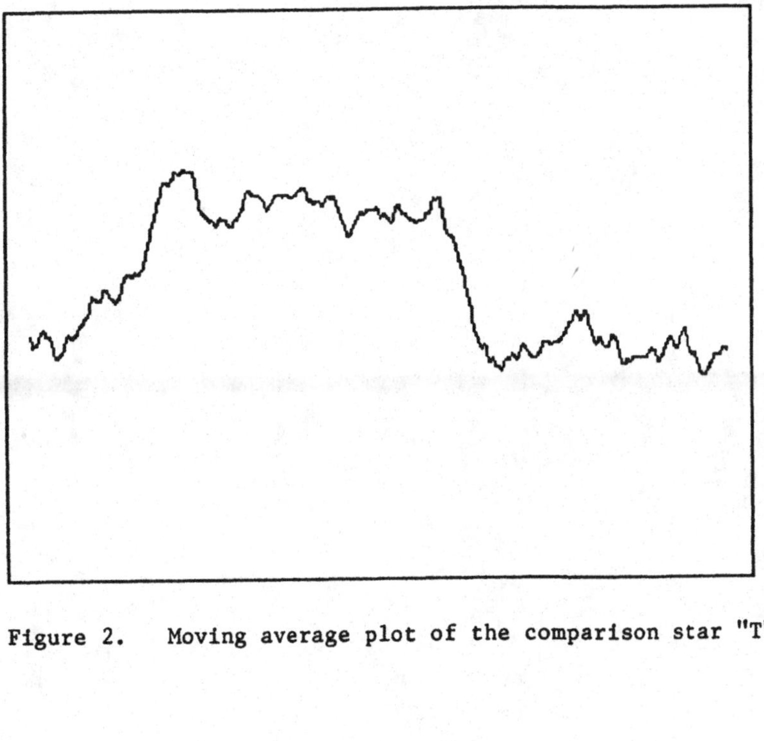

Figure 2. Moving average plot of the comparison star "T".

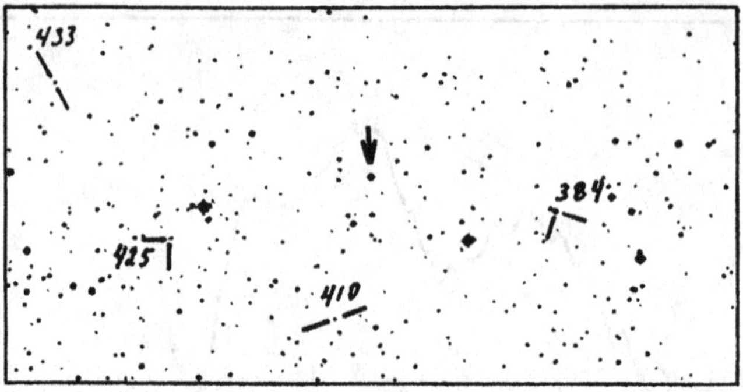

Chart 1. Enlarged part of the Map 14. published in G. Haro et al. (1982). The arrow points to the suspected new flare star.

DATA FILTERING IN STATISTICAL STUDIES OF FLARES RECORDED IN SKY FIELDS OF STELLAR AGGREGATES - DEMONSTRATED WITH THE EXAMPLE OF THE PLEIADES

G. SZÉCSÉNYI-NAGY
Eötvös University of Budapest
Department of Astronomy
H- 1083 Kun B. tér 2.
Hungary

ABSTRACT. Photometric data on flares of cluster flare stars published by different observers were - as a rule - collected during decades by different instruments and amidst various conditions. The uncritical use of these data in global statistical studies is usually unfair since such a collection is an indefinite mixture of high and low quality results. In order to make these investigations more reliable even the best and most complete photometric catalogues have to be filtered.

The first step of the procedure described in the paper is rejecting all flare events having too small amplitudes - seeing that they are underrepresented in the catalogue as opposed to medium and large amplitude flares. Then the photographic coverage of the aggregate has to be checked and all events observed in partially covered regions have to be excluded. Next any non-randomness in the time distribution of flares of each active object is to be unveiled and flares of the doubtful stars have to be omitted. Finally some objects are to be rejected for their individual characteristics.

The usefulness of the method is demonstrated with the example of the Pleiades, one of the richest open clusters known. More than 3000 hours of effective observing time has been devoted to flare photometry in fields centered upon Eta Tauri and the outcome of the programme the CPFS - the most complete catalogue of flare stars of the region - contains data of 519 flare stars and their 1532 flare events. Earlier versions of this catalogue or subsets of its data have been used to estimate the probable number of flare stars in the Pleiades region. Results of these attempts range from 600 to 2700 partly owing to the unreliability and incompatibility of the data sets. The method presented here is able to give much more reliable estimates but of course in the limited largest common field (LCF) only. As a result of the filtering the LCF contains 77% of the objects and 67% of the flares listed in the CPFS. Time sequence analysis of the 1026 flare events observed on 402 flare stars and statistical modelling of the phenomena suggested in an earlier paper fit each other almost perfectly.

The conclusion of the study is that the most likely number of flare stars in the reach of the telescopes used in this programme is between 520 and 600 in the LCF of the Eta Tauri fields. Their average flare frequency is about 1/2400h and the CPFS lists 70% of these objects. In order to be able to compare the Pleiades field flare stars and those of other

L. V. Mirzoyan et al. (eds.), Flare Stars in Star Clusters, Associations and the Solar Vicinity, 71–76.

aggregates the method of filtering will be carried out throughout the analysis of the most extended data sets available.

1. PHOTOMETRY OF CLUSTER FLARE STARS

1.1. The Instruments

Since stellar aggregates containing a large number of known flare stars are in the 0.1-1.0 kpc distance range their flare-active members are very seldom brighter than m(U)=15. For that reason and the very low frequency of large amplitude flares wide field photographic cameras (mainly Schmidt telescopes) are used in flare patrol observations. In order to secure the most complete time-coverage the method of multiple exposures is employed as a rule. More details of the photographic flare photometry were discussed by Szécsényi-Nagy (1985).

1.2. Evaluation of the Plate Material and Publication of the Results

Photometric plates are visually checked and brightness changes noticed on them are estimated or measured by iris- or microphotometers. Observers usually report the position and minimum brightness of the active stars found and the amplitudes of their flares.

For a reliable statistical investigation of flare stars and flares the unambigous identification of these objects is essential. As data on flare stars and their flares in a field of about 30-35 square degrees containing the open cluster M45 were compiled and published with detailed maps of the region (The Catalogue of Pleiades Flare Stars, the CPFS, Haro et al. 1982) and since this is the home ground of many Northern hemisphere observers, it was obvious to choose the Pleiades to demonstrate the use of the method.

The CPFS lists photometric data of 1532 outbursts with amplitudes ranging from 0.3 to 9.0 magnitudes.

1.3. Reliability of the Published Flare Amplitudes

The number distribution of flare ups versus amplitude was computed with a resolution of 0.1 magnitude - the greatest one allowed by the published results (Fig.1.). The plot demonstrates plainly that observers tend to report - of course inadvertently - much more integer and half-integer (e.g. 1, 1.5, 2, 2.5, etc.) amplitude outbursts than their expected number might be. This serious selection effect is even more obvious if one compares the content of the integer (or half-integer) bins and that of their immediate neighbours. There are always local minima at these latter amplitudes.

Another characteristic feature of the amplitude distribution is that the frequency of outbursts decreases with increasing amplitudes. It is self-evident and well known from flare statistics of nearby flare stars or that of the sun that low amplitude flares are much more numerous than the high amplitude ones. It is clear from the plot of Fig.1. that the flare sample listed in the CPFS must be incomplete in the lowest amplitude bins. At least one third of the outbursts is missing in the 0.5 ± 0.2 mag range.

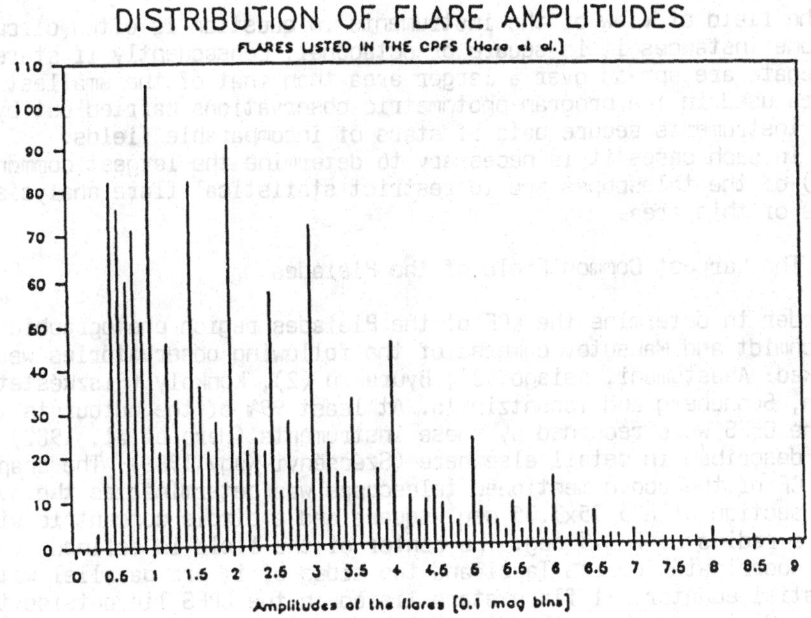

Figure 1. Low amplitude flares of stars of the Eta Tauri fields are considerably underrepresented in the CPFS because most of the observers are unable to detect these outbursts or quite simply ignore them.

The number of 0.8 mag outbursts is significantly higher than that of 0.7 magnitude ones but there is only a slight difference between the multitude of 0.6 and 0.5 mag flares. This fact and a comparison between the ratios of the length of 0.6/0.5 and that of 1.6/1.5 magnitude bars have to convince everybody of the incompleteness of the flare surveys at these low amplitudes. Although the 0.7 mag bin is surely much more complete than the half magnitude one for maximum safety all 237 flares of the 0.0-0.7 mag amplitude range were rejected. This move reduced the total number of accepted outbursts to 1295.

2. THE LARGEST COMMON FIELDS OF THE AGGREGATES

2.1. Uneven Photometric Coverage of the Fields

Photographic flare patrol observations have been carried out in at least eight observatories using nine or more different instruments. Unfortunately not only the optical system the aperture and the focal length of the cameras are different but their fields of view too (Szécsényi-Nagy 1983). In order to illustrate this problem it is enough to compare the larger Schmidt telescope of the Byurakan AO and the smaller one of the Asiago AO. While plates taken with the former cover an area of 14.14 square degrees those taken with the latter show more than 2.5 times larger field (approximately 37 square degrees). What is more the shape of the field of view of the instruments in question is often circular but in some instances it is square or octagonal. Consequently if stars of an aggregate are spread over a larger area than that of the smallest field camera used in the program photometric observations carried out by different instruments secure data of stars of incomparable fields.

In such cases it is necessary to determine the largest common field (LCF) of the telescopes and to restrict statistical flare analysis to stars of this area.

2.2. The Largest Common Field of the Pleiades

In order to determine the LCF of the Pleiades region photographic fields of Schmidt and Maksutov cameras of the following observatories were cross-checked: Abastumani, Asiago (2), Byurakan (2), Konkoly (Piszkéstető), Rojen, Sonneberg and Tonantzintla. At least 99% of the outbursts listed in the CPFS were recorded by these instruments (Haro et al. 1982) which were described in detail elsewhere (Szécsényi-Nagy 1986). The shape of the LCF of the above-mentioned telescopes was determined as the overlapping section of a 3.75x3.75 deg^2 square and a circle concentric with that with a radius of 2.375 deg. The center of the field is defined by the actual coordinates of Eta Tauri and two sides of it are parallel with the celestial equator. 71 flare stars listed in the CPFS lie outside the LCF of the Pleiades. These objects (which produced 121 observed outbursts) were wholly listed by Szécsényi-Nagy (1986). As a result of this spatial filtering the total number of accepted flare events dropped to 1174 and that of the accepted flare stars to 414.

3. TIME-DISTRIBUTION ANOMALIES OF FLARES OF THE MOST ACTIVE STARS

3.1. The Assumed Random Distribution of Stellar Flares

Flare occurrence studies of stars of the solar vicinity have concluded that outbursts of these objects are randomly distributed in time and consequently can be modelled by Poisson-distributions (Oskanyan and Terebizh 1971). For the definite physical similarity between cluster and solar vicinity flare stars the flare contribution of stars of aggregates has been analysed as the outcome of a group of random generators. Estimates of the flare star content of the Pleiades region based on these assumptions have ranged from 600 to 2700 (Szécsényi-Nagy 1986a and refs. cited therein).

3.2. Deviations From Random Distribution

For three flare stars of the solar neighbourhood Pazzani and Rodonó (1981) demonstrated that their flares were not randomly distributed in time but showed a tendency towards grouping. Some other flare stars seemed to change their mean flare frequency from year to year. Rodonó (1975) published markedly different values for the star II Tauri (1972:0.25h^{-1} and 1973:1.56h^{-1}) which very probably shows cyclic activity changes on the long run too (Szécsényi-Nagy 1986b and 1990). When the star is very active it produces 2-4 times more photographically detectable high amplitude flares than when it is quiet or least active. This particular object contributed 8% of all of the flares of the Eta Tauri fields consequently its variable activity should not be ignored.

3.3. Changing Activity of Stars of the Eta Tauri Fields

II Tau (also HII 2411 or CPFS 377) with its 120 flares listed in the CPFS seems to be the most powerful source of outbursts in the Eta Tauri fields. Despite its substantial contribution to the CPFS and for its above-mentioned cyclic activity variations (and its Hyades membership) it is better to reject all of its flares in a statistical study of the Pleiades region. In order to achieve the most reliable results flare distribution of all flare stars of the CPFS which produced at least 10 outbursts were checked. The method introduced and described by the author (Szécsényi-Nagy 1986 and 1989) led to the conclusion that besides II Tau 11 other flare stars of the CPFS (Nos 91, 143, 150, 194, 211, 256, 298, 354, 453, 454 and 477) must be omitted. This last move reduced the number of the stars involved to 402 and that of the flares to 1026. The effect of the filtering upon the relative contribution of flare stars of different activity to the total is illustrated by Fig.2. It demonstrates that changes due to filtering are really significant in the higher frequency bins and that implication of those objects which contributed 7 or more outbursts into the statistics may conclude to divergent solutions. To avoid this difficult problem models attributing lower statistical weights to these stars should be used in preference to any other.

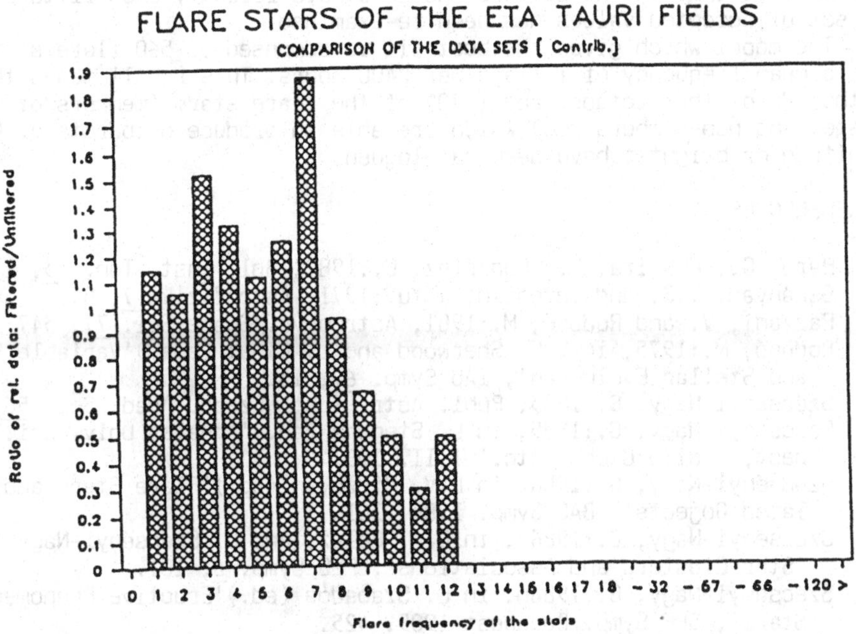

Figure 2. Filtering significantly diminished the relative flare contribution of the higher frequency (9...12) bins to the total of the M45 field.

4. THE MOST LIKELY NUMBER OF FLARE STARS IN STELLAR AGGREGATES

4.1. Restrictions on Statistical Estimations of Flare Star Populations

Statistical studies of flare stars and their flares in fields of aggregates have been aimed at the estimation of the amount of these objects. Unfortunately most of the papers dealing with this topic ignore some very important limiting conditions. First of all simple statistical methods are unable to distinguish between cluster member and field flare stars. Consequently each result refers to both kind of objects. Secondly the estimates refer to the given region of the sky and to those latent objects only which are able to produce outbursts with high enough amplitudes to be detectable by the photometric techniques employed. Flare stars of the foreground may therefore significantly differ from those of the background of the aggregate.

4.2. New Values for the LCF of the Eta Tauri Fields (Conclusions)

It is well known that members of the Pleiades cluster are spread over a field of 50 square degrees or more. The present work was aimed at the analysis of the LCF defined in paragraph 2.2. which covers only 1/4 of the Pleiades region. Previous studies in this field which were based on a new global statistical procedure (Szécsényi-Nagy 1986) concluded that the most likely number of flare stars which are able to produce 0.5 mag. amplitude or brighter outbursts is about 650 and that their average flare frequency is 1/3000h. As data of the CPFS have recently been filtered the new set of accepted values has been re-examined.

The model which gives the best fit is composed of 560 flare stars with a mean frequency of 1 flare per 2400 hours. This result means that in the LCF of the Pleiades about 70% of the flare stars (members of the cluster and non-members too) which are able to produce outbursts of 0.8 magnitude or brighter have been catalogued.

5. REFERENCES

Haro, G., Chavira, E., González, G.:1982, Bol. Inst. Ton., 3, 1, 3.

Oskanyan, V.S. and Terebizh, V.Yu.:1971, Astrofizika 7, 83.

Pazzani, V. and Rodonó, M.:1981, Astrophys. Space Sci. 77, 347.

Rodonó, M.:1975, in V.E. Sherwood and L. Plaut (eds.)'Variable Stars and Stellar Evolution', IAU Symp. 67, 15.

Szécsényi-Nagy, G.:1983, Publ. Astr. Inst. Czech. Acad. Sci. 56, 219.

Szécsényi-Nagy, G.:1985, in L. Stegena (ed.)'Annales Univ. Sci. Budapest, Sectio Geoph. etc.' I-II. 308.

Szécsényi-Nagy, G.:1986, in L.V. Mirzoyan (ed.)'Flare Stars and Related Objects', BAO Symp. 1984, 101.

Szécsényi-Nagy, G.:1986a, in B.A. Balázs and G. Szécsényi-Nagy (eds.) 'Star Clusters and Associations', PES Symp. 6, 101.

Szécsényi-Nagy, G.:1986b, in L. Szabados (ed.)'Eruptive Phenomena in Stars', SPE Symp. Budapest 1985, 425.

Szécsényi-Nagy, G.:1989, in B.M. Haisch and M. Rodonó (eds.)'Solar and Stellar Flares', IAU Coll. 104, 143.

Szécsényi-Nagy, G.:1990, in M. Vázquez (ed.)'New Windows to the Universe', IAU XI ERAM (in press).

ON TWO POSSIBLE GROUPS OF FLARE STARS IN PLEIADES

A.L.MIRZOYAN, M.A.MNATSAKANIAN
Byurakan Astrophysical Observatory
Armenian Academy of Sciences, USSR

ABSTRACT. The correlation between the magnitude of the proper motion and the mean frequency of flares for flare stars in the Pleiades cluster region is discussed.The magority of them are physical members of this star cluster.It is shown that there exist probably two groups of flare stars which differ each other by their parameters, in particular by the magnitude of proper motions. A possible explanation of this phenomenon is suggested.

The photographic observations of stellar flares in the Pleiades region show that for the flare stars having small proper motions the dispersion of the mean flare frequencies is very large meanwhile for the flare stars having comparatively large proper motions the mean frequencies of the flares are always small.

This is well seen in Fig.1 which represents the dependence between the quantity of the obsereved photographic flares k according to the chronological catalogue of the authors (see [1]) and, proper motions μ of the stars according to Jones [2] for the Pleiades flare stars. The total number of flare stars with the known proper motions in this system used by usis 230. They are comparatively bright stars.

One can notice on Fig. 1 that near the magnitude μ=2.5 really a gap is observed.

Jones [2] thinks that the flare stars of the Pleiades region having large proper motions are the stars of the general galactic field. However as was suspected [3,4] long time ago and has been shown by one of the authors [5] in the Pleiades the further from the centre of the cluster the smaller the number of flare stars irrespective of the magnitude of the proper motion i. e.practically all of these stars are members of the cluster.The number of flare stars belonging to the general galactic star field among them

L. V. Mirzoyan et al. (eds.), Flare Stars in Star Clusters, Associations and the Solar Vicinity, 77–80.

does'nt exceed 10% [6].

On the Hertzsprung-Rassell diagram of flare stars of the Pleiades cluster the mean frequency of the flares increases towards its right upper corner. This correlation can be presented by the formula

$$g = \frac{6}{7}(B-V) - \frac{1}{7}V + \text{Const} ,$$

where g is certain indicator of flare activity. Here B and V are stellar magnitudes.

In Fig.2 the correlations (k,g) and (μ,g) are presented,using the mean data on the considered parameters for five concentric regions having a thickness $0^{o}.5$ around the centre of the cluster.

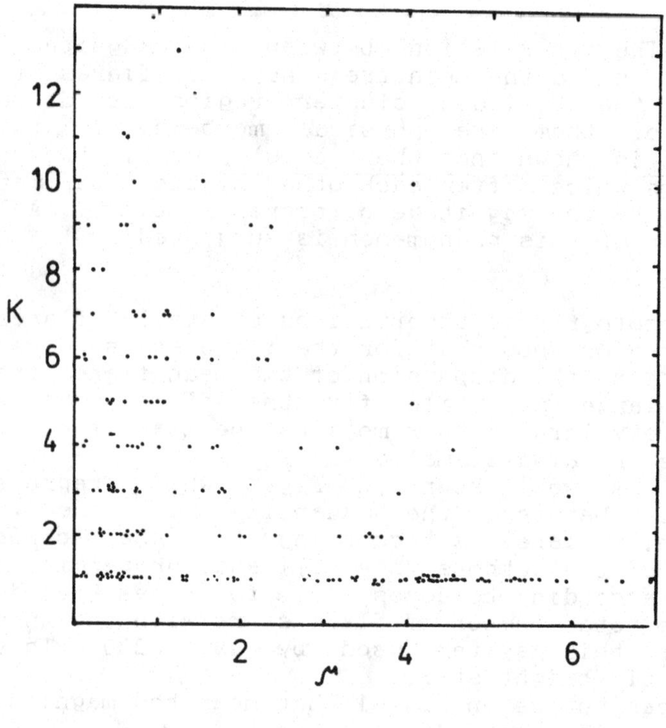

Figure 1. Observational data for flare stars on μ and k

A similar picture is observed on the correlation (g,Sp) constructed by us using the data on spectra of the Pleiades flare stars from the paper [7].

The correlation (k,g) shows that the number of observed flares k in average is increasing with the increasing of the magnitude g.

It means that really the magnitude g can be considered

as an indicator of the flare activity level of flare stars.

The second coreelation (μ,g) on contrary gives an evidence of the decreasig of the proper motion with the magnitude g .i.e. with the mean frequency of flares.

On the other hand ,both correlations (k,g) and (μ,g) show clearly that all flare stars in the Pleiades region can be devided into two separate groups which differ each other by their parameters (mean frequency of flares, proper motion and spectrum).

Thus the existence in the Pleiades cluster two groups of flare stars having different parameters is an observational fact.

It can be suggected different explanations of this unexpected result.

But the authors prefer the explanation ,according to which the flare stars having small proper motions are double and multiple ones, and the stars with large proper motions are single.

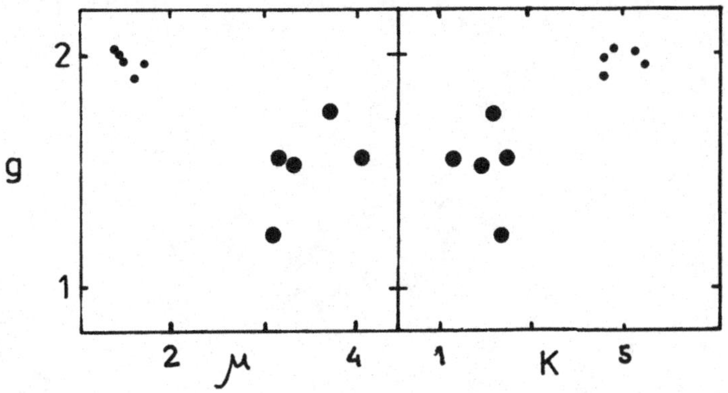

Figure 2. The dependence of the magnitude g from the number of observed flares- k and the proper motion- μ for two considered groups marked by small circles and large circles respectively.

It can be added that due to projection effect a part of flare stars which belong to the group of high-velocity stars can be observed when we use the proper motions as the members of low-velocity stars.

REFERENCES

1. Mnatsakanian,M.A. and Mirzoyan,A.L.(1989) 'Prediction of the flare activity in stellar aggregates', this volume.
2. Jones, B.F.,(1981) 'Proper-motion membership for Pleiades flare stars', Astron. J.,86,290-297.

3. Mirzoyan, L.V.(1976) 'Observational aspects in studing the early stages of the evolution of stars', in Kharadze E.K.(ed.).Stars and Galaxis from observational Points of view, Georgian Academy of Sciences, Tbilisi, pp.121-127.
4. Chavushian,H.C.(1979) 'A study of flare stars in the Pleiades aggregate region, Byurakan Astrophysical observatory.
5. Mirzoyan,A.L.(1983) 'On the proper motions and distribution of flare stars in the Pleiades'. Astrofizika, 19, 588-592.
6. Mirzoyan,L.V.,Hambarian,V.V.,Garibjanian,A.T.and Mirzoyan,A.L.(1988)'Statistical study of flare stars III.Flare stars in the general galactic star field', Astrofizika. 29, 531-540.
7. Parsamian, E.S. and Oganian, G.B., Astrofizika, in press.

FLARE ACTIVITY OF STARS IN THE TAURUS REGION

A.S.HOJAEV
High Energy Astrophisics Department
Astronomical Institute
Uzbek Academy of Sciences, USSR

ABSTRACT. Using the actual observational data on photometric and spectral stellar activity in the region of central area of the Taurus Dark Clouds complex, the question of flare activity of the stars and its connection with another kinds of activities is considered. The comparison of stellar flares in this region with that of in other associations and clusters (especially, Orion and Pleiades) is made and the youth of the group in Taurus is shown. The detection of flaring variables (including stars with T Tauri-type features) in Taurus region is the evidence in favour of evolutionary connection of these stages.

1. INTRODUCTION

It is well known, that flare stars are in the early stages of stellar evolution. The establishment of this regularity by Ambartsumian [1] and Haro [2] has stimulated the investigations in this field. Until now the regions of more than ten stellar systems were studied for search of flare stars (see, for example, [3]). One of them is T-association Taurus T3, connected with central area of the Taurus Dark Clouds (TDC).

2. DATA AND RESULTS

The first flare stars in the TDC were detected by Haro and Chavira [4]. A bit later some other flare stars in this region were found by Petit [5], Tsesevich [6] and Huang et al [7]. Systematic and extensive investigations of flare stars in the TDC were carried out at Byurakan astrophisical observatory [8-10]. Total exposition time of the TDS patrolling by chain method is equal to about thousand hours.

L. V. Mirzoyan et al. (eds.), Flare Stars in Star Clusters, Associations and the Solar Vicinity, 81–84.

Until 1987 in the TDC region 102 flare stars are found, on which 122 flares were fixed. 9 stars had 2, 4 stars 3 and 1 star 4 flares.

An important result is the detection of flares on 13 irregular Orion population variables [11], including at least 7 T Tauri stars. It turned out, that 21 flare stars had Hα-emission in spectrum. For 3 of them emission was detected by us [12].

3. GENERAL DISCUSSION

Two-dimensional distribution of flare stars in the TDS has shown, that it is not connected with the changes of average surface density of stars (i.e. with respective transparency of area). The preferable location of these flare stars is similar to the location of TDC's irregular variables.

An important result of our observations was the detection of flares on approximately 20% of irregular variables (Hojaev [11]). It turned out, that at least more than a half of these flare irregular variables are of T Tauri type.

They include the both components of visual bynary system FY/FZ Tau. FY Tau was more active, which showed 3 strong flares from 1980 till 1984. 2 flares with an interval of approximately a month was shown by another star of T Tauri type - VY Tau and one of them is evidently double (see Fig.1a).

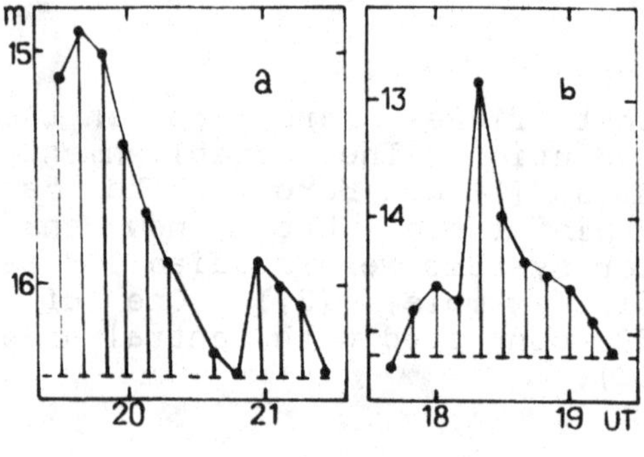

Figure 1. The light curves of /a/ VY Tau flash (15.11.83,U filter); /b/ CI Tau flare (03.02.83, without filter). The dotted line corresponds to the minimum, before and after flare.

Slow and of complex structure flare was demonstrated by T Tauri type star - CI Tau (Fig.1b), which obviously had preflare.

Among the flare stars of the TDC region the Hα-emission

was observed in a quarter of them. Relative number of them isclose to that of Orion, where approximately on 30% of flare stars the Hα-emission was observed.

The comparison of the distributions of flare stars by their luminosities in the TDC, Orion and Pleiades shows their similarity by form, as well as the shift of distribution's maximum towards lower luminosities with the increasing of the system's age (see [13]).

On Hertzsprung-Russel diagram of flare stars in the TDC region as well as flare stars in the Orion are situated in a broad width around ZAMS line, which is in accordance with youth of the TDC system.

The brightest flare stars according to our photometry had $V=12^{m}.5$, which agree with the earliest Sp of them near G9-KO.

On flare stars in the TDC region some "slow" flares by classification of Haro were fixed. The ratio of "slow" flares to "fast" ones in the TDC (= 4%) is close to analogous ratio for the Orion association (= 5%). In particular such persentage is observed among the flares of irregular variables. In this case 3 of 13 flares were "slow". Thus, due to major properties (the presence of irregular variables including doubtless T Tauri stars with the flares, Hα-emission flare stars, "slow" flares, the earliest Sp of the brightest flare star, as well as complex character of light curves of numerous flares and presence of extremely active flare stars) the TDC system is alike the young Orion association. But, the average rate of the flares in the TDC region (near 0.00025 flare per hour) is a bit lower than in Orion and NGC 2264 (0.0005 fl.per h.) and Pleiades (0.00035 fl.per h.). Though it contains also flare stars with respectively high rate of flares (approx. 0.0033 fl.per h.).

Based on all above mentioned data we can conclude about the youth of the subsystem of flare stars in the TDC region. The further study of the flare activity in the TDC region seems to be very perspective.

ACKNOWLEDGEMENTS

We are grateful to Prof. V.A.Ambartsumian and Prof. L.V.Mirzoyan for their fruitful discussions and generous support of this work.

REFERENCES

1. Ambartsumian,V.A.(1954) 'Continuous emission phenomenon and stellar energy sourses', Soob.Byurakan Obs.13,3-35.
2. Haro,G.and Morgan,W.W.(1953) 'Rapid variables in the Orion nebula',Astrophis.J. 118,16-17.

3. Mirzoyan,L.V.(1984) 'Flare stars', Vistas in Astronomy. 27,77-109.
4. Haro,G. and Chavira,E.(1955) 'Nuevas estrellas rafaga y la relacion spectro-rapidez de la variacion',Bol.Obs. Tonantzintla 12,3-16.
5. Petit,M.(1957) 'Stars with the flashes', Peremen.Zvezdy, 12,4-17.
6. Tsesevich,V.P.(1972) 'Flare star SVS 1849', Astron.Tsirk. 733,7.
7. Huang,C.-c.,Zhang,C.-s. and Wang,K.-m.(1974) 'A note on flare stars in the Taurus cloud region',Acta Astronomica Sinica, 20,329-332.
8. Hojaev,A.S.(1983). 'Flare stars in Taurus'. Commis.27 IAU IBVS 2412.
9. Hojaev,A.S.(1984). 'Stellar flares in Taurus'.ibid.2635.
10. Hojaev,A.S.(1984),'New flare stars in Taurus', ibid.2635
11. Hojaev,A.S.(1987). 'Flares of Orion population variables in the association Taurus T3',Astrofizika 27,207-217.
12. Parsamian,E.S. and Hojaev,A.S.(1985) 'New Hα emission stars in the region of the Taurus dark clouds', Astrofizika. 23,203-206.
13. Mirzoyan,L.V. and Hambarian,V.V.(1988) 'Statistical study of flare stars I. The UV Ceti stars of solar vicinity and the flare stars in clusters and associations', Astrofizika, 28,375-389.

GAHM: Are any of your flare stars in the Taurus-Auriga region identical to any of the post-T Tauris located by Walter et al. (1986) ?

HOJAEV: Some of these flare stars certainly are identical to obvious T Tauri type stars of the Taurus-Auriga complex, while it was no identification with known post T Tauri stars. Moreover, post T Tauri stars on the average are brighter than many flare stars so we believe it would be desireable to carry out observations with the specific purpose of detecting optical flares on these stars.

PARSAMIAN: VY Tau is a very interesting T Tauri star. The discovery of classical flares on this star is very important. Can you tell me what the flare amplitudes were for VY Tau?

HOJAEV: This star has shown two flares separated by about one month in our ultraviolet observations with the 40 inch Byurakan Schmidt. The first flare had an amplitude of 1.5 magnitude, the second had 1.1.

PARSAMIAN: What were the pre-flare and post-flare brightness levels for VY Tau?

HOJAEV: VY Tau is approximately 16.4 magnitude in U.

AN AUTOMATIC SEARCH FOR FLARE STARS IN SOUTHERN STELLAR AGGREGATES OF DIFFERENT AGES*

R. ANIOL, H. W. DUERBECK and W. C. SEITTER
Astronomical Institute of Muenster University, F.R.Germany

M. K. TSVETKOV[1]
Department of Astronomy, Academy of Sciences, Sofia, Bulgaria
and Astronomical Institute of Muenster University, F.R.Germany

ABSTRACT. Statistically relevant samples of flare stars in stellar aggregates can be used to specify the stellar mass-age-activity-relation when the ages of the aggregates are known from independent investigations. Associations and clusters of the southern sky are currently surveyed with the GPO astrograph of the European Southern Observatory. The plates are digitized with the PDS 2020 GM^{plus} microdensitometer and the data are reduced automatically at the Astronomical Institute of Muenster University. The programme package "FLARE" is described. First results from the Orion association are presented.

1. Introduction

Stellar activity as observed in flare stars (UV Cet stars) and related objects (T Tau, Hα emission, FU Ori stars) appears to play an important role during the early stages of stellar evolution (Ambartsumian 1969). The underlying physical processes and their decline with advancing age are, however, still not sufficiently well understood. Thus, it is of interest to study not only individual objects in detail but to improve the statistics of active stars and their degree of activity. The project described here aims at enlarging the number of known flare stars (currently 1414 verified plus 188 suspected UV Cet-type variables), and, in particular, to derive a mass-age-activity-relation from the frequency of flares and flare stars of different masses in stellar aggregates of various ages.
In addition to the general need for more data, the clear dichotomy of the data available in the northern and southern hemispheres requires additional surveys particularly in the south. The status of observations prior to the present study is shown in Tables 1 and 2 and in Fig. 1.

2. The Muenster Flare Star Project

The Muenster Flare Star Project started in 1985. The large demand on southern Schmidt telescope time for other surveys and atlases prompted us to work with the GPO astrograph of the European Southern Observatory, hoping that the smaller flare amplitudes in a blue band which does not extend into the ultraviolet would be compensated by the faster growth

*This paper, based on observations collected at the European Southern Observatory, La Silla, Chile, is dedicated to Academician V.A. Ambartsumian in honour of the 80th anniversary of his birthday.
[1]Research fellow of the Alexander von Humboldt-Stiftung Bonn-Bad Godesberg.

L. V. Mirzoyan et al. (eds.), Flare Stars in Star Clusters, Associations and the Solar Vicinity, 85–94.

TABLE 1. Flare data in northern stellar aggregates.

aggregate	no. of flare stars	total observing time (h)	references
Pleiades	539	3250	1,2,3
Orion M42/M43	499	1590	4,5,6,7
Taurus Dark Clouds	102	870	8
NGC 7000 - IC5068-70	75	1185	9
Praesepe	54	660	10,11,12
Monoceros - NGC 2264	42	100	13
Cassiopeia - W3, IC 1848	25	62	14,15,16,17
γ Cyg - IC 1318	17	300	9
Coma Open cluster	14	338	18
Cepheus - NGC 7023	14	50	19,20
total	1381	8404	

(1) Haro *et al.* 1982; (2) Chavushian 1979; (3) Tsvetkov *et al.* 1989; (4) Haro 1953; (5) Natsvlishvili 1989; (6) Konstantinova-Antova and Tsvetkov 1988; (7) Present work, see Table 4; (8) Hojaev 1986; (9) Tsvetkov and Tsvetkova 1989; (10) Jankovics 1975; (11) Haro *et al.* 1976; (12) Mirzoyan and Ohanian 1986; (13) Parsamian *et al.* 1985; (14) Pulakos 1976a; (15) Pulakos 1976b; (16) Pulakos 1977a; (17) Pulakos 1977b; (18) Erastova 1981; (19) Mirzoyan *et al.* 1968; (20) Parsamian 1982.

TABLE 2. Flare star surveys in southern fields.*

field	no. of flare stars	total observing time (h)	references
Sco-Oph	10	118.5	1,2
Coalsack	3**	27	3,4,5
R CrA	2	9	6,7
o Vel	1	5.5	8
total	16	160	

* without Orion association (M42/M43), which is included in Table 1.

** plus 140 suspected fast irregular variables with small amplitudes found by Andrews (1972) in the first automatic search for flare stars and irregular variables.

(1) Haro and Chavira 1974; (2) Sun and Tong 1988; (3) Andrews 1972; (4) Sanduleak 1968; (5) Sanduleak 1969; (6) Hardy and Mendoza 1968; (7) Duerbeck and Tsvetkov 1989; (8) MacConnell and Mermilliod 1984.

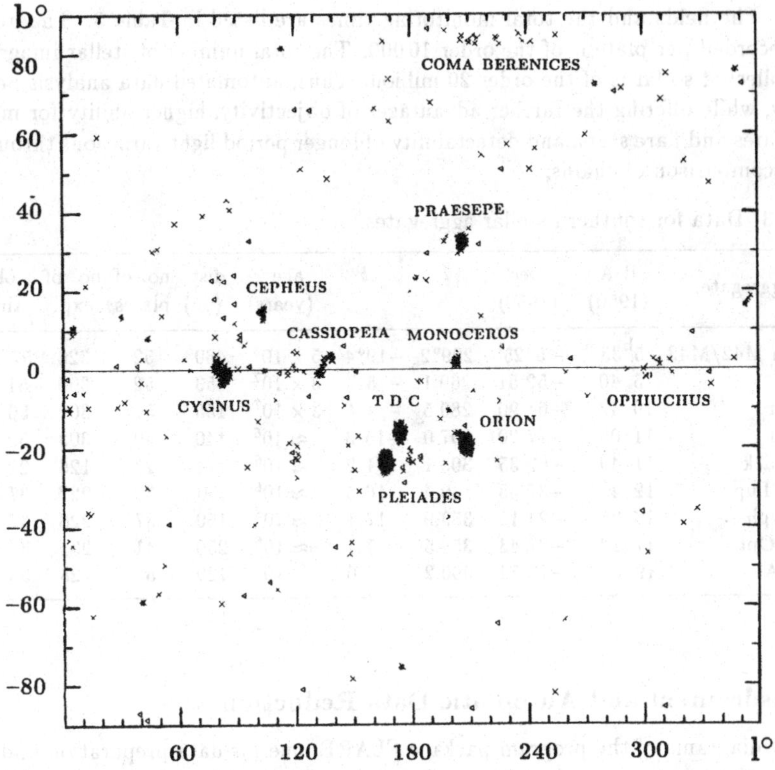

Fig. 1. Distribution of UV Ceti type stars in the Galaxy.

of the astrograph images. The field of the Orion association (M42/M43) which is not only well accessible from both hemispheres, but also one of the richest known regions of UV Cet and T Tauri stars, was chosen for tests of the detection power of the GPO and our reduction algorithms. The tests proved to be successful (Tsvetkov *et al.* 1986, Aniol *et al.* 1989).

3. Observations

The GPO astrograph has the following properties: 0.4 m aperture, focal ratio f:10, field size 4.2 square degrees, of those 2.3 square degrees practically coma-free. The complete fields are analyzed. With 10 min exposure time photographic magnitude 17.6 is reached on unsensitized Kodak IIa-O plates. Observations were carried out in the multiple exposure mode, usually with 5 to 7 observations per chain and ten minutes exposure time per image. To facilitate identifications of weak objects, deep plates were taken in the blue and red pass bands. So far, observations have been carried out during six observing periods in the southern summer and five observing periods in the winter; more observing time has been granted. The fields and the total monitoring times are listed in Table 3. The number of chains recorded per plate is of the order 10 000. The total number of stellar images on the plates collected so far is of the order 20 million. Thus, automated data analysis becomes a necessity, while offering the further advantages of objectivity, higher ability for measuring fainter flares and flare stars, and detectability of longer period light variations through plate to plate comparison of chains.

TABLE 3. Data for southern stellar aggregates.

aggregate	R A (1950)	Dec (1950)	l	b	age (years)	dist. (pc)	no. of plates	no. of exp.	obs. time
Orion M42/M43	5^h33^m	– 5°25′	209°.2	–19°.4	5×10^5	460	52	326	57^h40^m
o Vel	5 40	–52 51	269.1	– 5.7	3×10^7	153	52	309	51 33
Θ Car	10 42	–64 20	289.5	– 4.4	3×10^7	155	51	302	50 20
Cha T1	11 05	–77 20	297.0	–15.3	$\approx 10^6$	140	49	303	52 10
Coalsack	11 49	–62 35	303.1	1.2	$\approx 10^6$	174	21	126	21 50
B228 Lup	12 49	–37 38	340.4	16.5	$\approx 10^6$	150 :	38	227	37 50
Sco-Oph	16 25	–24 10	353.9	17.1	$\approx 10^6$	150	37	226	37 40
B 59 Oph	17 12	–27 43	358.5	7.1	$\approx 10^6$	250	41	231	36 16
R CrA	19 01	–72 54	000.2	–18.0	10^7	129	53	228	53 00

4. Measurement and Automatic Data Reduction

Two flow diagrams of the program package FLARE, one for data preparation and one for data reduction, summarize the steps leading from photographic images to candidate lists of flare stars and other variables. The flow diagrams are shown in Figs. 2a and b.

Digitization. In order to obtain the data in machine processible form, the plates are scanned with the PDS 2020 GM^{plus} microdensitometer of the Muenster Astronomical Institute. With 20μm aperture and the same step size the relative positional accuracies are 0″.07 to 0″.14, depending on stellar brightness. Relative magnitudes within a given chain differ by no more than $0^m.08$ to $0^m.12$, depending on brightness. The total scan time per plate is 4^h, supervised time (plate adjustment, initialization) is less than 0.3^h. Up to 19 plates are stored on a single 8 mm video tape.

Segmentation. For storage, the digitized plate data are 'segmented', *i.e.*, all images are stored in 'picture frames' of 15 pixel × 15 pixel each. This mode assures a minimum of storage space (1/3 that of a complete plate) and fast accessibility of all relevant information.

Image processing. This step includes the following procedures: Image reconstruction = object detection, determination of mean background density (averaged over several scan lines), local plate noise, positions in x and y, object radii, second order moments, and machine magnitudes for all objects.

Catalogue of image data. Through calibration with external catalogues, positions in α, δ, and photographic magnitudes are determined. The input catalogue for astrometric data is

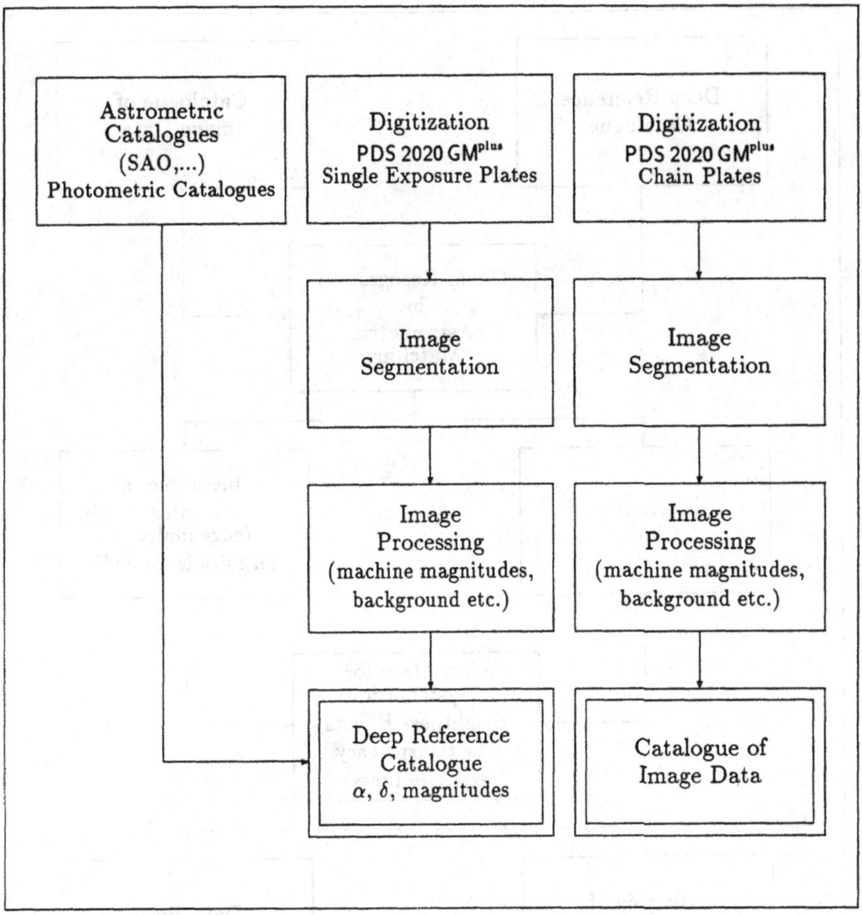

Fig. 2a. Preparation.

the SAO Catalogue, photometric data are taken from various sources. Catalogue positions are determined with an accuracy of 0.″5. External magnitude calibrations are strongly affected by the lack of suitable standards. Their accuracy is at present $0^m_.3$ to $0^m_.4$.

Deep reference catalogue. The catalogue contains the machine and calibrated positions and magnitudes of all objects found on the deep plates. It is used in the subsequent reduction process.

Chain recognition by matching. All image data files are matched with the deep reference catalogue using the eastern images of the brightest chains for x,y-coordinate transformation. Positional accuracies after matching are 0.″3. Also determined are the number of images per chain and the distances between chain images. The celestial coordinates from the deep plates are then assigned to all chains, including incomplete ones. The resulting lists of

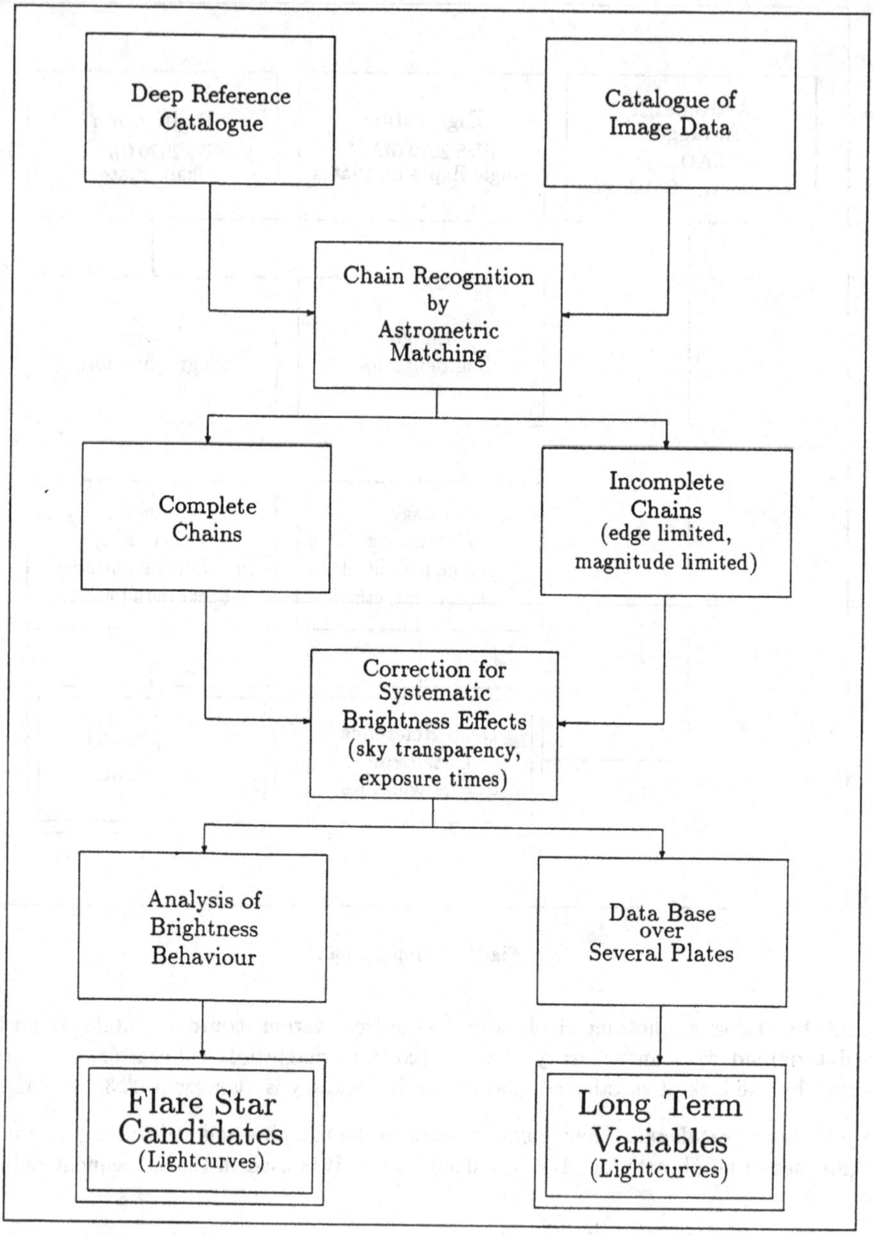

Fig. 2b. Main reduction.

complete and incomplete chains are used for further processing.

Correction for systematic brightness effects. Statistical analyses of image brightnesses (machine magnitudes) are used to correct for differences in exposure time and sky transparency within the chains.

Analysis of brightness behaviour. From the corrected magnitudes 'light curves' for all chains of a given plate are determined. In order to avoid magnitude effects, mean light curves for different magnitude intervals (generally 6 intervals over the total magnitude range) are obtained.

Flare star candidates. By comparison of all individual chains on a plate with the mean light curve of the proper magnitude range, the stars whose chains include at least one image brighter than a preset limit are classified as flare star candidates. Because relatively high accuracy internal plate magnitudes are used, the limit can be set as low as $0^{m}_{.}3$ (approximately 3σ). In true flares, generally more than one image deviates from the mean. High resolution tracings and two-dimensional images of the chains of all candidates are checked visually, and final assignments as flares take into account the shapes of the observed outbursts.

Data base for all plates. All corrected chains found on all plates for a given star are collected in a data base.

Long term variables. This list contains all stars with light variations larger than the preset limit in any of the available chains, including those which have no identified flare event. For the determination of long term light variations the individual plate magnitudes are reduced to a common internal system. The accuracy is sufficiently high to detect variations within $0^{m}_{.}5$. The majority of long term light curves found so far suggests irregular variables with time scales of days.

5. Results from the test region and conclusions

20 plates of the test region Orion M42/M43 were analyzed by the program FLARE. The results were compared with those obtained through visual inspection by one of us (M.K.T.) who has many years of experience in visual flare detection. The preset lower variation limit was 5σ following Oskanian and Terebizh (1971). The results are summarized in Table 4, which also includes one new flare star found by visual inspection of a section not included in the above test. Samples of the detected flare events are shown in Fig. 3. From the candidates of the automatic reduction program, 12 flare events could be verified, the others were included in the list of long term variables. Visual inspection of the plates without preselection yielded 8 flares. Besides the fact that automatic search increased the number of objects by 50%, it is interesting to note the characteristics of the additional flares: all are of low amplitude and some of them occurred on bright stars. They were verified by subsequent checks of the original plates. The results support our expectation that unbiased machine selection of flare candidates is particularly useful for objects which escape the strongly biased human eye. 10 of the observed stars are new flare stars. Six of them were previously known as variables (Kholopov 1985; Kholopov *et al.* 1985, 1987, 1989). This

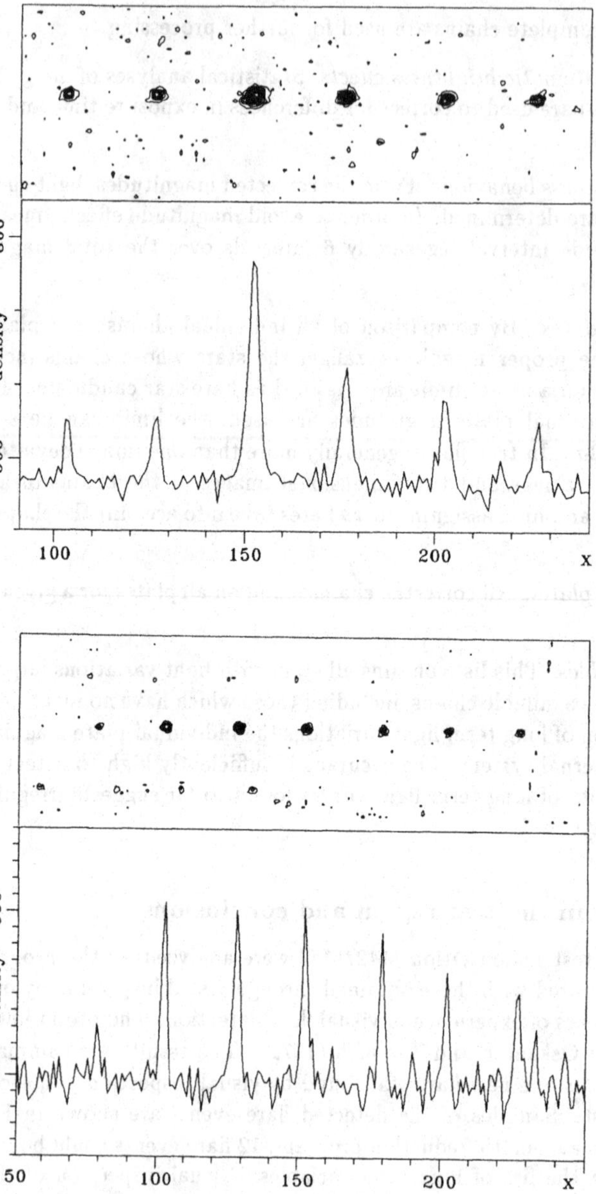

Fig. 3. Flare events in chains of LS 6 and LS 7 shown as high resolution two-dimensional contour plots and density tracings.

TABLE 4. New flare events and flare stars in the Orion M42/M43 association.

LS no.	ABC no.	GCVS CSV	R A (1950)	Dec (1950)	date (1985)	UT (max)	min. pg	max. pg	Δm (mag)	notes
1		AR	$5^h33^m28^s.4$	$-5°06'59''.3$	10 Jan	2^h43^m	$15^m.16$	$14^m.53$	$0^m.63$	1
2		V368	5 35 07.9	−5 13 45.1	11 Jan	1 26	15.84	15.20	0.65	
3	97	V714	5 30 45.9	−5 02 30.2	11 Jan	1 39	16.74	16.14	0.70	2
4*	416		5 32 22.7	−6 16 22.5	11 Jan	2 23	21.0:	16.5	4.5:	
5	344	6315	5 36 13.5	−5 50 39.7	13 Jan	4 02	16.92	16.32	0.59	3
6	145		5 34 02.4	−4 42 50.5	14 Jan	2 03	16.75	15.60	1.35	4
7	172		5 34 29.8	−5 00 45.8	15 Jan	1 42	16.89	16.07	0.82	3
8		AL	5 32 07.8	−5 46 39.9	17 Jan	2 38	16.11	15.28	0.83	
9			5 35 34.1	−4 54 56.9	21 Jan	1 53	16.84	15.79	1.15	
10	52		5 29 01.5	−5 39 43.3	26 Jan	2 45	16.46	15.34	1.12	5
11	450		5 36 11.1	−5 43 20.1	26 Jan	2 56	16.83	16.23	0.60	
12	453		5 36 23.9	−5 26 54.2	26 Jan	3 18	19.0:	16.32	1.7:	
13	239	KO	5 35 00.4	−4 33 05.8	26 Jan	3 29	15.84	14.24	1.65	1,3

* Found in a section outside the automatically reduced field.
(1) slow flare; (2) discovered as flare star T15 by Haro (1953); (3) Hα emission star according to Parsamian and Chavira (1982); (4) discovered on patrol plate with exposures of 20 min per image; (5) discovered as flare star T260 by Haro and Chavira (1969).

2% increase in the total number of known flare stars in Orion corresponds to a discovery rate of 0.5 per monitoring hour as compared to 0.3 for the same region from all previous surveys – not taking into account different field sizes and centers, and the pitfalls of small number statistics.

Acknowledgements

We thank the European Southern Observatory for observing time and assistance. Some of the reduction routines were kindly provided by H. Horstmann. M.K.T. acknowledges support from the Alexander von Humboldt Foundation. We also acknowledge travel grants from the International Astronomical Union (R.A.) and the Deutsche Forschungsgemeinschaft (W.S.).

References

Aniol, R., Duerbeck, H. W., Seitter, W. C. and Tsvetkov, M. K. (1988), Messenger **52**, 39.

Ambartsumian, V. A. (1969), *Stars, Nebulae, Galaxies*, Byurakan Symposium, Acad. Sci. Armenian SSR, Yerevan, p. 283.

Andrews, A. R. (1972), *Bol. Obs. Tonantzintla y Tacubaya* **6**, No. 38, 179.

Chavushian, H. S. (1979), Thesis, Yerevan.

Duerbeck, H. W., and Tsvetkov, M. K. (1989), in Mirzoyan, L. V., Pettersen, B. R., and Tsvetkov, M. K. (eds.) (1989), *Flare Stars in Star Clusters, Associations and Solar Vicinity*, IAU Symposium 137 (to be published).

Erastova, L.K. (1981), Thesis, Yerevan.
Hardy, E., and Mendoza, E. E. (1968), *Cordoba Bull.* **14**, 28.
Haro, G. (1953), *Astrophys. J.* **117**, 513.
Haro, G., and Chavira, E. (1969), *Bol. Obs. Tonantzintla y Tacubaya* **5**, No. 32, 59.
Haro, G., and Chavira, E. (1974), *Bol. Inst. Tonantzintla* **1**, No. 3, 189.
Haro, G., Chavira, E., and Gonzales, G. (1976), *Bol. Inst. Tonantzintla* **2**, No. 2, 95.
Haro, G., Chavira, E., and Gonzales, G. (1982), *Bol. Inst. Tonantzintla* **3**, No. 1, 3.
Hojaev, A. S. (1986), in Mirzoyan, L. V. (ed.), *Flare Stars and Related Objects*, Acad. Sci. Armenian SSR, Yerevan.
Jankovics, I. (1975), Thesis, Yerevan.
Kholopov, P.N. (ed.) (1985), *General Catalogue of Variable Stars*, Nauka, Moscow.
Kholopov, P. N., Samus, N. N., Kazarovets, E. V., and Perova, N. B. (1985), *Inf. Bull. Var. Stars* No. 2681.
Kholopov, P. N., Samus, N. N., Kazarovets, E. V., and Kireeva, N. N. (1987), *Inf. Bull. Var. Stars* No. 3058.
Kholopov, P. N., Samus, N. N., Kazarovets, E. V., Florov, M. S., and Kireeva, N. N. (1989), *Inf. Bull. Var. Stars* No. 3032.
Konstantinova-Antova, R., and Tsvetkov, M.K. (1988), *Inf. Bull. Var. Stars* No. 3190.
MacConnell, D. J., and Mermilliod, J.-C. (1984), *Inf. Bull. Var. Stars* No. 2633.
Mirzoyan, L. V., Parsamian, E. S. and Chavushian, H. S. (1968), *Soob. Byurakan Obs.* **39**, 2.
Mirzoyan, L. V., and Ohanian, G. B. (1986), in L. V. Mirzoyan (ed.), *Flare Stars and Related Objects*, Acad. Sci. Armenian SSR, Yerevan.
Natsvlishvili, R. Sh. (1989), *A Catalogue of Flare Stars in the Orion T2 Association* (to be published).
Oskanian, V. S., and Terebizh, V. Yu. (1971), *Astrofizika* **7**, 281.
Parsamian, E. S. (1982), Thesis, Yerevan, 1982.
Parsamian, E. S., and Chavira, E. (1982), *Bol. Inst. Tonantzintla* **3**, No. 1, 69.
Parsamian, E. S., Rosino, L., and Chavushian, H. S. (1985), *Astrofizika* **22**, 515.
Pulakos, C. (1976a), *Praktika Acad. Athen* **51**, 765.
Pulakos, C. (1976b), *Praktika Acad. Athen* **51**, 771.
Pulakos, C. (1977a), *Astron. Astrophys. Suppl.* **27**, 429.
Pulakos, C. (1977b), *Acta Astr.* **27**, 87.
Sanduleak, N. (1968), *Inf. Bull. Var. Stars* No. 275.
Sanduleak, N. (1969), *Astrophys. J.* **155**, 1121.
Sun, Y.-l., and Tong, J.-h. (1988), *Vistas Astr.* **31**, 385.
Tsvetkov, M. K., Seitter W. C., and Duerbeck, H. W. (1986), *Comm. Konkoly Obs.* No. 86, 429.
Tsvetkov, M. K., and Tsvetkova, K. P. (1989), in Mirzoyan, L. V., Pettersen, B. R. and Tsvetkov, M. K. (eds.) *Flare Stars in Star Clusters, Associations and Solar Vicinity*, IAU Symposium 137 (to be published).
Tsvetkov, M. K., Stavrev, K. Y., and Tsvetkova, K. P. (1989), *New machine-readable version of the Pleiades Flare Star Catalogue* (in preparation).

SPECTRAL OBSERVATIONS OF FLARE STARS

L.V.MIRZOYAN, V.V.HAMBARIAN, A.T.GARIBJANIAN
Byurakan Astrophisical Observatory
Armenian Academy of Sciences, USSR

ABSTRACT. Spectral observations of 6 flare stars in the Pleiades cluster are carried out which occupy different positions on the Hertzsprung-Russell diagram relative to the main sequence: above and below it. The spectral indices which are sensible to luminosity or temperature of the star photosphere are determined. Significant differences between indices of the stars belongings to these two groups are not detected.

1. INTRODUCTION

It is well known that red dwarf stars of young population on the HR diagram are located around the main sequence. For example, according to Haro and Chavira [1] flare stars in star clusters and associations are distributed on the HR diagram on both sides of the main sequence.

Meanwhile current theories of stellar evolution predict the location of young stars only above the main sequence.

For the study of possible differences between spectra of stars belonging to these two groups spectral observations of young red dwarf stars are begun. Here we present some preliminary results.

2. OBSERVATIONS

The spectral observations of 6 flare stars in the Pleiades are obtained with 6-m telescope of the Special Astrophysical Observatory of the USSR Academy of Sciences in North Caucasus.

Of observed flare stars 3 are located above and others-below the main sequence on the HR diagram.

The obtained spectra cover the range 3600-6800A. In all spectra strong emission lines are observed. The

L. V. Mirzoyan et al. (eds.), Flare Stars in Star Clusters, Associations and the Solar Vicinity, 95–98.

registrograms of spectra for two flare stars are presented on Fig. 1. To compare the obtained spectra the spectral indices introduced by Stauffer [4] and Stauffer and Hartmann [5] are used.

3. RESULTS

The results of our spectral observations are presented in Table 1 where the following data are given: flare star number [6], location of star relative to the main sequence [2,3], spectral indices of molecular bands and spectral class. The latters are determined by intrinsic colours R-I

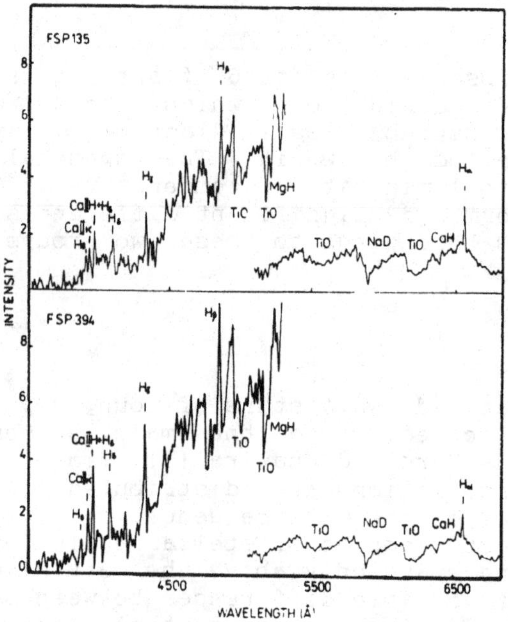

Figure 1. Registrograms of spectra obtained for flare stars located below (FSP 135) and above (FSP 394) the main sequence

using the tabulation (R-I,Sp) suggested by Joy and Abt [7]. The magnitudes R-I are derived from the TiO band strengths.

4. DISCUSSION

The simple comparison of the mean spectral indices (Table 1) determined for flare stars situated above and below the main sequence on the HR diagram shows that there are no significant differences between them.

For more detailed comparison of the obtained data the correlations between indices sensible to temperature and gravity are used. It was expected that flare stars having different location relative to the main sequence should show different correlations.

Indeed they show that for the same value of the TiO spectral indices wich are sensible to star photospheric temperature considerable differences observed in spectral indices CaH and NaI'D'-sensible to star gravity. However , one can not be sure that these differences are conditioned by different correlations for the considered two groups of stars.

TABLE 1 Physical parameters of flare stars

Flare star	Location on HR diagram	Spectral indices							Sp
		TiO							
		D_{54}	D_{59}	D_{61}	D_{65}	CaH	Na'D'	MgH+ Mg'b'	
263	Above	0.237	0.290	0.088	0.210	4.8	11.0	-9.9*	M3
313	"	0.247	0.591	0.166	0.257	2.6	5.5	132.9	M5.5
394	"	0.231	0.611	0.159	0.257	1.7	5.5	139.9	M5.5
	Mean	0.238	0.497	0.138	0.241	3.0	7.3	87.5	
79	Below	0.223	0.663	0.154	0.258	5.6	16.8	132.8	M6
124	"	0.289	0.652	0.206	0.381	3.2	3.9	48.2	M6
135	"	0.223	0.615	0.187	0.279	3.0	6.1	90.7	M6
	Mean	0.245	0.643	0.182	0.306	3.9	8.9	90.6	

* - High-noise

Probably this uncertain result is due to the small number of flare stars of each group and possible errors in position determinations of studied flare stars on the HR-diagram.

We intend to discuss the considered problem after obtaining the spectra of a large number of stars located on both sides of the main sequence.

REFERENCES

1. Haro,G. and Chavira,E.(1966) 'Flare stars in stellar aggregates of different ages',Vistas in Astronomy,8,89-107.

2. Stauffer,J.R.(1982) 'Observations of low-mass stars in the Pleiades: has a pre-main sequence been detected?', Astron. J., 87, 1507-1514.
3. Stauffer,J.R.(1984) 'Optical and infrared photometry of late-type stars in the Pleiades',Asrophys.J.,280,189-20
4. Stauffer,J.R.(1982) 'The faint end of the Hyades main sequence', Astron. J., 87, 899-905.
5. Stauffer,J.R. and Hartmann,L.W.(1986) 'The cromospheric activity, kinematics, and metallicities of nearby M dwarfs', Astophys. J., Suppl.Series, 61, 531-568.
6. Haro,G., Chavira,E. and Gonzalez,G.(1982) 'A catalog and identification charts of the Pleiades flare stars', Bol. Inst. Tonantzintla, 3, No.1, 3-68.
7. Joy,A.H. and Abt,H.A.(1974) 'Spectral types of M dwarf stars', Astrophys. J., Suppl.Series, 28, 1-18.

LANG: How certain can you be that only a few stars are below the main sequence? You assume they belong to the same cluster and lie at the same distance. If your two stars were foreground objects they would be brighter than supposed and actually be on or above the main sequence.

MIRZOYAN: The existence of young stars located below the main sequence in the HR-diagram can probably not be doubted. For individual stars it is difficult to be sure, which is why a study of this problem should be made statistically for a large number of young stars. One possible data set would be flare stars that belong to a given system of stars.

SHAKHOVSKAYA: In the solar vicinity there are flare stars known to be members of binaries with the other component being a white dwarf. Other flare stars are members of old groups of stars. What do you think about this?

MIRZOYAN: The duplicity and multiplicity of flare stars cannot explain the appearance of stars below the main sequence in the HR-diagram.

GIAMPAPA: Can you discuss the flaring properties, and the flare characteristics, for M dwarf stars that are not dMe stars? In other words, what are flares like on M dwarfs without H-alpha emission? Is there a difference in the physics of the flare event between dMe and dM stars?

MIRZOYAN: The spectral classes of flare stars in star clusters and associations are unknown for the majority of the stars. in a few cases, when observed with high enough resolution, the spectra are like those of the UV Ceti stars in the solar vicinity. For example, all spectra obtained by us show strong emission lines. The other physical properties are similar to those of the UV Ceti stars. Taking into account these circumstances one can assume that the differences between dMe and dM stars as far as flare events in star clusters and associations are concerned, must be like those for the red dwarf stars in the solar vicinity. All differences between them are conditioned by the differences in their ages.

NTT SPECTRA OF THE FLARE STARS HM1 AND HM2 IN THE R CORONAE AUSTRINAE AGGREGATE

H. W. DUERBECK[1,2] and M. K. TSVETKOV[2,3,4]

[1] *European Southern Observatory, La Silla, Chile*

[2] *Astronomisches Institut der Universität Münster, F.R. Germany*

[3] *Department of Astronomy and National Astronomical Observatory, Bulgarian Academy of Sciences, Sofia, Bulgaria*

[4] *Research Fellow, Alexander von Humboldt Foundation, Bonn-Bad Godesberg.*

ABSTRACT. Spectroscopic observations of the flare stars HM1 and HM2 (V667 CrA) in the direction of the R Coronae Austrinae aggregate are presented. The respective spectral classes are dM1e and dM6e. Both stars are at the distance 40 pc, much closer than the R CrA association.

The R CrA region is a well-known star forming region, investigated by a variety of observational techniques. The interest in the region is due to the presence of the R CrA T association, containing the bright reflection nebulae NGC 6726 and NGC 6729, Herbig Ae/Be stars, several well established T Tau type stars, Hα emission stars, two Herbig-Haro objects and many rapid irregular variables of unknown type. The distance to the R CrA dark cloud is about 130 pc, according to Marraco and Rydgren (1981).

Among the young stellar objects two flare stars were discovered by Hardy and Mendoza (1968) on plates taken with the Curtis Schmidt telescope. During multi-exposure observations in September 1968, the ultraviolet brightness of both stars increased by two magnitudes in less than two hours. The data of the stars and flare events are summarized in Table 1.

TABLE 1. Data of the flare stars HM1 and HM2

Star design.	GCVS design.	RA (1950)	Dec (1950)	JD (max) 2440114 ...	mag pg	Duration min	Sp. type
HM1	Anon	$18^h54^m08^s$	$-36°38'.6$	$0^d.518$	$13^m.6$	90	dM1e
HM2	V667 CrA	18 57 54	−37 00 .7	0.550	18	60	dM6e

The spectra of both stars were taken on 1989 September 8 with the Faint Object Spectrograph and Camera EFOSC 2, attached to the 3.5 m New Technology Telescope of ESO. Exposure times in three different spectral regions were of the order of a few minutes each.

L. V. Mirzoyan et al. (eds.), Flare Stars in Star Clusters, Associations and the Solar Vicinity, 99–100.

The spectra were flatfielded, wavelength- and flux-calibrated by means of standard objects (lamp, planetary nebula, spectrophotometric standard), and combined into one spectrum from three (HM 1) and two (HM 2) different grism settings. Because accurate guiding was not possible, the stars slowly drifted from the slit, and the resulting spectra (Fig. 1) are not absolutely flux-calibrated.

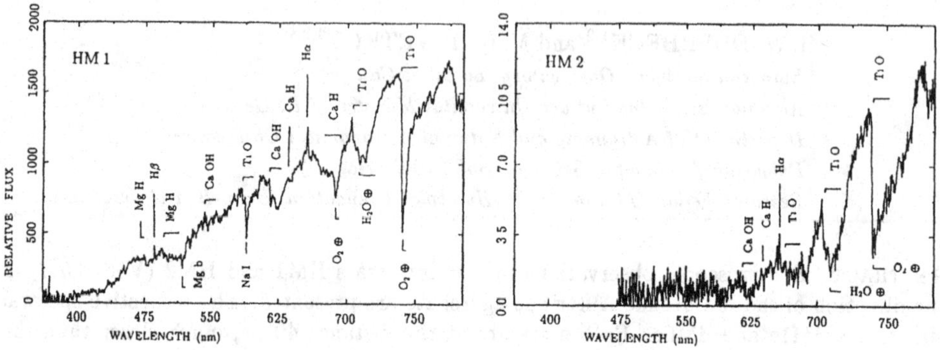

Fig. 1. Spectra of HM1 and HM2, observed with EFOSC 2 at the ESO 3.5 m NTT.

A spectral classification, using Turnshek et al.'s (1985) Atlas and Pettersen's and Hawley's (1989) spectral tracings, gave the results of Tab. 1. In HM1, Hα and Hβ emission, in HM2, only Hα emission is visible.

With the spectral types and the photographic magnitudes of Hardy and Mendoza, the distances of HM1 and HM2 are found to be about 40 pc, much nearer than the distance of the R CrA association. The stars are thus foreground objects or the R CrA association reaches much nearer to us than previously assumed.

Acknowledgements

We thank W. C. Seitter for useful suggestions and comments. H. W. D. acknowledges support from the European Southern Observatory, and M. K. T. acknowledges support from the Alexander von Humboldt-Foundation.

References

Hardy, E., and Mendoza, E. (1968), *Cordoba Bol.* No. 14, 28.
Marraco, H. G. and Rydgren, A. E. (1981), *Astrophys. J.* **86**, 62.
Pettersen, B. R. and Hawley, S. L. (1989), *Astron. Astrophys.* **217**, 187.
Turnshek, D. E., Turnshek, D. A., Craine, E. R. and Boeshaar, P. C. (1985), *An Atlas of Digital Spectra of Cool Stars*, Western Research Company, Tucson, Arizona.

CATALOG OF FLARE STARS IN ORION NEBULA REGION

R.SH.NATSVLISHVILI
Abastumani Astrophysical Observatory
Georgian Academy of Sciences, USSR

ABSTARACT. A Catalog of flare stars of the region around the Orion Trapezium is compiled. The results of photographic patrol observations of flare stars in this region carried out at the observatories: Abastumani, Asiago, Byurakan, La-Silla, Rojen, Tonantzintla and Uppsala, up to 1986 year are used. The Catalog contains infomation for 491 flare stars and their 654 flares. It is compared with the Catalog of flare stars in the Pleiades region.

1. INTRODUCTION

The association T2 Ori (a stellar complex around the Orion Trapezium) is the richest stellar system by abundance of flare stars. This system is one of the youngest ones containing flare stars.

The first flare stars in the association T2 Ori were discovered in 1953 by Haro and Morgan [1]. Since that time extensive photographic observations on searching and studying the flare stars in stellar systems were begun. It was established by Haro [2,3] and Ambartsumian [4,5] that the stage of a flare star is an evolution stage, one of the early ones in the life of red dwarfs.

2. CATALOG

Our catalog is based on the results of patrol observations of the flare stars in the Orion region performed at the Abastumani, Asiago, Byurakan, La-Silla, Tonantzintla and Uppsala Observatories till 1986 [6]. It contains 491 flare stars known by the beginning of 1986.

For each flare star the catalog provides: number in the catalog, designation according to the determination of the observatory where it was discovered, number according to the

L. V. Mirzoyan et al. (eds.), Flare Stars in Star Clusters, Associations and the Solar Vicinity, 101–104.

General Catalog of Variable Stars [7], right ascention and declination for 1950.0 (mainly by our determinations), data of discovery, magnitude in the light minimum, amplitude of the first flare, quantity of registered flares, number of the remark, number of the identification chart and references.

The data available on the spectrum of the star and the number in Parenago's Catalog [8] are given in the notes.

3. THE SUMMARY DATA ON STELLAR FLARES IN THE ORION REGION

The summary data on the flare stars and stellar outbursts found in the Orion region up to early 1986, based on the catalog data, are listed in the Table 1, showing the number of single flare stars and occurence of their outbursts: total and separately according to the observatories.

The successive columns of the Table 1 give: name of the observatory where the flare stars in the Orion region were discovered, number of flare stars having k (k=1,2,3...) flares, total number of flare stars discovered in the given observatory, total number of registered flares.The last two lines of the Table 1 list the summary numbers of the corresponding columns.

4. COMPARISON WITH THE CATALOG OF FLARE STARS IN THE PLEIADES REGION

Comparison of the present catalog of flare stars in the Orion region with that by Haro et al [9] for the Pleiades region is of a certain interest. as in these cases we deal with two complexes of flare stars of essentially different ages.

Comparison shows that the luminosities of the Orion flare stars, in average, are significantly higher than in the Pleiades, inlike these latter objects they coexist with T Tauri type stars and diffuse matter. A mean frequency of "slow" flares seems to be higher [10] etc.

It should be noted that when speaking on the whole complex of flare stars, we assume that all flare stars discovered in the Orion and Pleiades regions, in their overwhelming majority, are physical members of the appropriate systems. There are reasonable grounds for assuming this. The observations indicate that a relative number of flare stars in the galactic field, which could be found during photographic observations both in Pleiades and Orion, is small-about 10% [11].

Between the complexes of flare stars under study there are also defferences of other nature. A total time of patrol photographic observations of the Pleiades region exceeds

3000 hours [12], meanwhile for the Orion region it is about 1600 hours [6]. The complex of flare stars in the Orion is at least twice as rich as the complex in the Pleiades [6, 12]. The Orion association is four times farther than the Pleiades cluster.

In spite of the fact that the Orion region was observed half as much as the Pleiades region, the number of flare stars found in both regions differs slightly. This can be explained due to the fact that the mathematical expectation of observing a flare in the Orion is twice as large: there are more flare stars in this system, the distance effect is compensated by the fact that the luminosities of the flare stars are higher there (by $\propto 2^m$) than in the Pleiades.

Table 1.Number of Orion Flare Stars and Incidence of Flare-Ups

Observatory \ No. of Flares	1	2	3	4	5	6	7	86*	Total No.of F.S.	Total No.of Flares
Tonantzintla	178	54	19	6	1	1	2		261	392
Abastumani	113	10	2						125	139
Asiago	43	6	1						50	58
Byurakan	20	4		1					25	32
Uppsala	9	2	1					1	13	16
Rojen	10								10	10
La-Silla	7								7	7
Total Number of Flare Stars	380	76	23	7	1	1	2	1	491	
Total Number of Flares	380	162	69	28	5	6	14			654

* These Flares refer to the star FS461=TZ Ori detected in Uppsala are not included into the last column of the Table 1

Based on the comparison of the two catalogs it can be concluded as well that the frequency functions in the Orion and Pleiades systems differ significantly. For example, in the Orion region the number of flare stars showing repeated flares is markedly less than in Pleiades.

The author expresses his sincere thanks to Prof. L.V. Mirzoyan for his constant care and assistance in preparation of the present paper.

REFERENCES

1. Haro,G. and Morgan,W.W.(1953)'Rapid variables in Orion nebula', Astrophys.J.,118,16-17.
2. Haro,G.(1957),'Stars of T Tauri and UV Ceti types,and the phenomenon of continuous emission', in G.H.Herbig (ed.), Non-Stable Stars, Cambridge University Press, Cambridge,p.26.
3. Haro,G. (1976),'An observational approach to stellar evolution.-I.Flare stars and related objects', Bol.Inst. Tonantzintla, 2, 3-54.
4. Ambartsumian,V.A.(1969),'On the statistics of flaring objects',in V.V.Sobolev (ed.),Stars, Nebulae, Galaxies, Armenian Academy of Sciences, Yerevan, pp.283-292.
5. Ambartsumian,V.A. and Mirzoyan,L.V. (1971),'Flare Stars', in: New Directions and New Frontiers in Variable Star Research, Veroff Bamberg,9,No.100,pp.98-108.
6. Natsvlishvili,R.Sh.(1987),'Flare Stars in Regions Orion and Pleiades', Byurakan Astrophysical Observatory.
7. Kholopov,P.N. et. al.(1985),General Catalog of Variable Stars, Vol.2, Nauka, Moscow.
8. Parenago.P.P.(1954),'A study of stars in the Orion nebula region', Trudy Astron.Inst.Stenberga,25,3-546.
9. Haro,G.,Chavira,E. and Gonzalez,G.(1982),'A catalog and identification charts of the Pleiades flare stars', Bol.Inst.Tonantzintla,3,No.1,3-68.
10. Mirzoyan,L.V. and Melikian,N.D. (1986), 'Some characteristics of stellar flares',in L.V.Mirzoyan (ed.),'Flare Stars and Related Objects',Armenian Academy mof Sciences, Yerevan, pp.153-161.
11. Mirzoyan,L.V., Hambarian,V.V., Garibjanian,A.T. and Mirzoyan,A.L.(1988)'Statistical study of flare stars II Flare stars in the general galactic star field', Astrofizika, 29, 531-540.
12. Mirzoyan, L.V. and Oganian,G.B.(1986),'Flare stars in stellar clusters and associations', in L.V.Mirzoyan (ed.), Flare Stars and Related Objects, Armenian Academy of Sciences,Yerevan,pp.68-78.

A CATALOGUE OF FLARE STARS IN THE CYGNUS REGION

MILCHO K. TSVETKOV* and KATYA P. TSVETKOVA
Department of Astronomy and National Astronomical Observatory, Bulgarian Academy of Sciences, Lenin Blvd. 72, Sofia-1784, Bulgaria
and
Astronomical Institute of Muenster University, Wilhelm-Klemm-Str. 10, 4400 Muenster, F.R. Germany

ABSTRACT. A catalogue of known flare stars in the region of the emission nebulae NGC 7000 – IC 5068-70 and IC 1318 a,b,c in Cygnus is presented. The UBV magnitudes and spectral types, as far as available, for 96 flare stars and their 144 flares were collected. Patrol observations from the Asiago, Byurakan, Konkoly, Rozhen and Tonantzintla observatories, covering a total of 1500 hours effective observing time were used.

1. Introduction

Flare star observations, which started systematically in the early nineteen-sixties, showed the importance of flare stars as a separate class of variables for our understanding of processes of star formation and evolution (Ambartsumian and Mirzoyan 1977). Today, a large number of data from flare star observations in different stellar aggregates is available. For the reduction of existing observations and the preparation of future patrol observations, as well as for statistical investigations of flare stars, a compilation of the available data is needed. Up to now two catalogues exist – for the Pleiades region (Haro et al. 1982) and for the aggregate in Orion (Natsvlishvili 1988). In the present paper a new catalogue is described, containing 96 flare stars in Cygnus and their 144 observed flare events. The complete version of the catalogue will be published later because of the limited space available here.

A machine-readable version of the catalogue is available upon request.

2. The Catalogue

The catalogue lists the results of flare star observations in Cygnus (photographic and photoelectric) amounting to about 1500 hours effective observing time. Systematic observations of the region around the emission nebulae NGC 7000 – IC 5068-70 (Cyg T1 association) were conducted mainly at Byurakan Observatory, these of IC 1318 a,b,c (Cyg T2 association) mainly at Rozhen Observatory. A few additional flare stars in the vicinity of those fields are also included. The list of the 96 flare stars in Cygnus, given in the catalogue, is presented in Table 1.

*Research Fellow of the Alexander von Humboldt-Stiftung

L. V. Mirzoyan et al. (eds.), Flare Stars in Star Clusters, Associations and the Solar Vicinity, 105–108.

TABLE 1. List of flare stars in the catalogue of the Cygnus region.

Cyg No	Fl Des	GCVS	RA (1950)	Dec (1950)	Cyg No	Fl Des	GCVS	RA (1950)	Dec (1950)
1	Ct	V 1581	$19^{h}52^{m}3$	44°17′	49	B 22	V 1599	$20^{h}50^{m}6$	42°48′
2	Cr	V 1513	20 03.9	54 18	50	T 6	V 1494	20 50.6	43 24
3	R# 6	V 1772	20 15.5	42 28	51	B 25	V 1600	20 50.7	44 25
4	A	V 1381	20 19.1	41 10	52	B 31	V 1601	20 51.0	43 05
5	R#12	V 1777	20 23.0	42 23	53	B 33	V 1602	20 51.1	41 45
6	R# 7	V 1778	20 23.6	41 55	54	B 26	V 1603	20 51.3	42 26
7	R# 8	V 1779	20 24.6	40 18	55	K 4	V 1709	20 51.6	44 53
8	R# 9	V 1780	20 25.4	41 32	56	B 43	V 1710	20 51.7	41 33
9	R#10	V 1781	20 25.6	41 09	57	A 2	V 1924	20 52.0	44 09
10	B# 2	V 1750	20 26.2	40 06	58	Bj	V 1926	20 52.9	42 48
11	R# 1	V 1752	20 27.3	40 16	59	B 55	V 1795	20 53.0	40 17
12	R# 2	V 1753	20 30.1	43 01	60	T 7	V 1495	20 53.4	44 38
13	B# 3	V 1754	20 30.4	43 06	61	B 18	V 1604	20 53.7	43 23
14	B 28	V 1586	20 32.0	44 13	62	B 30	V 1605	20 53.9	40 45
15	B# 1	V 1695	20 32.5	43 30	63	B 3	V 1536	20 53.9	44 09
16	R# 3	V 1755	20 33.3	42 21	64	B 5	V 1537	20 54.0	43 31
17	R#11	V 1785	20 33.8	41 01	65	B 37	V 1606	20 54.5	42 49
18	R# 4	V 1756	20 34.3	39 51	66	B 49	V 1713	20 54.5	43 03
19	R# 5	V 1757	20 34.5	40 37	67	B 48	V 1712	20 54.5	43 05
20	B 32	V 1587	20 39.6	41 38	68	B 47	V 1927	20 55.0	43 53
21	B 16	V 1522	20 40.1	40 03	69	A 4	V 1928	20 55.3	41 50
22	B 40	V 1588	20 40.3	44 08	70	B 24	V 1607	20 55.5	40 53
23	B 19	V 1589	20 41.0	41 11	71	B 44	V 1714	20 55.5	42 58
24	B 35	V 1590	20 41.4	40 07	72	B 8	V 1538	20 55.6	43 39
25	B 38	V 1591	20 41.4	44 08	73	B 17	V 1608	20 55.6	44 32
26	B 27	V 1592	20 41.7	43 08	74	A 5	V 1929	20 55.8	43 46
27	B 34	V 1593	20 42.3	42 13	75	B 12	V 1539	20 56.2	43 41
28	K 1	V 1698	20 43.1	44 33	76	B 29	V 1609	20 57.1	42 41
29	B 53	V 1789	20 43.3	42 26	77	T 2	V 1496	20 57.3	42 26
30	B 6	V 1526	20 43.9	42 41	78	B 50	V 1717	20 57.5	41 44
31	B 41	V 1699	20 44.1	43 38	79	Cr	V 1396	20 58.1	39 53
32	B 46	V 1700	20 46.2	42 49	80	T 5	V 1497	20 58.2	43 20
33	B 7	V 1528	20 47.1	41 02	81	B 45		20 58.6	43 04
34	B 39	V 1594	20 47.2	42 10	82	Bj	V 1930	20 58.6	44 19
35	R 1	V 1790	20 48.3	43 14	83	B 15	V 1544	20 58.8	42 10
36	B 10	V 1595	20 48.5	41 33	84	B 56	V 1796	20 59.0	43 58
37	B 11	V 1529	20 48.7	41 46	85	B 51	V 1797	20 59.5	41 45
38	B 36	V 1596	20 48.7	43 15	86	B 57	V 1798	20 59.6	40 29
39	B 1	V 1530	20 48.8	40 43	87	B 58	V 1799	20 59.9	42 31
40	B 54	V 1791	20 48.8	40 50	88	B 2	V 1545	21 00.0	42 26
41	K 2		20 48.8	44 25	89	T 3	V 1498	21 00.5	44 54
42	B 59		20 49.0	43 46	90	T 1	V 1424	21 00.7	42 08
43	K 3		20 49.0	44 08	91	B 20	V 1611	21 01.3	44 22
44	B 9	V 1597	20 49.2	44 04	92	B 21	V 1612	21 01.6	40 59
45	B 14	V 1598	20 49.3	44 00	93	T 4	V 1499	21 01.7	42 03
46	B 42		20 49.5	40 43	94	B 23	V 1613	21 02.3	43 24
47	B 52	V 1793	20 49.9	43 26	95	B 13	V 1550	21 04.5	44 25
48	B 4	V 1534	20 50.5	41 26	96	B	V 1758	21 11.1	48 44

The columns in the table are: catalogue number of the flare star; original designation of first registered flare event; designation according to the General Catalogue of Variable Stars (Kholopov 1985) and subsequent Namelists of Variable Stars; equatorial coordinates. In addition to the information given in Table 1, the original catalogue contains the following data: catalogue number of flare event; dates of observed flare events, and apertures of the telescopes used; minimum and maximum magnitudes of the flare stars in U– or pg–light; values of V, B–V, U–B in quiescence from photographic photometry and corresponding spectral type. Flare stars not satisfying the criterion $\Delta m \geq 5\sigma$ (Oskanian and Terebizh 1971) are marked.

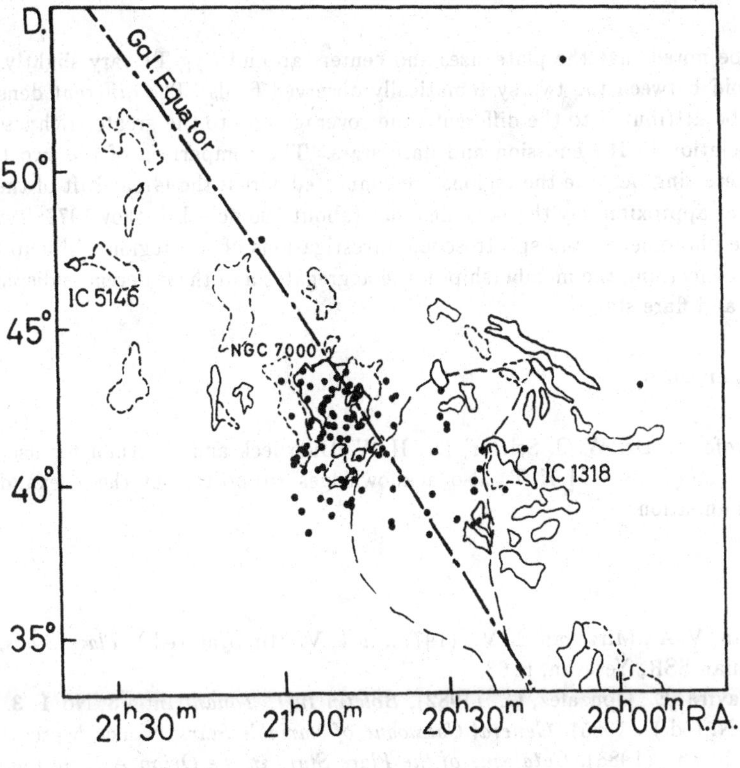

Figure 1. Distribution of known flare stars in Cygnus

In Fig. 1 the distribution of all known flare stars in Cygnus is shown. They were mostly found in the two patrol fields, around the emission nebulae NGC 7000 – IC 5068-70 with more than 1000 hours effective observing time, and around IC 1318 a, b, c with 300 hours.

In Table 2 the frequency distribution of flare events during the time of observations is given.

TABLE 2. Numbers of Cygnus flare stars and incidence of flare events

Number of flare events	1	2	3	5	7	10	Total fl. stars	Total fl. events
Number of flare stars	78	12	1	1	2	2	96	144

It should be noted that the plate sizes and centers around Cyg T1 vary slightly. There is no clear void between the two systematically observed fields. The different densities of objects may be attributed to the different time coverage and to the greater richness of the Cyg T1 association in Hα emission and flare stars. The comparison of the two fields in Cygnus is interesting because the regions are connected across the Great Rift of the Milky Way and are at approximately the same distance (about 500 pc – Tsvetkov 1977; Tsvetkova 1986). Future photometric and spectroscopic investigations of this region will help to clarify questions concerning the membership in the aggregates and the differences displayed by Hα emission and flare stars.

Acknowledgements

We thank Professor Dr. W. C. Seitter, Dr. H. W. Duerbeck and R. Aniol for useful suggestions and comments. M. K. T. also acknowledges support from the Alexander von Humboldt-Foundation.

References

Ambartsumian, V. A., Mirzoyan, L. V. (1977), in L. V. Mirzoyan (ed.), *Flare Stars*, Acad. Sci. Armenian SSR, Yerevan, p. 63.

Haro, G., Chavira, E., Gonzalez, G. (1982), *Boletin Inst. Tonantzintla* **3**, No. **1**, 3.

Kholopov, P. N. (ed.) (1985), *General Catalogue of Variable Stars*, Nauka, Moscow.

Natsvlishvili, R. Sh. (1988), *Catalogue of the Flare Stars in the Orion Association* (to be published).

Oskanian, V. S., Terebizh, V. Yu. (1971), *Astrofizika* **7**, 281.

Tsvetkov, M. K. (1977), in L. V. Mirzoyan (ed.), *Flare Stars*, Acad. Sci. Armenian SSR, Yerevan, p. 78.

Tsvetkova, K. P. (1986), *Star Clusters and Associations*, Publ. Eötvös Univ., Budapest, No. **8**, p. 129.

SLOW FLARES IN STELLAR AGGREGATES AND SOLAR VICINITY

E.S.PARSAMIAN, G.B. OHANIAN
Byurakan Astrophysical Observatory
Armenian Academy of Sciences, USSR

ABSTRACT. The study of slow flares in star clusters, associations and Solar Vicinity is carried out. The dependances of flare amplitude from the inverse velocity of flare increasing time in U,B,V bands are obtained. It is shown, that strong flares frequently take place in deep layers of stellar photosphere.

The first attempt to find the reason of the difference between slow and fast flares was made by Ambartsumian [1]. According to Ambartsumian, slow flares take place under photospheric layers. First slow flares was found by Haro [2].

1. In work [3], based on existing observational data, attempt was made to classify slow flares by the form of brightness curve :

1. Slow increase and slow decrease.
2. Slow increase, continuous maximum and decrease.
3. Combination of two flares:slow and fast and vice versa.

At present, distribution of slow flares in stellar clusters by types is the following:

TABLE 1. A distribution of slow flares in stellar clusters

Aggr\Type	I	II	III
Orion	15	7	1
Pleiades	28	2	5
Preasepe	2	–	–

Comparison with similar table from work [3] shows, that insrease in the quantity of flares took place due to flares of type I and flares of type II are again rare. The flat maximum of brightness curve of slow flares of type II can be

L. V. Mirzoyan et al. (eds.), Flare Stars in Star Clusters, Associations and the Solar Vicinity, 109–112.

considered as a result of superposition of several slow flares, which took place in the same layer at small time interval, perhaps having the same source.

As for flares type III, they take place both, under photosphere and over it, being divided by small time interval.

2. Based on new observational data, an attempt was made to find the relationship between amplitude in brightness maximum and inverse velocity of the brightness increase $t_m = t_B/\Delta m$, where t_B – the time of increase till to maximum, Δm – amplitude, for Orion association and Pleiades cluster.

1. Orion:

$$\ln\Delta m_U = -0.05\ t_m + 2.3 \quad (1)$$

$$\ln\Delta m_B = -0.02\ t_m + 1.5 \quad (2)$$

2. Pleiades:

$$\ln\Delta m_U = -0.04\ t_m + 1.9 \quad (3)$$

$$\ln\Delta m_B = -0.05\ t_m + 1.6 \quad (4)$$

The obtained relationships confirm the results received before, that flare amplitude depends on depth of layer where the flare occured, as t_B depends on depth [3].

Slow flares occure not only in stellar aggregates, but also on flare stars of Solar Vicinity. However only two stars AD Leo and EV Lac show flare with t_B about 20 min [4,5]. In other cases t_B<10 min. For that case, when t_B varies in the range of 5–10 min we get:

$$\ln\Delta m_U = -0.17\ t_m + 1.27 \quad (5)$$

Let us note, that slow flares of Solar Vicinity can also be classified by types I,II,III. If photoelectric observations of flare stars in stellar aggregates were made, it is doubtless , that slow flares with small t_B, similar to those in Solar Vicinity can be found.On the other hand, we can surely say, that probability to find slow flares with t_B>20 min in Solar Vicinity is small and depends on evolution stage of these stars.

3. Observational data do not let us to get immediate connection between maximal amplitude and t_B, as flares of different energy can take place on the same depth. That is why, having an idea of flare energies distribution with depth only, we can consider the known slow flares in order of increasing of t_B, i.e.with depth, where slow flares occur.

Fig.1 shows dependance of $\ln\Delta m_B$ on t_m for Pleiades cluster of different intervals of t_B:

$$\ln\Delta m_B = -0.07\ t_m + 1.58, \qquad t_B = 16 - 25\ \text{min}, \quad (6)$$

$$\ln \Delta m_B = -0.06\, t_m + 1.86, \qquad t_B = 27 - 36 \text{ min.} \qquad (7)$$

the angle coefficients show that the diffusion of radiation takes place by the same law. The increase of constant in (6),(7) shows that the deeper the flare, the stronger it in average. The same is true for Orion association.

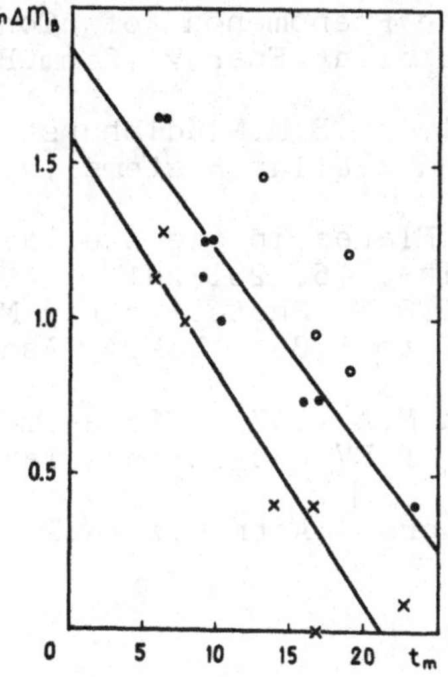

Figure 1. Dependance of $\ln \Delta m_B$ on t_B for Pleiades cluster of different intervals of t_B:

crosses	16 - 25 min,
dark circles	27 - 36 min,
light circles	>36 min.

4. According to Ambartsumian [6] the probability of fast and slow flare appearances must depend on width of corresponding layer. The photosphere width is of the order of 10^2 km. The small dispersion values of t_B confirms, that the layers, where slow flares take place are relatively narrow. The width of the layers where fast flares occure is about 10^4-10^5 km. The ratio of the numbers of fast and slow flares must be proportional to the width of the layer where the flares occur and is about 10^2-10^3. Let us compare this value with the results of observations in Orion and Pleiades. From the observed data it is seen that the ratio of fast and slow flares is in order of 10^2. In this case fast flares ($\Delta m < 1$) which can not be observed as slow, if

they occured under photospheric layers, because of small amplitude is taken into account. Consequently, the ratio of the numbers of slow and fast flares must be more than 10^{-2}. Thus, in the photosphere the flares with great energy happen more frequently than they are thought to be, on the base of ratio of numbers of slow and fast flares.

REFERENCES

1. Ambartsumian,V.A.(1954) 'The Phenomenon of Continuouse Emission and Sources of Stellar Energy',Comm.Byurakan obs.,13,3-35.
2. Haro,G.(1968) 'Flare Stars',in .B.M.Middlehurst and L.H.Aller (eds.),Stars and Stellar Systems ,v.7, Chicago, pp.141-166.
3. Parsamian,E.S.(1980) 'Slow Flares in the Stellar Aggregates.II', Astrofizika, 16, 231-241.
4. Ichimura,K.and Shimizu,Y.(1981) 'Photoelectric Monitoring of Flares Stars from 1977 to 1980', Tokyo Astr .Bull.. No.264, 2999-3013.
5. Grigorian,K.M. and Eritsian,M.A.(1971) 'Polarimetric and Photometric Observations of EV Lac during Flares', Astrofizika,7,303-306.
6. Ambartsumian,V.A.(1971) 'Fuors', Astrofizika,7, 557-572.

A PREDICTION OF THE FLARE ACTIVITY OF STELLAR AGGREGATES

M.A.MNATSAKANIAN, A.L.MIRZOYAN
Byurakan Astrophysical Observatory
Armenian Academy of Sciences, USSR

ABSTRACT. The problem of the statistical prediction of flare activity of a group of flare stars in stellar aggregates is considered.

1. INTRODUCTION

Two important problems were put forward by Ambartsumian [1,2] connected with the existence of flare stars in stellar aggregates: on the estimation of the complete quantity of flare stars, and the determination of the distribution of flare stars according to flare frequency. Both problems are certainly connected with of incomplete data, and being reverce problems they are somewhat incorrect, so their accurate solutions seem actually impossible.

Thus we turn here to another more correct (and at the same time more general) statement of the problem - a prediction into future as much as possible of the flare activity of stellar aggregates. Namely, the prediction for the future time number $n_k(t)$- the quantity of aggregate's flare stars showing k flares in time t.

2. PREDICTION FORMULAE

We suppose that the flare activity of stellar aggregate is stationary in time and that observed flares for each star and different stars take place irrespective of each other. None other supposition for the distribution function is made.

If for the present time T of the observations we have a statistics $n_k(T)$, the theoretical formulas defining the behaviour $n_r(t)$ are given by expressions [3-5]:

L. V. Mirzoyan et al. (eds.), Flare Stars in Star Clusters, Associations and the Solar Vicinity, 113–116.

$$n_r(t) = \sum_{k=r}^{\infty} n_k(T) C_k^r \left(\frac{t}{T}\right)^r \left(1-\frac{t}{T}\right)^{k-r}, \quad r=1,2,\ldots \qquad (1)$$

where C_k^r are the binomial coefficients.

The complete number n(t)of the aggregate's flare stars discovered for the moment of time t, is equal to

$$n(t) = \sum_{k=1}^{\infty} n_k(T)\left[1-\left(1-\frac{t}{T}\right)^k\right] \qquad (2)$$

It is of interest that these formulae give an exact analytical behaviour of the statistics both for the past time and for the future time according to the data available for the present moment T only. The problem is completely coming to a possible more exact definition of the data $n_k(T)$.

3. THE METHOD OF LINEAR REGRESSION

As far as the observational numbers $n_k(T)$ have errors of natural fluctuations, for their more exact definition we use the known chronological behaviour $n_r(t)$ for the past time of observations ($t<T$). Putting this observational chronology into the left part of the formulae (1-2), we consider these formulas as linear algebraic equations relative to numbers $n_k(T)$ and use the classical method of linear reggression.

4. COMPARISON OF THEORY WITH OBSERVATIONS

The problem put forward by us still continues to remain incorrect enough and the time interval of the prediction is quite restricted. It can be proved that principally the prediction cannot be made for times, exceeding the double time of the observations 2T.

In the Figure 1 the observational values for the Pleiades [6] and Orion [7] aggregates are marked by different symbols and the continuous curves are constracted according to the theoretical formula (1-2). If the data be changed slightly, the prediction curves greatly begin to diverge from one to another.

The comparison of the theoretical and observational curves $n_k(t)$ shows that though they are in good agreement with one another, there exists a systematical divergence, i.e. a selection of observational data. The reason for such divergence may be in disturbance of the suggestions on stationarity of the aggregate activity or the independance of the flares from one another. However, similarity of the behaviour of these divergenses in time for Orion and Pleiades aggregates and our study showed that there is the selection made by the observers, who had a nonstandard approach towards detection of flares and flare stars at different observational times. It seems to us that it is necessary to make an inspection of the plates to remove selection factors for further corrections of the prediction.

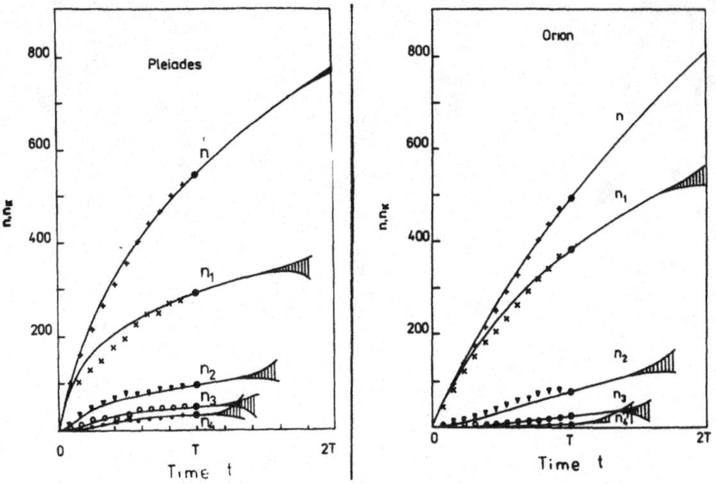

Figure 1.The observation values of the numbers n(t) and $n_k(t)$ for Pleiades and Orion (different symbols) for the past time (t<T), and the theoretical behaviour (continuous curves) of them for the past and future (the prediction) times.

REFERENCES

1. Ambartsumian.V.A.(1968) 'On the statistics of flare stars' in V.V. Sobolev (ed.), Stars, Nebula, Galaxies, Armenian Academy of Sciences, Yerevan, pp. 283-289.
2. Ambartsumian,V.A.(1978) 'On the derivation of the distribution of the frequences of stellar flares in a stellar aggregate', Astrofizika, 14, 367-377.
3. Mnatsakanian.M.A.(1986) 'On the question of the distribution of frequences of stellar flares in stellar aggregates', Astrofizika, 24, 621-623.
4. Mnatsakanian,M.A.(1987) 'The renormalization group analogies in astrophisics' in Contributions of 'Conference "Renormalization Group-86", D-2-87-123, Dubna, pp.375-393.

5. Mnatsakanian.M.A. and Mirzoyan,A.L.(1988) 'A prediction of the flare activity of stellar aggregates I. Theory'. Astrofizika, 29, 32-43.
6. Haro,G., Chavira,E and Gonzales.G.(1982) 'A catalog and identdification charts of the Pleiades flare stars', Bol.Inst.Tonantzintla, 3, No.1, 3-68.
7. Natsvlishvili.R.Sh.(1987) Flare Stars in Regions Orion and Pleiades, Byurakan Astrophysical Observatory.

FLARE STARS AS AGE INDICATORS IN OPEN CLUSTERS

L.PIGATTO
Osservatorio Astronomico
Vicolo dell'Osservatorio, 5
35122 Padova
Italy

ABSTRACT. A new set of isochrones derived from overshooting evolutionary models, allows one to give a homogeneous age determination for open clusters with solar chemical composition and in a range of age $4\ 10^7 \div 1.5\ 10^9$ yr. In this context is derived an empirical new age calibration - absolute magnitude of the supposed brightest flare star in a cluster versus log age - . The evolutionary implication in finding or not finding flare stars in candidate open clusters is also examined.

1. Introduction

The main tools in clusters age determination are the isochrones, the same time-lines, computed from the evolutionary models of the same chemical composition. In these last years the evolutionary models have come to reach an higher accuracy because of new and up-to-date physical inputs: but some controversies are still open among modelers about the best evolutionary models capable of explaining some features in the clusters HR diagram not understandable in the classical evolutionary scheme. We think that the best models are those which are able to explain the greatest number of evolutionary problems as a whole. In this context the overshooting evolutionary models and the derived isochrones (Bertelli et al. 1986) were able to account for several questions about the HR diagram characteristics in open clusters (Mazzei and Pigatto, 1988, 1989 and references therein).

2. Age determination in open clusters

Let us concentrate on the problems of the young and intermediate-age open clusters. Mazzei and Pigatto (1988) put into evidence the necessity of a correct method in the clusters age determination by using not only the pure best-fit technique, but also comparing the observed and theoretical luminosity functions, finally deriving from the isochrones synthetic clusters which are able to mimic the morphology of the observed HR diagrams. The age of the Pleiades derived by this method is of $1.5\ 10^8$ yr (Mazzei and Pigatto,1989). The uncertainty in the Pleiades age determination is of about 50%; but a better determination of about 10% of uncertainty can be derived for intermediate-age open clusters (Mazzei and Pigatto,

L. V. Mirzoyan et al. (eds.), Flare Stars in Star Clusters, Associations and the Solar Vicinity, 117–120.

1988) where the presence of the red giants clump provides an additional constraint in the age calibration. In the same way the age of the open cluster NGC 7092 is here derived by using the overshooting isochrones for solar chemical composition. The choice of this cluster will be discussed in the next section. Figure 1a shows the colour-magnitude diagram of NGC 7092 in which, for the assigned age, the 6 10^8 yr old isochrone is overimposed.

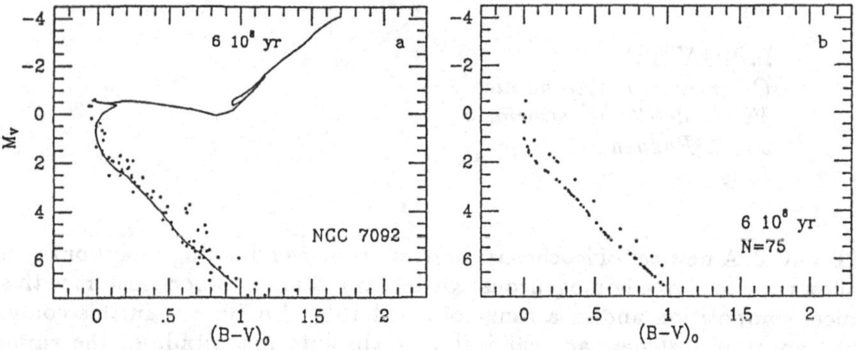

Figure 1. The colour-magnitude diagram of NGC 7092 (a) and overimposed the overshooting isochrone of 6 10^8 yr with solar chemical composition and a synthetic cluster (b) from the same isochrone for N=75 objects.

To support the reliability of this age determination, a synthetic cluster chosen for the best morphological resemblance from a group of samples computed from the 5, 6 and 7 10^8 yr old isochrones respectively is shown in fig. 1b.

3. Flare stars as age indicators

Usually ages in clusters are derived by the analysis of the brilliant Main-Sequence (MS) portion just above and around the turnoff region, but it is also interesting, from the evolutionary point of view, to examine the fainter MS portion, i.e. the turnon region, where the stars are just arriving in the MS after the contraction phase. The flaring activity is generally associated with this evolutionary phase, and the relation with the clusters age is well demonstrated by many observational facts such as the earliest spectral type and/or the brightest magnitude of the flare stars in a cluster. Parsamian (1976, 1985) first put into evidence the dependence on the age of the absolute magnitude M_f of the brightest flare star in a cluster or aggregate and the possible monotonous behaviour of this relation. Figure 2 shows the relation between the brightest flare star (Haro,1968 and references therein) versus log t, in three well known clusters as NGC 2264, the Pleiades and the Praesepe (full squares). Ages are derived from the turnoff overshooting isochrones. Because of the new age and because of a distance modulus of 5.88 mag by Pérez et al.(1988), NGC 2264 is correctly located in the linear correlation M_f, log t. The reason for this empirical and rather uncertain calibration, is based on a first attempt at listing open clusters in which flare stars could be looked for. In this calibration one fact is evident: in the range of age $10^8 \div 10^9$ yr, from the Pleiades to the Praesepe, no open clusters with known flare stars are present.

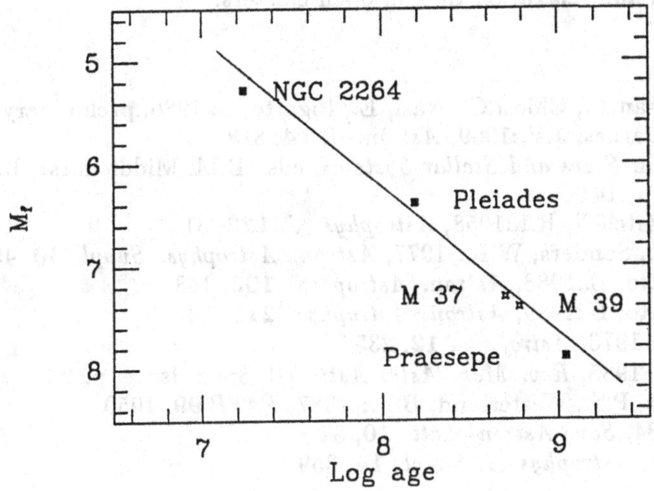

Figure 2. Absolute magnitude of the brightest flare star versus log age. The ages are derived from the overshooting isochrones for solar chemical composition.

The first reason can refer simply to the fact that a detectability of a flare is connected with its amplitude and with the cluster distance. Another reason can be due to the fact that in a cluster not very young and with very few stars, the probability of finding flare stars is very low. In fact, the flare activity is going to reduce and then disappears in the oldest clusters where even the low mass stars have already arrived on the MS. From the calibration the supposed apparent magnitude of the brightest flare stars for NGC 2099 and NGC 7092 has been derived taking into account the reddening and the distance modulus (photometric data are from West, 1967 and Platais, 1984 respectively). The not very faint value of this magnitude (14.5 mag) for NGC 7092 suggests that this cluster can be a good candidate for the search of flare stars. NGC 7092 (M 39) was believed peculiar by Mc Namara and Sanders (1977) because of the lack of faint stars in its HR diagram. A more recent research by Platais (1984) has shown stars in the low MS down to the 15th mag. If the flare activity is an obligatory step in the evolutionary path of the stars in a cluster, flare stars can be expected in NGC 7092, accounted for the dependence on the number of the cluster stars and on the observation time. On the other hand the lack in finding flare stars in the cluster area could give interesting information on the mass segregation and on the cluster relaxation time.

4. Conclusions

To discover flare stars in open clusters can signify answering many still open questions such as: 1) turnon identification and age determination from pre-main-sequence models, 2) rotation, binarity and/or multiplicity and their influence on the flare stars position in the HR diagram, 3) eventual difference between the turnoff and turnon age in a cluster as indicator of a possible bimodal star formation, 4) mass segregation and relaxation time in open clusters.

References

Bertelli,G., Bressan,A., Chiosi,C., Nasi, E., Pigatto, L.:1986, preliminary draft

Crawford,D.L., Barnes, J.V.:1969, *Astron. J.* **74**, 818

Haro, G.: 1968, in *Stars and Stellar Systems*, eds. B.M. Middlehurst, L.H. Aller, Chicago, **7**, p. 141

Johnson, H.L., Mitchell, R.I.:1958, *Astrophys. J.* **128**, 31

Mc.Namara, B.J., Sanders, W.L.: 1977, *Astron. Astrophys. Suppl.* **30**, 45

Mazzei, P., Pigatto, L.:1988, *Astron. Astrophys.* **193**, 148

Mazzei, P., Pigatto, L.:1989, *Astron. Astrophys.* **213**, L1

Parsamian, E.S.: 1976, *Astrofisika* **12**, 235

Parsamian, E.S.: 1985, *Rev. Mex. Astr. Astr.* **10**, Spec. Issue, p. 221

Pérez, M.R., The, P.S., Westerlund, B.E.: 1987, *PASP* **99**, 1050

Platais, I.K.: 1984, *Sov. Astron. Lett.* **10**, 84

West, R.F.: 1967, *Astrophys. J. Suppl.* **14**, 359

LANG: What are the uncertainties in your age determinations? Why aren't there any young stars above or below the main sequence in your HR-diagram?

PIGATTO: The uncertainties are 50 per cent. Only the brightest stars were plotted in my HR-diagrams; they show no stars off the main sequence, so such behavior only applies to the low luminosity stars.

THE RELATIVE NUMBER OF FLARE STARS IN SYSTEMS OF DIFFERENT AGES

V.V.HAMBARIAN, A.T.GARIBJANIAN,L.V.MIRZOYAN,
A.L.MIRZOYAN
Byurakan Astrophisical Observatory
Armenian Academy of Sciences, USSR

ABSTRACT. The problem of flare activity frequency among red dwarf stars is discussed.The observational data on stars in Orion, Pleiades and solar vicinity are used. It is shown, that relative number of flare stars among all red dwarf stars in the considered samples is increasing towards the lower luminosity stars. The flare stars are found in the considered samples begining with a definite luminosity, this limiting luminosity being decreased with the aging of the system which includes flare stars. A possible explanation of the observed phenomenon is given.

1. INTRODUCTION

An observational approach to the evolution of flare stars has shown that the flare activity stages is one of the early stages in stellar evolution [1-3].

On the other hand flare stars of different ages belong to the same class of non-stable stars possessing flare activity and the observed differences between them are due to their different ages [4].

Conseqently, one can expect that there is certain dependence of the flare activity upon the star age. In this report the frequency of flare activity phenomenon among red dwarf stars of different ages is estimated.

2. RELATIVE NUMBER OF FLARE STARS AMONG RED DWARF STARS

The relative number of flare stars is estimated for three samples of different ages:in the Orion association,Pleiades cluster and solar vicinity.

As criteria of star membership to these samples besides proper motion some physical parameters, in particular the flare activity are used, according to catalogues by

L. V. Mirzoyan et al. (eds.), Flare Stars in Star Clusters, Associations and the Solar Vicinity, 121–124.

Hertzsprung et al [5], Parenago [6] and Gliese [7,8]. For each sample the number of flare stars for corresponding magnitude range is determined as a sum of the numbers of known and potential ones. The latter number is estimated by Ambartsumian's formulae [1].

The results of these estimates are presented in Table 1, including M_{Pg} range of absolute photographic magnitudes, N - number of red dwarf stars in a given sample, N_F - number of flare stars among them and N_F/N - relative number of flare stars among all red dwarf stars in the sample.

Table 1 shows, that the relative number of flare stars- N_F/N, is increasing while the luminosities are decreasing for all three samples.

However, there is one significant difference between the considered samples, namely the magnitude ranges of flare star luminosities.

Both these regularities can be explained by different ages of these samples. Indeed, according to generally accepted proposition the rates of stellar evolution depend on star mass (luminosity). As a result the flare stars of higher luminosities lose their flare activity earlier than the lower luminosity flare stars.

Table 1 testifies in favour of this conclusion. In each of the samples the relative number of flare stars is increasing to lower luminosities. At the same time the older the sample the lower luminosities of flare stars meet in it.

TABLE 1. Relative number of flare stars among red dwarf stars [9].

M_{Pg}	N			N_F			N_F/N		
	I	II	III	I	II	III	I	II	III
4.4-5.5	73	-	-	2	-	-	0.03	-	-
5.5-6.5	115	-	-	10	-	-	0.09	-	-
6.5-7.5	143	-	-	26	-	-	0.18	-	-
7.5-8.5	104	34	50	79	14	1	0.76	0.41	0.02
8.5-9.5	*	55	100	*	37	3	*	0.67	0.03
9.5-10.5	*	64	131	*	40	7	*	0.63	0.05
10.5-11.5	*	73	67	*	58	10	*	0.79	0.15
>11.5	*	*	113	*	*	47	*	*	0.42

I - Orion, II - Pleiades, III - Solar vicinity

Another important result following from Table 1 is the decreasing of the relative number of flare stars with the age of the sample.

This result can also be interpreted by the dependence of stellar evolution rates from star mass.

It should be added that the data presented in Table 1 are only qualitative. Besides, they are not complete for Orion sample: in some intervals of absolute photografic magnitude (M_{pg}) the data on the numbers of not flaring red dwarf stars are absent.

However, even these qualitative data, which are trustworthy enough seem to be important for further study of the flare activity phenomenon in evolution of red dwarf stars.

3. CONCLUSION

The analysis of the observational data for three samples of flare stars in Orion, Pleiades and solar vicinity having different ages shows that relative number of flare stars among red dwarf stars is increasing to lower luminosities. It is the direct consequence of the fact that the duration of flare activity phenomenon depends on the masses (luminosities) of corresponding stars. Due to this dependence the older is the flare star system the lower are luminosities of corresponding flare stars.

REFERENCES

1. Ambartsumian, V.A. (1969) 'On the statistics of flaring objects', in V.V.Sobolev (ed.) Stars, Nebulae, Galaxies, Armenian Academy of Sciences, Yerevan, pp.283-292.
2. Ambartsumian, V.A. and Mirzoyan, L.V. (1981) 'Flare stars', in: New Directions and New Frontiers in Variable Star Research, Veroff, Bamberg 9, Nr.100, 98-108.
3. Mirzoyan, L.V. (1984) 'Flare stars', Vistas in Astronomy, 27, 77-109.
4. Mirzoyan, L.V. and Hambarian, V.V. (1988) 'Statistical study of flare stars I. The UV Ceti stars of solar vicinity and the flare stars in clusters and associations', Astrofizika, 28, 375-389.
5. Hertzsprung, E. et al (1947) 'Catalogue de 3259 etoiles dans les Pleiades', Ann.Leiden Obs., 19, No.1A 1-85.
6. Parenago, P.P. (1954) 'A study of stars in the Orion nebulae region', Trudy Astron.Inst.Sternberga, 25, 3-546.
7. Gliese, W. (1969) 'Catalogue of nearby stars', Veroff, Astron.Rechen. Inst.Heidelberg, No.22, 1-116.
8. Gliese, W and Jahreiss, H. (1979) 'Nearby star data published 1969-1978', Astron.Astrophys.Suppl.Series, 38, 423-448.

9. Mirzoyan.L.V.,Hambarian.V.V.,Garibjanian.A.T.and Mirzoyan, A.L.(1989) 'Statistical study of flare stars IV. Relative number of flare stars in the Orion association, Pleiades cluster and in the solar vicinity',Astrofizika, 31, 258-269.

LADA: In your comparative study of Orion, Pleiades, and the solar vicinity, did you take into account differing observing times for each group?

Of the stars listed in your Table 1, how many were actually observed to flare and how many were included as a result of correction or normalization due to different amounts of observing time? In other words how many stars did you add to account for the fact that each group of stars was monitored for different amounts of time?

HAMBARIAN: Of course you are right. The observational times for the three samples listed in Table 1 are quite different. But the relative number of flare stars among all red dwarf stars (N_F/N) do not depend on the observational time because N_F includes not only the known flare stars, but also the potential ones which we have estimated from the formula by Ambartsumian, $n_o = (n_1^2/2n_2)$, where n_1 is the number of stars seen to flare once and n_2 are those seen to flare twice.

RODONO: Did you take into account the different detection thresholds that favour the detection of faint flare stars or do your data refer to original uncorrected observations?

HAMBARIAN: No. But this circumstance cannot explain the observed regularity.

FLARE STARS AT RADIO WAVELENGTHS

KENNETH R. LANG
Department of Physics and Astronomy
Tufts University
Medford, MA 02155
U.S.A.

ABSTRACT. The radio emission from dMe flare stars is discussed using Very Large Array and Arecibo observations as examples. Active flare stars emit weak, unpolarized, quiescent radio radiation that may be always present. Although thermal bremsstrahlung and/or thermal gyroresonance radiation account for the slowly-varying, quiescent radio radiation of solar active regions, these processes cannot account for the long-wavelength quiescent radiation observed from nearby dMe flare stars. It has been attributed to nonthermal gyrosynchrotron radiation, but some as yet unexplained mechanism must be continually producing the energetic electrons. Long-duration (hours), narrow-band ($\Delta\nu/\nu < 0.1$) radiation is also emitted from some nearby dMe stars at 20 cm wavelength. Such radiation may be attributed to coherent plasma radiation or to coherent electron-cyclotron masers. Impulsive stellar flares exhibit rapid variations (< 100 msec) that require radio sources that are smaller than the star in size, and high brightness temperatures $T_B > 10^{15}$ K that are also explained by coherent radiation processes. Quasi-periodic temporal fluctuations suggest pulsations during some radio flares. Evidence for frequency structure and positive or negative frequency drifts during radio flares from dMe stars is also presented.

1. INTRODUCTION

Pioneering single-dish observations in the 1970s showed that dwarf M flare stars occasionally emit radio bursts with extremely high flux densities and brightness temperatures of $T_B > 10^{12}$ to 10^{15} K if the radio source is comparable to the star in size (see Lang and Willson (1986b) and Kundu and Shevgaonkar (1988) for some historical details). Such powerful radio flares are extremely rare, sporadic and brief, however, leading many to suspect that they might be confused with terrestrial interferance. Moreover, identification by correlation with optical flares could not be relied on, for different radiation mechanisms often dominate in the two spectral domains.

L. V. Mirzoyan et al. (eds.), Flare Stars in Star Clusters, Associations and the Solar Vicinity, 125–138.

Interferometric observations with the Very Large Array (VLA) have unambiguously differentiated stellar radio emission from terrestrial interferance, and the large collecting area of both the VLA and the Arecibo Observatory have enabled detection of the relatively weak radio flares that are presumably more frequent than the more powerful ones. One survey, for example, indicates that flaring emission can be detected for about 40% of flare stars nearer than 10 parsecs and visible with the VLA (White, Kundu and Jackson, 1990). Under the assumption that the radio source size is equal to the stellar radius, brightness temperatures of $T_B = 10^8$ to 10^{10} K and $T_B = 10^9$ to 10^{11} K have been inferred for the detected stars at 6 cm and 20 cm wavelength, respectively.

Although the observed radio luminosity from stellar flares is only about one thousandth the luminosity observed at X-ray or optical wavelengths, the radio emission is still thousands of times more powerful than solar radio flares, and it serves as an important diagnostic tool for studies of stellar coronae. Such studies have been carried out in detail for the most active dwarf M flare stars listed in Table 1. They are all nearby dMe stars that show evidence for chromospheric activity in the form of Ca II, Mg II and Hα emission lines.

Table 1. Accurate 6 cm positions, spectral type, Sp, quiescent flux density, S_{Q6}, at 6 centimeters wavelength, peak flaring flux density, S_{F20}, at 20 centimeters wavelength, distance, D, in parsecs, and the logarithm of the quiescent X-ray luminosity, log L_x, for radio-active flare stars.

Star	R.A.(1950.0) h m s	Dec.(1950.0) ° ′ ″	Sp	S_{Q6} (mJy)	S_{F20} (mJy)	D (pc)	log L_x (erg s^{-1})
L 726–8A**	01 36 33.314	− 18 12 23.20	dM5.5e	1.0	20	2.7	27.5
UV Ceti**	01 36 33.404	− 18 12 21.56	dM6e	3.2	100	2.7	27.5
YY Gem	07 31 25.691	+31 58 47.23	dM1e	0.4	1	14.5	29.5
YZ CMi	07 42 02.962	+03 40 30.39	dM4.5e	0.5	20	6.0	28.5
AD Leo	10 16 52.604	+20 07 17.59	dM3.5e	1.1	100	4.9	29.0
Wolf 630A,B	16 52 46.455	− 08 15 13.715	dM4.5e	0.9	3	6.2	
AT Mic†	20 38 44.4	− 32 36 49.5	dM4.5e	3.6	6	8.8	29.3
AU Mic	20 42 04.558	− 31 31 17.50	dM0e	0.8	26	8.8	29.8
EQ Peg A††	23 29 20.910	+19 39 41.11	dM5e	0.3	25	6.4	

*Adapted from Kundu et al. (1987), and Jackson, Kundu and White (1989) for the 6 cm data, with positions accurate to 0″.1 or better.

**The separation and position angle of L 726–8A and L 726–8B (UV Ceti) is 2″.080 ± 0″.080 at 38°.0 ± 1°.9.

†Southern component of a fully resolved binary whose components are both active radio emitters; the nothern component lies about 3″.6 away at a position angle of about 15°.

††Quiescent emission from both component of EQ Peg A, B has been previously detected with respective 6 cm fluxes of 0.7 and 0.4 mJy and an angular separation of about 3″.0 (see Gary (1985)).

This review will focus on these radio-active dMe stars. They include the very few cases in which we can detect the relatively weak (a few mJy), quiescent radio flux of the star (see Section 2). Long-duration, narrow-band radio flares (Section 3) have also been observed from several of these stars; they are unlike anything observed on the Sun.

Powerful (up to 200 mJy), impulsive (a few minutes) radio flares are also emitted by these stars. Such flares exhibit rapid variations (Section 4.1), quasi-periodic fluctuations (Section 4.2), both narrow-band and broad-band features (Section 5) and positive and negative frequency drifts (Section 5).

Figures 1 and 2 illustrate such flares for the dMe star EQ Pegasi. They are often up to 100% circularly polarized. Successive oppositely polarized flares have been detected for EQ Pegasi (Fig. 2) and AD Leonis (Willson, Lang and Foster (1988)), suggesting the presence of both magnetic polarities; but YZ Canis Minoris always exhibits left-handed circular polarization that remains the same over a wide range of wavelength (6 cm to 90 cm), suggesting a global, dipolar magnetic field that is viewed pole-on (Kundu and Shevgaonkar (1988)).

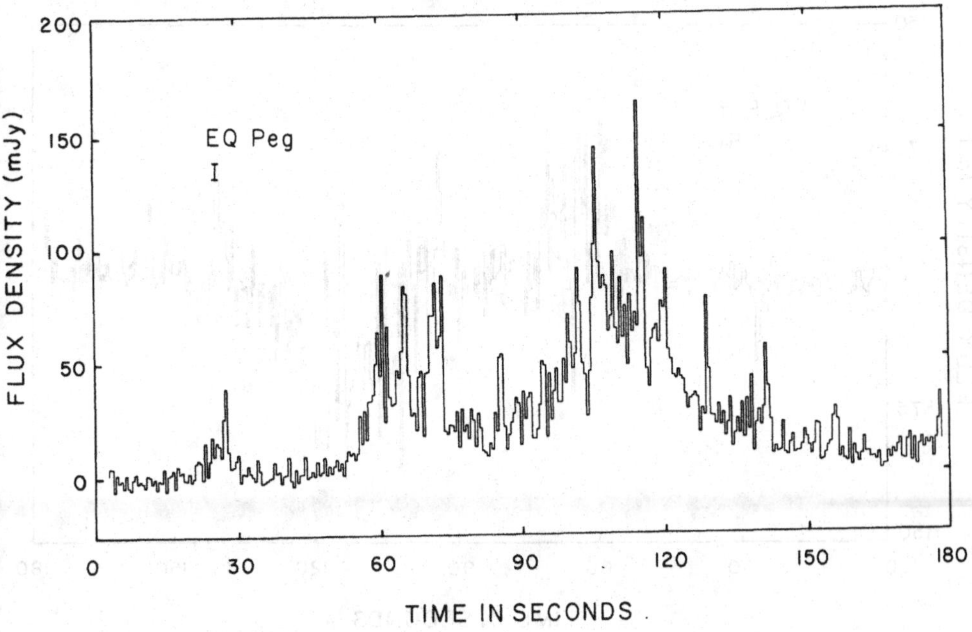

Figure 1. A previously unpublished VLA observation of the total intensity, I, at 1420 MHz (21 cm) from the dwarf M star EQ Pegasi. The flaring emission has a total duration of minutes, with components on shorter time scales of seconds or less. The radiation is up to 100 percent circularly polarized (see Fig. 2).

2. QUIESCENT RADIO EMISSION FROM dMe FLARE STARS

Unpolarized radio radiation that is nearly always present, and shows only slow variations with time, has been termed quiescent radio emission to distinguish it form the highly variable, brief radio flares that are often highly circularly polarized. The quiescent emission has flux densities of a few mJy at 6 cm wavelength (see Table 1 and Gary (1985)). It was at first attributed to thermal radiation from the same electron population that gives rise to the stellar X-ray emission (also see Table 1). If this were the case, it would be consistent with the Sun's slowly varying radiation at centimeter wavelengths; the solar radiation is due to either thermal bremsstrahlung or the thermal gyroresonance radiation of hot (10^6K) electrons trapped within coronal loops that radiate strongly at X-ray wavelengths.

Nevertheless, thermal emission from the stellar coronae observed in X-rays cannot easily account for the quiescent radiation. Thermal bremsstrahlung of the X-ray plasma is so optically thin at 6 cm wavelength that its flux density is one or two orders of magnitude below the detection limit of the VLA - even when the X-ray plasma covers the entire surface of the nearest flare star.

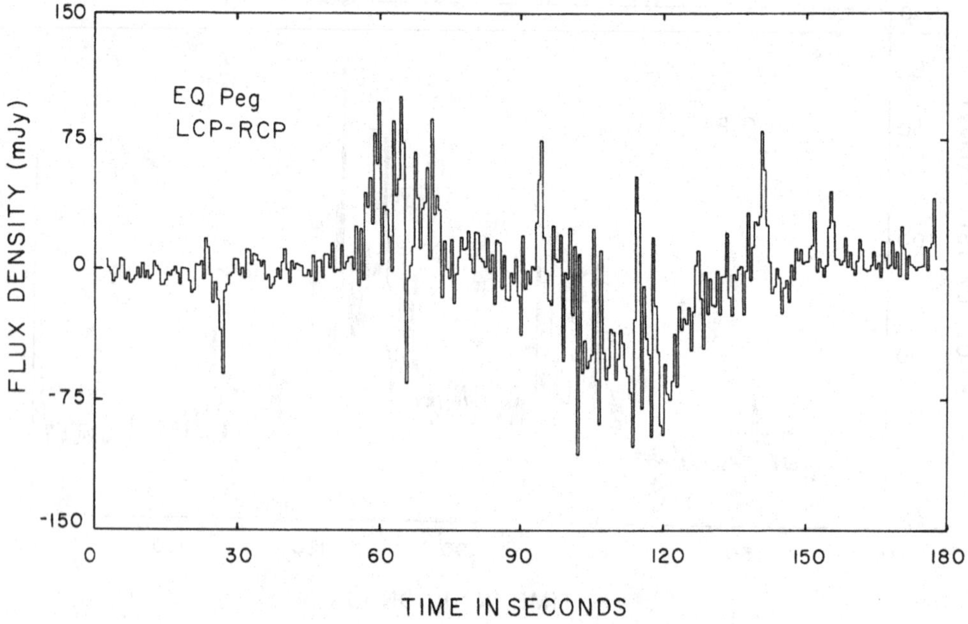

Figure 2. The difference between the left-hand circularly polarized (LCP) and right-hand circularly polarized (RCP) radiation from the dwarf M star EQ Pegasi at 1420 MHz (21 cm). Up to 100 percent circularly polarized radiation is emitted with opposite senses, or directions, for successive bursts (courtesy of Robert F. Willson).

The optical depth can be enhanced for the gyroresonance of thermal electrons in relatively strong magnetic fields. Short-wavelength < 6 cm) quiescent emission might then be **explained** by optically thick gyroresonant radiation of the high-temperature tail of the X-ray emmitting plasma (Gudel and Benz, 1989). In this case, the radio data require electrons that are at a higher temperature than the average X-ray emitting electrons, and they are probably emitted from a source that is much larger than the star. The maximum observed flux density, S, is given by the Rayleigh-Jeans law, and the radius of the emitting source is therefore given by:

$$R^2 = 10^{13} \frac{SD^2}{\nu^2 T} \quad cm^2 ,$$

where S is the source flux denisty in Jy, D is the distance in cm, T is the temperature in K, and the observing frequency is $\nu = 2.8 \times 10^6$ nH Hz for the n th harmonic in a magnetic field of strength H. Radii comparable to those of the dwarf M stars are only obtained for low flux densities, $S = 1$ mJy, short wavelengths, $\lambda < 6$ cm, high temperatures $T > 10^7$ K, and nearby dMe flare stars, $D < 10$ pc.

Thermal gyroresonance radiation cannot explain the long-wavelength, $\lambda > 20$ cm, quiescent radiation where higher flux densities of S = 2 to 20 mJy have been observed (Lang and Willson, 1986 a, Bastian and Brokbinder, 1987, Gudel and Benz, 1989). The radio source would have to be tens to hundreds of times larger than the star with implausibly intense magnetic fields at these remote locations. This long-wavelength radio emission has been attributed to nonthermal gyrosynchrotron radiation.

The gyrosynchrotorn hypothesis requires a hotter corona with smaller sizes and lower densities than the gyroresonance model, but there must be a currently - unexplained, steady source of energetic electrons. Both models have been discussed by Kundu et al. (1987); because the gravitational scale height is comparable to the height of the stellar coronae, magnetic structures are required to confine the radio-emitting plasma. The coronae of nearby dMe flare stars therefore bear a closer resemblance to the Earth's magnetosphere than to the Sun's corona. We do not know if the relevant magnetic structures on the flare stars are due to several small active regions or to a global dipolar field, and we do not know how the radio-emitting electrons interact with the X-ray emitting plasma. The long-wavelength quiescent **radiation might alternatively be due to** continued, low-level, narrow-band, coherent radiation that resembles radio flares from these stars.

3. LONG-DURATION, NARROW-BAND EMISSION

Relatively intense(S = 100 mJy), narrow-band ($\Delta\nu/\nu < 0.1$) radiation lasting for several hours has been observed at 20 cm wavelength in several flare stars (Lang and Willson, 1986; White, Kundu and Jackson, 1986; Kundu et al., 1987; Lang and Willson, 1988). These long-duration events are slowly variable, so they might be more energetic version of the process that accounts for the quiescent radiation. However, the long-duration, narrow-band radiation is highly circularly polarized, so its polarization bears a closer resemblance to the stellar flares than to the unpolarized quiescent radiation.

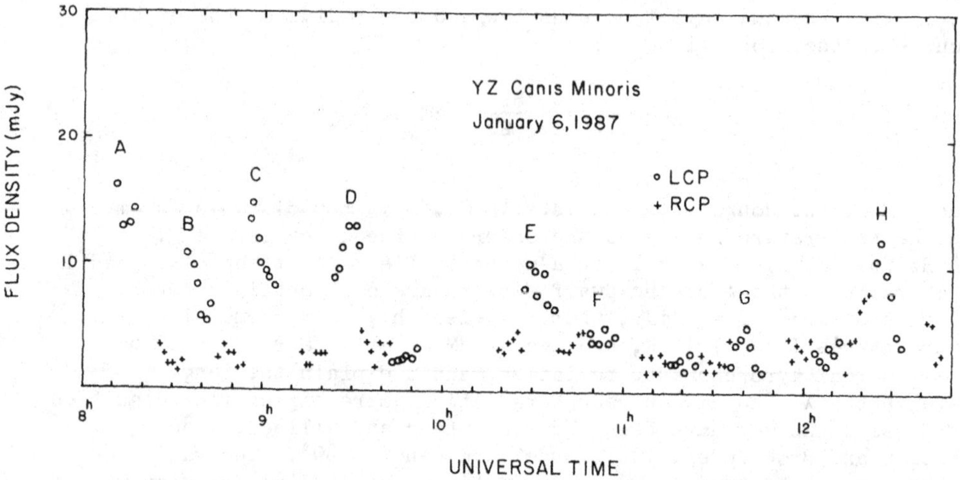

Figure 3. A five-hour VLA observation of the total intensity of the radiation from the dwarf M star YZ Canis Minoris in a 50 MHz bandwidth centered at 1464.9 MHz. **The radiation is 100 percent** left-hand circularly polarized. Its narrow-band frequency structure is illustrated in Fig. 4. (adapted from Lang and Willson (1986a)).

The long-duration, narrow-band highly polarized radiation (see Figures 3 and 4) is unlike any flares observed on the Sun, and cannot be easily explained using the solar analogy. The energy release mechanism lasts at least an order of magnitude longer than solar flares, and is difficult to understand if magnetic reconnection is the source of energy for the stellar flares (White, Kundu and Jackson, 1986). The narrow-band structure cannot be attributed to continuum emission processes such as thermal bremsstrahlung, thermal gyroresonant radiation, or nonthermal gyrosynchrotron radiation; it may be due to coherent mechanisms like electron-cyclotron masers or coherent plasma radiation. Both mechanisms require a magnetic field to produce the high circular polarization - either at the site of radiation production or during subsequent propagation of initially unpolarized radiation.

The coherent radiation processes provide constraints on the physical conditions in the coronae of flare stars (Lang, 1986). An upper limit to the electron density in the source is given by the requirment that the observing frequency must be greater than the

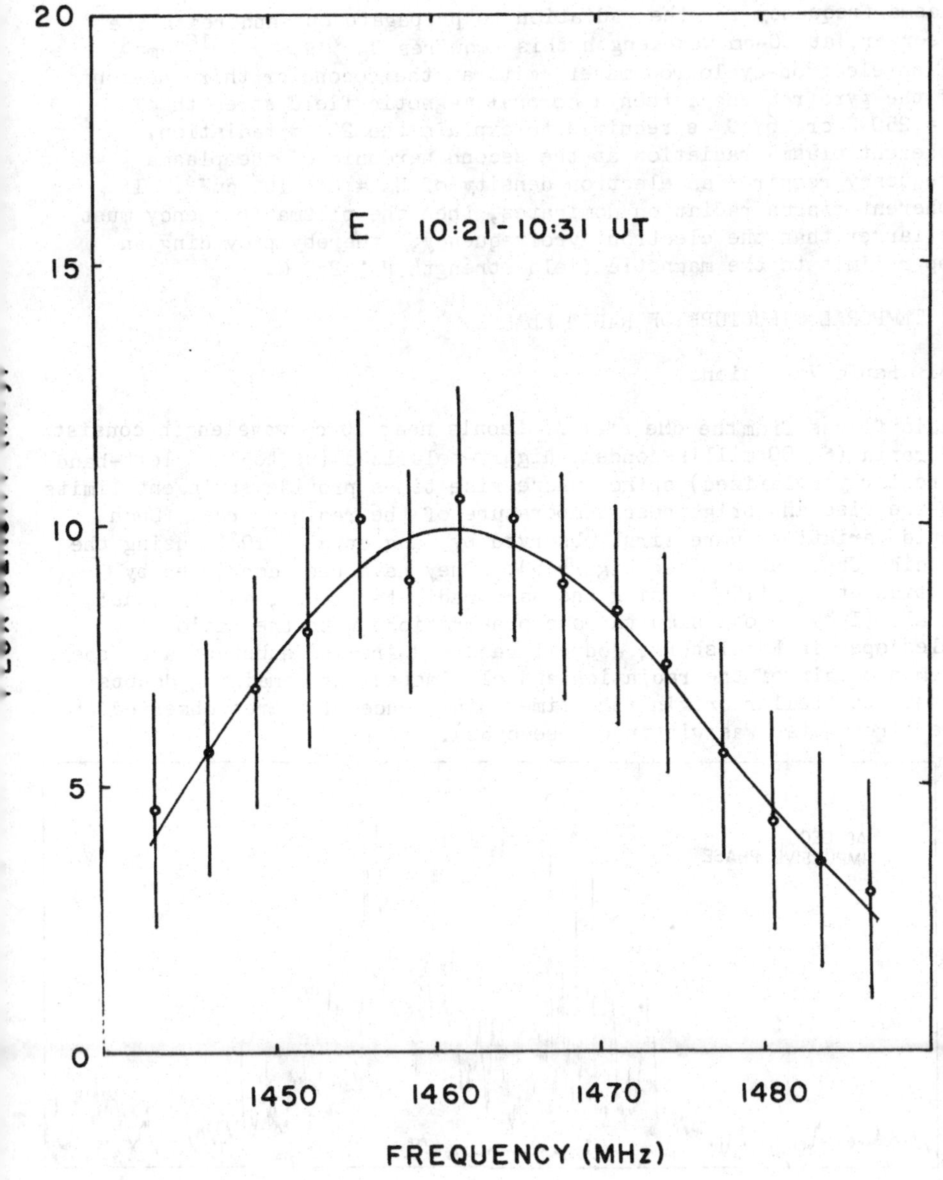

Figure 4. Frequency spectrum of the left circularly polarized radiation from YZ Canis Minoris for the interval marked E in Figure 1. Here the total intensity is plotted for 15 contiguous channels, each 3.125 MHz wide, for the 10-second interval. These data, as well as those for other intervals lettered in Fig. 3, show evidence for narrow-band radiation with a bandwidth $\Delta\nu < 30$ MHz, or $\Delta\nu/\nu = 0.02$. (Adapted from Lang and Willson (1986a)).

plasma frequency for the radiation to propagate out and reach the observer; at 20-cm wavelength this requires $N_e < 2.5 \times 10^{10}$ cm^{-3}. If an electron-cyclotron maser emits at the second or third harmonic of the gyrofrequency, then a coronal magnetic field strength of H = 250 G or 167 G is required to explain the 20 cm radiation. Coherent plasma radiation at the second harmonic of the plasma frequency requires an electron density of $N_e = 6 \times 10^9$ cm^{-3}. If coherent plasma radiation dominates, then the plasma frequency must be larger than the electron gyrofrequency, thereby providing an upper limit to the magnetic field strength H < 250 G.

4. TEMPORAL STRUCTURE OF RADIO FLARES

4.1 Rapid Variations

Radio flares from the dMe star AD Leonis near 20 cm wavelength consist of rapid (< 100 milliseconds), highly-polarized (up to 100% left-hand circularly polarized) spikes whose rise times provide stringent limits to the size and brightness temperature of the radio source. Such rapid variations were first observed by Lang et al. (1983) using the Arecibo Observatory (see Figure 5). They have been confirmed by Bastian et al. (1990) using the same radio telescope, and by Gudel, et al. (1989) whose simultaneous observations with the radio telescopes in Effelsberg, Jodrell Bank and Arecibo substantiated the common origin of the radiation and eliminated any remaining doubts about its stellar origin (the time coincidence of flares observed at the three sites was within 0.4 seconds).

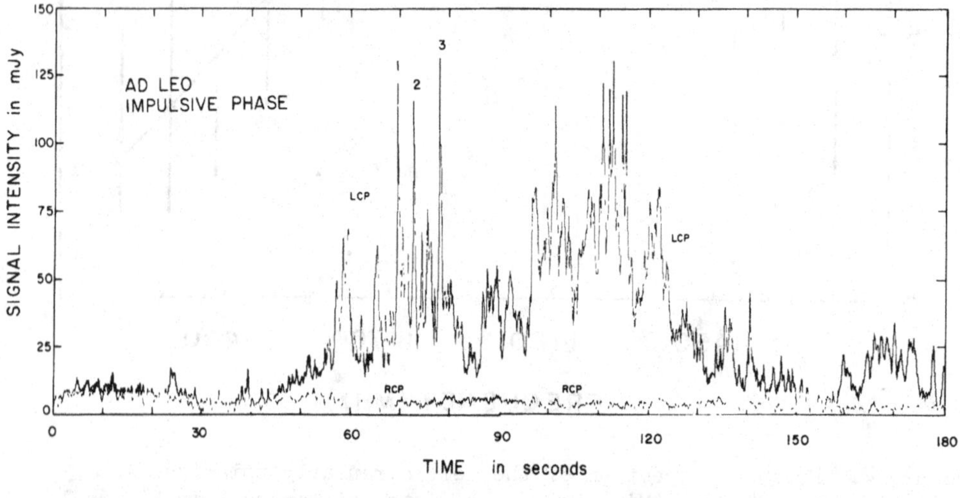

Figure 5. These observations, taken at 1400 MHz (21 cm) with the Arecibo Observatory, indicate that highly left-circularly polarized (LCP) radiation from the dwarf M star AD Leonis consists of rapid spikes whose duration τ < 100 milliseconds. The emitting source must be much smaller than the star in size.(Adapted from Lang et al. (1983)).

The **spikes** labeled 1,2 and 3 in Figure 5 have rise times $\tau < 200$ milliseconds, and upper limits of $\tau < 20$ milliseconds have been observed. An upper limit to the linear size, L, of the emitting region is provided by the distance that light travels in time, τ, or $L < c \times \tau$. A light-travel time of 20 milliseconds indicates $L < 6{,}000$ km, which is less than 1% of the stellar diameter. If the burst emitter is symmetric, it has an area less than 0.0003 of the stellar surface area, and the brightness temperature, T_B, is $T_B > 10^{15}$ K.

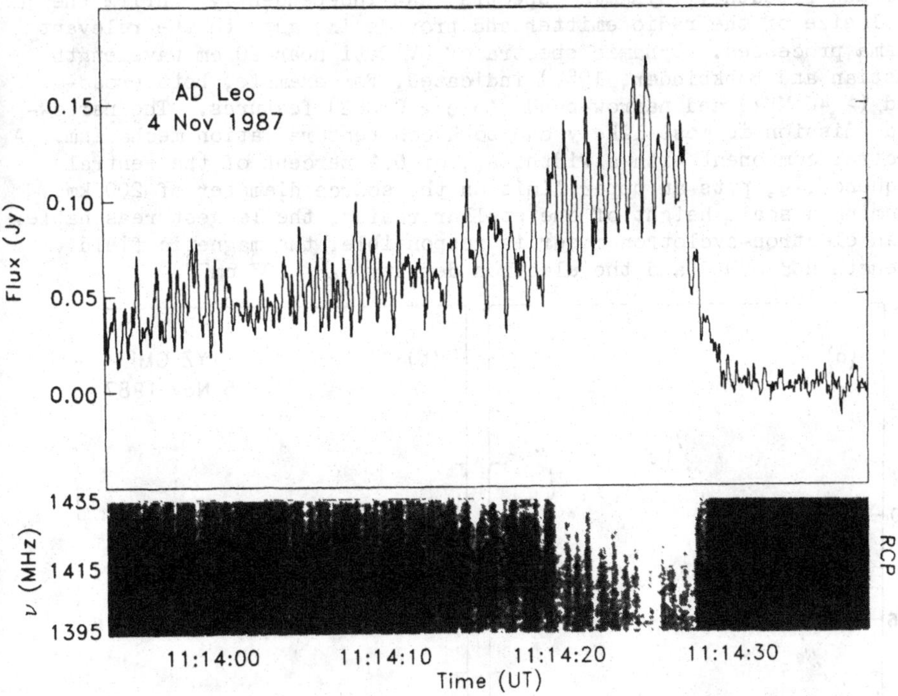

Figure 6. Observations of AD Leonis at 1415 MHz (21 cm) with the Arecibo Observatory indicate quasi-periodic pulsations with an amplitude modulation of ≈ 50 percent and a period of ≈ 0.7 seconds (top). The dynamic spectrum (bottom) indicates that the pulsations are broad-band with bandwidths $\Delta\nu > 40$ MHz. (Adapted from Bastian et al. (1990)).

4.2. Quasi-Periodic Fluctuations

Radio flare emission from AD Leonis has also been resolved into a multitude of broad-band, quasi-periodic fluctuations called pulsations. Such pulsations have been reported by Lang and Willson (1986), Gudel et al. (1989) and Bastian et al. (1990); and example is shown in Figure 6. The typical interval between pulses is about 1 second, which is comparable to that of solar decimetric pulsations. The AD Leo pulsations are up to 100% circularly polarized; they are coherent across the observing bandwidths of up to 100 MHz. They might be attributed to oscillations in a coronal loop with dimensions of about 5,000 km, which is close to the upper size limit inferred from the light-travel time.

5. FREQUENCY STRUCTURE OF RADIO FLARES

Observations of the radio radiation intensity as a function of both time and frequency (dynamic spectra) can independently confirm the small size of the radio emitter and provide insights to the relevant plasma processes. Dynamic spectra of UV Ceti near 20 cm wavelength (Bastian and Bookbinder, 1987) indicated, for example, both broad-band (> 40 MHz) and narrow-band ($\Delta\nu/\nu$ < 0.002) features. The narrow-band emission is most likely due to a coherent radiation mechanism. A spectral component with a width, $\Delta\nu$, of 0.2 percent of the central frequency, ν, puts an upper limit on the source diameter of 200 km, assuming a scale height of one stellar radius, the largest reasonable. If an electron-cyclotron maser is responsible, the magnetic field strength $H \approx 250$ G and the electron density $N_e < 10^9$ cm^{-3}.

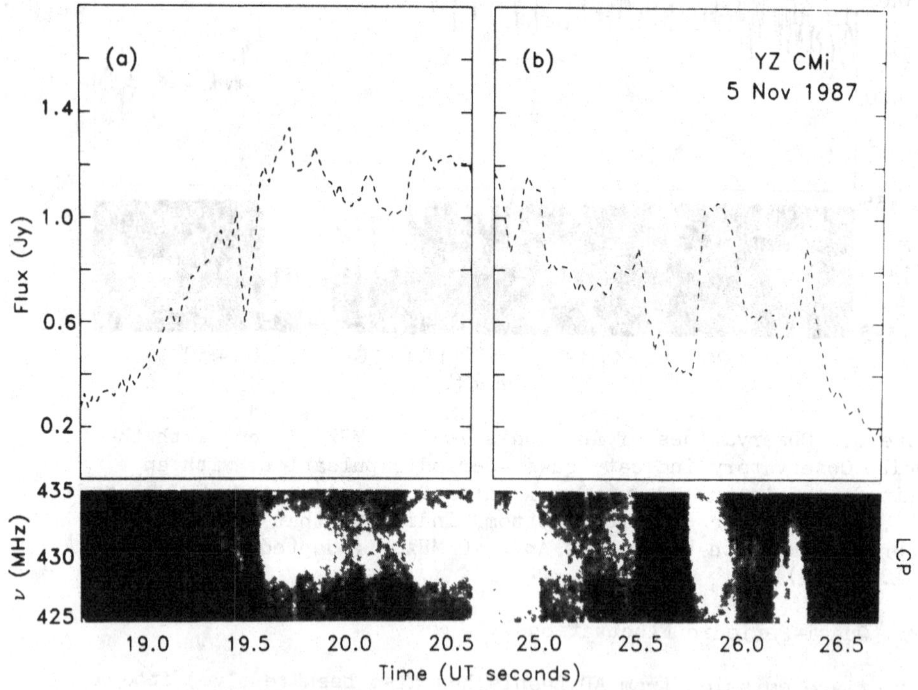

Figure 7. These Arecibo observations of the dynamic spectra (lower panel) of the 430 MHz (70 cm) radiation from the dwarf M star YZ Canis Minoris indicate a sudden reduction feature, in (a), with a drift in frequency of 250 MHz/s from high to low frequencies, as well as other narrowband and drifting features in (b). (Adapted from Bastian et al. (1990)).

Dynamic spectra of the radio radiation from the dMe flare stars AD Leo, YZ Cmi and UV Ceti show considerable complexity, with both narrow-band and broad-band features, and both positive and negative frequency drifts (Bastian and Bookbinder, 1987; Jackson, Kundu, and White, 1987; Bastian et al., 1990). The example shown in Figure 7 has a negative drift of 250 MHz per second from high to low frequencies. Such a negative drift is commonly observed in the Sun (eg. type II and type III bursts) suggesting electron beams or shock waves that propagate outwards in the stellar corona. But positive frequency drifts have also been observed for flare stars, suggesting a distrubance that propagates downward in the stellar corona and progressively excites plasma radiation at higher frequencies (larger electron densities).

Thus, there is clear evidence for apparent frequency drifts and narrow-band features in radio flares from dMe stars, but their interpretation is currently open to question. They could be due to the progagation of an exciter, group delays or some other cause.

6. DISCUSSION

To sum up, the quiescent radio radiation from dMe stars might be due to exceptionally hot thermal electrons, nonthermal electrons, or near continual coherent flaring. The high brightness temperatures, strong circular polarization and narrow frequency extent of long-duration radio events require a coherent plasma process, as does the more impulsive stellar radio flares. The cyclotron maser could explain many aspects of the flaring emission (Dulk, 1985), but several other coherent radiation processes might be involved (Kuijpers, 1989; Mullan, 1989). When the correct radiation mechanisms are identified, perhaps as the result of future observations with broader bandwidths, we can accurately specify the physical parameters in the stellar coronae.

Different processes probably dominate at different wavelengths, as they do on the Sun, and both solar and stellar flares must be related to magnetic fields. Past theoretical studies of solar radio radiation can therefore provide a useful background for exploring plausible radiation mechanisms. However, the direct analogy of the Sun as a radio flare star is probably a mistake; the Sun is the wrong spectral type, and its radio flares are so weak that they would be undetectable at the distance of the nearest star. In addition, solar flares have near-simultaneous signatures at optical, radio and X-ray wavelengths, while flaring radio emission from the dMe stars is often undetectable in other regions of the electromagnetic spectrum (see for instance Kundu et al. (1989)). The available evidence therefore indicates that radio flares from dwarf M flare stars are physically very different from those occuring on the Sun.

7. ACKNOWLEDGMENTS

Radio astronomical studies of the Sun and other nearby active stars at Tufts University are supported under grant AFOSR-89-0147 with the Air Force Office of Scientific Research. Related solar observations are supported by NASA grant NAG 5-501. The Arecibo Observatory is part of the National Astronomy and Ionosphere Center, which is operated by Cornell University under contract with the National Science Foundation (N.S.F.). The Very Large Array is operated by Associated Universities, Inc., under contract with the N.S.F.

8. REFERENCES

Bastian, T.S. and Bookbinder, J.A. (1987) 'First dynamic spectra of stellar microwave flares', Nature, 326, 678-680.

Bastian, T.S., Bookbinder, J., Dulk, G.A. and Davis, M. (1990) 'Dynamic spectra of radio bursts from flare stars', Astrophysical Journal, April 10, 1990 issue.

Dulk, G.A. (1985), 'Radio emission from the sun and stars', Annual Review of Astronomy and Astrophysics 23, 169-224.

Gary, D.E. (1985) 'Quiescent stellar radio emission', in R.M. Hjellming and D.M. Gibson (eds.), Radio Stars, D. Reidel, Boston, pp. 185-196.

Güdel, M. and Benz, A.O. (1989) 'Broad-band spectrum of dMe star radio emission', Astronomy and Astrophysics Letters, 211, L5-L8.

Güdel, M., Benz, A.O., Bastian, T.S., Fürst, E., Simnett, G.M. and Davis, R.J. (1989) 'Broadband spectral observation of a dMe star radio flare', Astronomy and Astrophysics Letters 220, L5-L8.

Jackson, P.D., Kundu, M.R. and White, S.M., (1987) 'Dynamic spectrum of a radio flare on UV Ceti', Astrophysical Journal Letters 316, L85-L90.

Jackson, P.D., Kundu, M.R. and White, S.M. (1989) 'Quiescent and flaring radio emission from the flare stars AD Leonis, EQ Pegasi, UV Ceti, Wolf 630, YY Geminorum, and YZ Canis Minoris', Astronomy and Astrophysics 210, 284-294 .

Kuijpers, J. (1989) 'Radio emission from stellar flares', Solar Physics 121, 163-185.

Kundu, M.R., Jackson, P.D., White, S.M., and Melozzi, M. (1987) 'Microwave observations of the flare stars UV Ceti, AT Microscopii, and AU Microscopii', Astrophysical Journal 312, 822-829.

Kundu, M.R., Pallavicini, R., White, S.M., and Jackson, P.D. (1985) 'Co-ordinated VLA and EXOSAT observations of the flare stars UV Ceti, EQ Pegasi, YZ Canis Minoris and AD Leonis', Astronomy and Astrophysics 195, 159-171.

Kundu, M.R. and Shevgaonkar, R.K. (1988) 'Detection of the dMe flare star YZ Canis Minoris simultaneously at 20 and 90 centimeter wavelengths', Astrophysical Journal 334, 1001-1007.

Lang, K.R. (1986) 'Radio wavelength observations of magnetic fields on active dwarf M, RS CVn and magnetic stars', Advances in Space Research 6, No. 8, 109-112.

Lang, K.R., Bookbinder, J., Golub, L. and Davis, M.M. (1983) 'Bright, rapid, highly-polarized radio spikes from the M dwarf AD Leonis', Astrophysical Journal (Letters) 272, L15-L18.

Lang, K.R. and Willson, R.F. (1986a) 'Narrow-band, slowly varying decimetric radiation from the dwarf M flare star YZ Canis Minoris', Astrophysical Journal (Letters) 302, L17-L21.

Lang, K.R. and Willson, R.F. (1986b) 'Millisecond radio spikes from the dwarf M flare star AD Leonis', Astrophysical Journal 305, 363-368.

Lang, K.R. and Willson, R.F. (1988) 'Narrow-band, slowly varying decimetric radiation from the dwarf M flare star YZ Canis Minoris II', Astrophysical Journal 326, 300-304.

Mullan, D.J. (1989) 'Solar and stellar flares: questions and answers', Solar Physics 121, 239-259.

White, S.M., Jackson, P.D., and Kundu, M.R. (1989) 'A VLA survey of nearby flare stars', Astrophysical Journal Supplement, December 15, 1989.

White, S.M., Kundu, M.R., and Jackson, P.D. (1986) 'Narrow-band radio flares from red dwarf stars', Astrophysical Journal 311, 814 - 818.

Willson, R.F., Lang, K.R., and Foster, P. (1988) 'VLA observations of dwarf M flare stars and magnetic stars', Astronomy and Astrophysics 199, 255-261.

RODONO: Can you spend a few words on the correlation between microwave and hard X-ray flares on the Sun and stars?

LANG: Although the solar analogy would predict a correlation between microwave and hard X-ray emission, such a correlation will not exist on many flare stars because their radio emission is not continuum synchrotron radiation. The radio emission of dwarf M flares is often narrowband ($\Delta f/f<0.02$), high brightness temperature (T_B>10**15 K), and sometimes of long total duration (hours). However, simultaneous observations across the electromagnetic spectrum from X-ray to optical to radio should be encouraged to search for correlations and to increase the net observational time on these stars.

PALLAVICINI: I would like to comment on the question just raised by Dr. Rodono. On the Sun, there is a very good correlation between X-ray and radio microwave emission both during the impulsive phase and during the gradual phase of flares. This apparently is not the case for flare stars. We have carried out a detailed comparison of EXOSAT and VLA observations of 4 flare stars (cf. Kundu, Pallavicini, White and Jackson 1988, Astron. Ap. 195, 159), and we have found virtually no correlation between the two data sets. This shows that there are other processes going on on stars as Dr. Lang just pointed out.

BROMAGE: This is a comment. As Professor Rodono said, it is necessary for investigating the impulsive phase, to observe hard X-rays, greater than 20-30 keV. There has not been any possibility of observing such flares so far.

LANG: The GRO satellite will provide such an opportunity after it is launched next June. I would expect a possible correlation of hard X-rays with short centimeter radiation or millimeter radiation.

BENZ: The energy radiated in radio emission by a typical (dMe) radio burst is at most 10**26 erg. If it is a coherent emission process such as cyclotron maser, the electron energy may be as low as 10**28 erg. This is a small number compared to the observed energies in the optical and soft X-ray regimes. Thus the absence of correlation is less surprising.

LANG: I agree. The energy emitted during radio flares is less than that radiated at optical and soft X-ray wavelengths, and this difference may become less for coherent emission, but the difference in any event does support a lack of correlation, and different radiation mechanisms for the radio and other spectral regions.

BROADBAND SPECTRAL RADIO OBSERVATIONS OF FLARE STARS

A.O. BENZ[1], M. GÜDEL[1], T.S. BASTIAN[2], E. FÜRST[3], G.M. SIMNETT[4], L. POINTON[5]

[1]Institut für Astronomie, ETH-Zentrum, CH-8092 Zürich, Switzerland
[2]National Radio Astronomy Observatory, P.O.Box O, Socorro, NM, USA
[3]Max-Planck-Institut für Radioastronomie, Auf dem Hügel 69, D-5300 Bonn, Federal Republic of Germany
[4]Department of Physics and Space Research, University of Birmingham B 15 2TT, U.K.
[5]The Nuffield Radio Astronomy Laboratories, Macclesfield, Cheshire SK11 9DL, U.K.

Summary

Several nearby flare stars have been observed with spectrometers on various large telescopes. The experience of several attempts to record stellar radio flares is summarized including an event on AD Leo that was detected by three widely separated large telescopes. The event is analogue to solar decimetric pulsations. An interpretation in terms of coherent cyclotron emission (maser) is given and the source size, exciter drift velocity and source magnetic field are estimated.

1. Introduction

The interest in stellar radio burst spectra stems from the solar experience. Different types of solar radio bursts, caused by different emission mechanisms, can easily be distinguished in a time-frequency representation of radio flux density (known as spectrogram). It is therefore an important first step to know the spectra of stellar radio bursts and compare them with those observed on the Sun. This may help to identify the emission mechanism, which allows the determination of coronal parameters of the stellar source.

The first stellar flare radio spectra were taken with the Very Large Array with a bandwidth of 44 MHz and a time resolution of 5 s by Bastian and Bookbinder (1987). Bastian et al. (1990) have since expanded upon the initial work by using an autocorrelator with the 305 m telescope at Arecibo. However, further progress on the interpretation of stellar radio bursts can only be achieved by broadband multifrequency observations at high time resolution.

In this paper, the first broadband spectrometer observation of a stellar event is re-analyzed and discussed. The data were originally published in Güdel et al. (1989), but since publication additional timing information has become available which now allows us

L. V. Mirzoyan et al. (eds.), Flare Stars in Star Clusters, Associations and the Solar Vicinity, 139–144.

to make a more detailed analysis.

2. Instruments

We began these observations with the *Jodrell Bank* 76m telescope using an acousto-optic spectrometer operating over a frequency range of 100 MHz. This bandwidth was divided into 128 channels and the spectrum was integrated over 2 s. To discriminate against the terrestrial interference discussed in the next section, we later added the 100m telescope in *Effelsberg* operating at a center frequency of 1665 MHz with total bandwidth of 25 MHz divided into 32 channels and having a time resolution of 0.125 s. For one observing run in November 1987 it was possible to coordinate these observations with the 305m telescope in Arecibo operating at 1415 MHz center frequency with a 40 MHz correlator and an effective time resolution of 0.2 s. More details of the instruments are given in Güdel et al. (1989).

Since the appearance of that publication, two unexpected developments allow us to time the common observations of all three telescopes with much higher accuracy than previously reported. First, recent checks of the clock of the Jodrell Bank spectrometer revealed that it runs fast by 8.44 s/day. Since the clock is manually set for each observation run, an additional timing error of order ± 1 second may result, which is within the time resolution of the spectrometer. For the work reported here, the first error has been corrected. Second, a check of each scan of the Arecibo data in succession revealed that data gaps of variable duration occured. Once corrected, an absolute timing uncertainty of a few milliseconds was achieved. The timing at Effelsberg was set by the atomic clock of the observatory and is accurate to within a few milliseconds. In practice, we estimate the timing error to be a few tens of milliseconds between Arecibo and Effelsberg.

The timing is now of sufficient accuracy to allow a detailed correlation of the observations in the different frequency bands.

3. Need for Confirmation of Stellar Origin

Solar radio astronomers usually can easily recognize terrestrial interference by its structure in the spectrogram. Radio interferometers effectively reject waves from weak terrestrial sources. However, for stellar flare observations with single telescopes the situation is much more difficult and has lead to erroneous results in the past. The problem is that terrestrial interference at the low intensity level of stellar radio flares may resemble the signals we are expecting and can easily be mistaken for them.

An early idea to discriminate terrestrial signals by on-source and off-source measurements turned out to be inefficient if not impossible. We found many similar events on and off source. The type of stellar event shown in the next paragraph occurred only one other time on-source and never off-source. However, the confirmation of its stellar origin by on/off observations would be very time consuming and would always be useful only in a statistical sense. An other idea to discriminate terrestrial signals by off-source monitoring has also proved to be inadequate (see e.g. Bastian et al. 1990). The strategy adopted for the observations reported here serve two purposes: 1) observations with three widely separated telescopes have provided unambiguous confirmation of the stellar origin of the event because there is no source of terrestrial interference that could interfere with all three telescopes simultaneously; 2) the use of three telescopes has allowed us to effectively increase the bandwidth used to observe the event.

4. First Broadband Observation

For the reasons given in the previous paragraph it is more efficient and in fact necessary to observe spectra of stellar flares with widely separated telescopes. After some short and unsuccessful observing sessions we finally recorded a flare on AD Leo with the three telescopes. In the previous report of this event by Güdel et al. (1989) the data of the different telescopes has been shifted in time for the presentation in their figures 1 and 2. Here we can present the much more accurate timing which basically confirms the previous best fits.

In the case of the Arecibo data, the correction in timing brings the flare observed in Arecibo within 0.4 s of the best fit with the Effelsberg data (on a different frequency, however). The time correction of the Jodrell Bank observation of the flare brings the observations to within 1.6 s of the best fit with the Effelsberg data, which is within the time resolution of the Jodrell Bank spectrometer. The corrected timing of the Arecibo and Jodrell Bank observations confirm the common origin of the radiation and eliminate any remaining doubts of the stellar origin of the flare emission.

The full display of data can be seen in the article by Güdel et al. (1989). Figure 1 shows a detailed comparison of the strongest parts of the event. Whereas some peaks seem to correlate, others do not or may be shifted between the Arecibo observation at 1415 MHz and the Effelsberg observation at 1665 MHz. We have not found any drifting structures in the single spectrograms of the three telescopes below the instrumental resolution of 200 MHz/s in Arecibo and Effelsberg. These lower limits are compatible with the shifts possibly observed in Fig.1. The cross-correlation, presented in Figure 2 for the same time intervals, has a (> 95 %) significant peak at zero lag only in the second 5 s data segment presented in Figure 1. If all sections are combined, the significance for correlation is less than 90 % . Other peaks at lags different from zero are more significant. This does probably not express a delay of one frequency but quasi-periodicity, as can be seen from the auto-correlation (Fig.3) having a broad peak at 1.1 s. The results indicate that the correlation between the emissions at 1415 MHz and 1665 MHz is weak or absent. Figure 1 also clearly shows that the broadband spectra of the pulses are not similar and most are of smaller bandwidth than the 250 MHz separation.

5. Discussion

Having shown that stellar radio bursts can be recorded over broad spectra and may be identified unambiguously as stellar in origin, we now obtain rough estimates of stellar source parameters. From the observed rise times, Δt, the light dimension of the source of individual pulses, Δs, may be estimated on the basis of light travel-time arguments: $\Delta s < c \cdot \Delta t$, where c is the speed of light. There have been pulses with rise times faster than the highest time resolution (0.1255 s for the Effelsberg data, cf.Fig.1), thus $\Delta s \lesssim 4 \cdot 10^9$ cm.

Another estimate of the source size can be derived from the bandwidth $\Delta\nu$ of an individual pulse. In all coherent solar bursts the emission frequency is related to height by the emission process depending either on the plasma density or, more likely in our case, on the magnetic field. For the latter case, which will be discussed below, we have for the extent of the source in height

$$\Delta h \approx \frac{\Delta \nu}{\nu} H_B \qquad (1)$$

where H_B is the magnetic scale height. Our observations suggest $\Delta\nu \approx 300$ MHz at $\nu = 1500$ MHz, thus $\Delta h \approx 0.2 H_B$. As an example (and upper limit) we may take a scale height of one stellar radius to estimate $\Delta h \approx 2 \cdot 10^9$ cm.

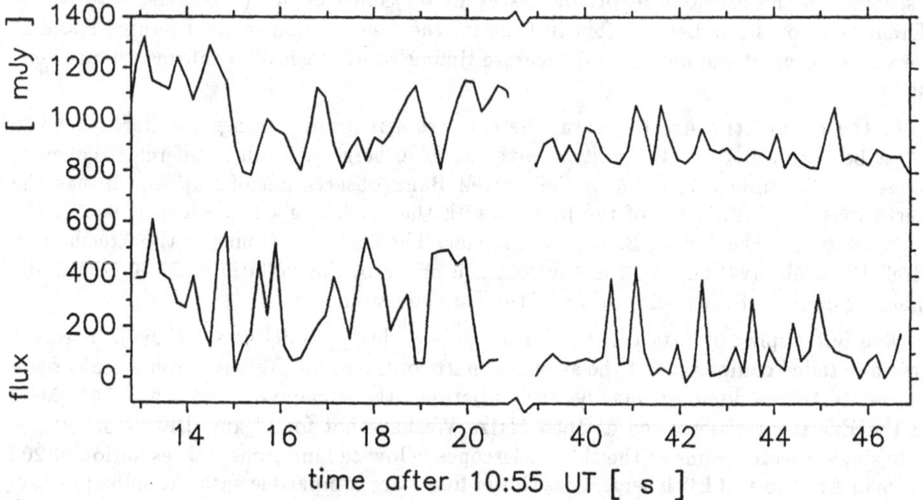

Figure 1: Detailed time profiles of the emission of a radio flare of AD Leo on November 7, 1987 at 1415 MHz integrated over 40 MHz for Arecibo (top) and at 1665 MHz integrated over 25 MHz for Effelsberg (bottom).

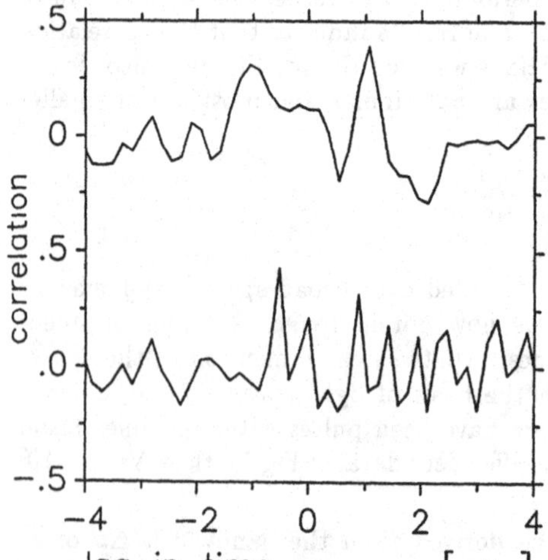

Figure 2: The cross-correlation of the 1415 and 1665 MHz flare emission in the time intervals shown in Fig. 1 (top: first section, bottom: second section).

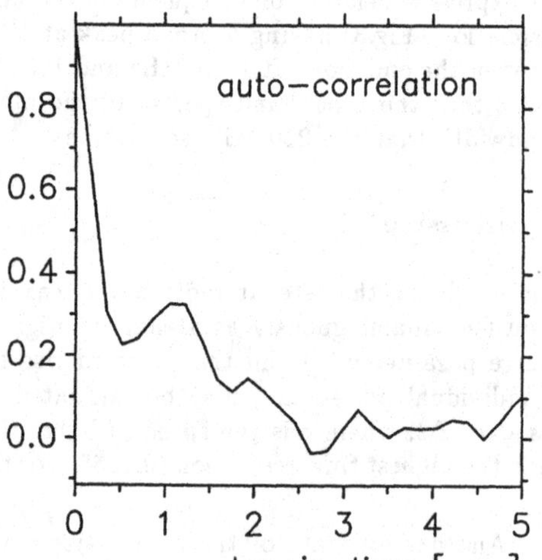

Figure 3: Auto-correlation of the 1415 MHz emission of the radio flare of AD Leo observed at Arecibo.

The observed flux density F and the estimated source size yield the brightness temperature defined from the Rayleigh-Jeans law

$$T_B = \frac{F \cdot c^2}{2 \cdot \nu^2 \cdot K}\left(\frac{D}{\Delta s}\right)^2 \qquad (2)$$

where K is Boltzmann's constant and D is the distance to the star. With an observed peak flux of 600 mJy and taking the light dimension as a more reliable estimate we obtain $T_B > 10^{14}$ K. This may not be the maximum reported in the literature, but it is the first thoroughly confirmed case.

The AD Leo radio flare is somewhat analogous to solar decimetric pulsations in the 300 – 1000 MHz band (Güdel and Benz, 1988). Solar decimetric pulsations have bandwidths in the range $\Delta\nu/\nu = 0.3 - 0.5$ and rise times of individual pulses of 0.1 – 0.5 s. Its quasi-periodic nature (Kurths and Herzel, 1989), high degree of circular polarization (Aschwanden, 1986), and high drift rate (Aschwanden and Benz, 1986) all agree well with the observed stellar parameters. Solar decimetric pulsations are generally attributed to trapped energetic (non-thermal) electrons in magnetic loops having a loss-cone velocity distribution. A model has been developed by Aschwanden and Benz (1988) interpreting the emission by the electron cyclotron instability (generally called 'maser') and the pulsations by relaxation oscillations of the same process.

Adopting the cyclotron maser model suggested by the solar analogy, we can constrain the magnetic field strength in the source if the ratio of plasma frequency to cyclotron frequency is approximately known. The high brightness temperature hints at the most efficient mode of the maser process where this ratio is below unity and the emission is in the fundamental mode. With this assumption a magnetic field strength of 600 G (resp. 300 G, if the emission is at the harmonic) is derived in the source.

Acknowledgements

The project received partial support by the Swiss National Science Foundation (Grant Nr. 2000-5.499). We are grateful to the staff of the Jodrell Bank Radio Telescope for making the facility available to us for these observations. The Arecibo Observatory is part of the National Astronomy and Ionosphere Center, which is operated by Cornell University under contract with the US National Science Foundation.

References

Aschwanden, M.J. (1986) *Solar Phys.*, **104**, 57.

Aschwanden, M.J. and Benz, A.O. (1986) *Astron.Astrophys.*, **158**, 102.

Aschwanden, M.J. and Benz, A.O. (1988) *Astrophys.J.*, **332**, 466.

Bastian, T.S. and Bookbinder, J. (1987) *Nature*, **326**, 678.

Bastian, T.S. and Bookbinder, J., Dulk, G.A., and Davis, M. (1989) *Astrophys.J.*, in press (10 April 1990 issue).

Güdel, M. and Benz, A.O. (1988) *Astron.Astrophys.Suppl.Ser*, **75**, 243.

Güdel, M., Benz, A.O., Bastian, T.S., Fürst, E., Simnett, G.M., and Davis, R.J. (1989), *Astron.Astrophys.*, **220**, L5.

Kurths, J. and Herzel, H. (1986) *Solar Phys.*, **107**, 39.

LANG: Why are some of the spikes not seen simultaneously at different radio telescopes?

BENZ: The time profiles of the telescopes observing the AD Leo flare at the same frequency completely agree. The profiles at 1415 and 1665 MHz are not well correlated in < 1 second details. The main reason seems to be that some of the single pulses have a smaller bandwidth. Another reason may be drifts which seem to be both positive and negative. These characteristics are similar to decimetric pulsations if one takes two widely separated frequencies.

LANG: What are the error bars in your radio spectrum of UV Ceti? If they are one sigma you could fit almost any theory to the high frequency data.

BENZ: The possibility to interpret the high frequency part of the quiescent radio spectrum of UV Ceti by another synchrotron emission has been discussed by Gudel and Benz (1989). A more likely interprertation is a low-frequency part of a thermal component. We have shown that a possible candidate is the soft X-ray emitting plasma of UV Ceti using the observed emission measure. The high frequency radio emission may thus correlate with quiescent soft X-ray emission.

LANG: What is the size of the radiating source in your thermal model?

BENZ: It is compatibel with the lower end of the temperature allowed by soft X-ray observations and the size of the star,or a higher temperature and a smaller source.

LANG: The quiescent radio emission that has been detected is all about 1 mJy, but with two orders of magnitude variation in X-ray luminosity, so it is unlikely the radio and X-ray plasmas are related.

RODONO: I would invite radio astronomers to advise and be advised by other observers at other wavelengths in order to organize simultaneous coverage at several wavelengths. This would also help in selecting genuine flares.

BENZ: The identification of stellar origin of a radio burst by a more or less coincident optical flare has lead to serious misinterpretation in the past. Contrary to interferometers, single dish observations are seriously affected by terrestrial interference. If radio astronomers do not make sure that the observed flare emission is stellar beyond doubt, they cannot start to compare it to optical flares. The association rate between optical and radio flares strongly varies for different solar radio burst types. The situation is unclear for flare stars and needs careful study. We are very interested in collaboration.

RADIO FLARE ON η GEMINI STAR

P.M.HEROUNI, V.S.OSKANIAN
Radiophysical Measurment Institute
Armenian Academy of Sciences, USSR

The first at the world Radio-Optical Telescope ROT-32/54/2,6 was mounted on the southern slope of Mount Aragats in Armenia at 1700 m above sea level. The Large Antenna of ROT with the unmovable hemisperical main mirror of 54 m in diameter and movable small correcting mirror is the extremely accurate and shortwave (down to 1 mm). Using aperture is 32m. The diameter of the Optical Telescope is 2.6 m. General view of ROT is shown in Fig.1.

During the very first test of the Antenna in summer 1985 at the 20 cm range an unexpected result was obtained. On June 23 at 9.23 UT a powerful radio flare was first registered on η Gemini [1]. This is a tripple system (two red giants of spectral class M3III and a G0IV star) classified as an irregular variable [2].

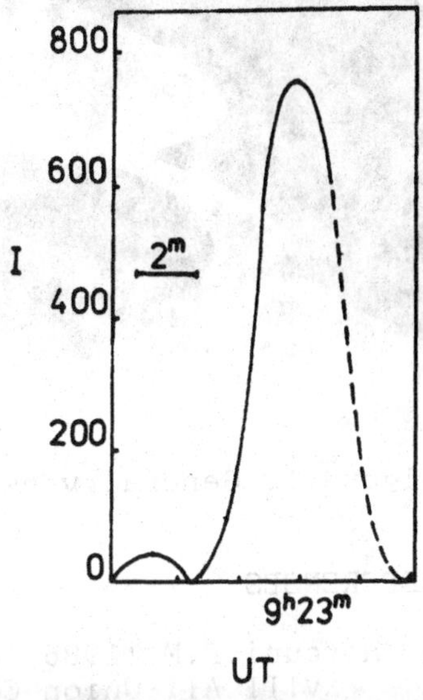

Figure 2.

The flare was registered, at the frequency 1.5 GHz with bandwidth 14 MHz, sensitivity 0.2 K at the time constant 1sec

Its duration exceeded 12 min, but was not longer than one hour. The flux density from the flare was about 800 units and the power density of about 3×10^{14}W per Hz.

The flare was detected while observing the radio-source 3C157 (a supernova remant) placed nearly the same direction, with a difference of about 2.2 min in time.

In Fig.2 the registered curve the η Gem flare is shown.

L. V. Mirzoyan et al. (eds.), Flare Stars in Star Clusters, Associations and the Solar Vicinity, 145–146.

Figure 1. General view of ROT.

REFERENSES

1. Herouni,P.M.(1986)'Radiooptical Telescope ROT-32/54/2,6', XVIII All-Union Conference of Radioastronomy, Proc.1, Irkutsk, pp. 5-7.
2. Kholopov,P.N. et.al.(1985)'General Catalogue of Variable Stars',Vol.2, Nauka, Moscow.

X-RAY EMISSION FROM SOLAR NEIGHBOURHOOD FLARE STARS

R. PALLAVICINI[1], L. STELLA[2] and G. TAGLIAFERRI[3]
1) Osservatorio Astrofisico di Arcetri, Firenze, Italy
2) Dipartimento di Fisica, Università di Roma, Italy
3) EXOSAT Observatory, ESTEC, Noordwijk, The Netherlands

ABSTRACT. A brief summary is given of a comprehensive analysis of EXOSAT observations of solar neighbourhood flare stars. Special attention is devoted to the discussion of quiescent X-ray emission and time variability.

1. Introduction

During its operational lifetime (May 1983-April 1986) the European satellite EXOSAT obtained many valuable observations of UV Ceti-type flare stars at X-ray wavelengths. The EXOSAT sample comprises 23 different objects from the flare star catalogue of Pettersen (1976). A number of them were observed on several occasions. If we also include a few other flare stars detected serendipitously, the sample comprises more than 40 separate observations for a total monitoring time of nearly 300 hours. The most significant aspect of these data is that about one half of them were long continuous observations extending over periods of many hours, without the data gaps that were usually associated to previous low-orbit satellites. This allowed us to study quiescent and flaring X-ray emission from UV Ceti-type flare stars in a much better way than it had been possible before (see Haisch 1983, Agrawal et al. 1986, Ambruster et al. 1987 for previous *Einstein* observations)

We have carried out a comprehensive analysis of EXOSAT observations of flare stars, using data from the Low Energy (LE: 0.05-2 KeV) and Medium Energy (ME: 1-10 KeV) experiments on EXOSAT. A full account of this study is in press in *Astronomy and Astrophysics*; here we give only a short summary of our work and present the most significant results.

2. Quiescent emission

The quiescent X-ray luminosities for the stars in our sample span a range of nearly a factor 500 (from 1.4×10^{27} erg s^{-1} for Prox Cen to 6.7×10^{29} erg s^{-1} for YY Gem), in spite of the fact that these stars do not differ much in spectral type. We have not found any obvious correlation between the quiescent X-ray luminosity and the rotation rate of the star, in contrast with what could be expected from simple dynamo models. Instead, we have found a very good correlation between X-ray quiescent luminosity and bolometric luminosity. This is shown in Fig. 1. The correlation coefficient is 0.95 for 24 data points and the best fit relationship is

$$\log L_x = -9.83+1.21 \log L_{bol} \quad (1)$$

This dependence of coronal activity on bolometric luminosity suggests a dependence on the stellar radius as would be expected if activity in these stars occurs in a "saturation" regime.

In contrast to the large variations observed between different objects, the quiescent X-ray emission of any given star appears to change little from one observation to the other. For stars observed repeatedly by EXOSAT, the observed variations were always less than a factor ≈ 2, and more typically did not exceed amplitudes of ≈ 20-30%. The time variations observed in the course of any individual observation were

L. V. Mirzoyan et al. (eds.), Flare Stars in Star Clusters, Associations and the Solar Vicinity, 147–151.

also typically small (less than ≈ 50%), except during flares. The latter show a wide range of amplitudes with respect to the quiescent level, from less than a factor 2 to more than one order of magnitude.

Twenty stars in our sample were observed previously with *Einstein*. We find a very good correlation between the EXOSAT and *Einstein* average fluxes, the difference being less than a factor ≈ 2 for most sources. This is comparable to the typical error on the determination of EXOSAT to *Einstein* flux ratios. For only 3 stars we found somewhat larger differences (a factor 3 to 4). This suggests that long-term variations of coronal X-ray emission (as might be expected from activity cycles) are probably small for these stars, at least in the spectral bands accessible to EXOSAT and *Einstein*.

3. Short-term variability

We have carried out a detailed time analysis of EXOSAT LE observations in search of low-amplitude short-term variability as might be expected from "microflaring" activity. Time variability has been investigated by us by applying several different techniques, including variance, correlation and power spectra analysis. For most stars in our sample we find substantial variability over a variety of different times scales (from a few minutes to hours). The observed variability appears in the form of either individual (usually sporadic) flares or of more gradual variations (on times scales of tens of minutes to hours). This result is consistent with what found previously from *Einstein* observations (Ambruster et al. 1987).

By contrast, we do not find evidence in the EXOSAT data for continuous low amplitude variability on time scales shorter than a few hundred seconds. More specifically, we do not confirm the low-level "microflaring" variability, consisting of a succession of events lasting from tens of seconds to several minutes and with characteristic energies of ≈ 2×10^{30} erg, that was reported by Butler et al. (1986).

In the case of UV Ceti, the availability of four different EXOSAT observations allows a more detailed investigation of its variability properties. To this end, we have combined the four observations (after excluding a strong flare that was present in one of them) and we have computed an average power spectrum for the entire data set. This is shown in Fig. 2, where the power spectrum has been normalized in such a way that the counting statistics noise corresponds to a power of 2. As shown in the figure, significant power above the noise level is detected only between frequencies of ≈ 3×10^{-5} and ≈ 2×10^{-3} Hz. No evidence for excess power at higher frequencies is found. This implies that the shorter variability time scales of UV Ceti revealed by the EXOSAT observations are on the order of a few hundred seconds. This is in agreement with what found independently by Collura et al. (1988) using a different variability analysis.

4. Flaring activity

A few dozens flares have been observed by EXOSAT from UV Ceti-type stars. For a few of them, ME data were obtained in addition to the LE ones. This allowed spectral analysis of the flare and determination of relevant physical parameters such as coronal temperatures and emission measures.

The observed flares cover a broad range of total X-ray energies (from ≈ 3×10^{30} erg to ≈ 1×10^{34} erg) and have a variety of different time scales (from a few minutes to hours). There is evidence in the EXOSAT data for at least two different types of stellar flares, i. e.: a) *impulsive flares*, which are reminiscent of compact flares on the Sun; and b) *long-decay flares*, which are reminiscent of long-duration solar two-ribbon flares.

Temperatures derived from spectral analysis of stellar flares are in the range from ≈ 2×10^7 K to ≈ 4×10^7 K, similar to the typical temperatures found for solar X-ray flares. The emission measures are in the range from ≈ 1×10^{51} cm^{-3} to ≈ 1×10^{53} cm^{-3}, much larger than for solar flares. There is evidence that the plasma is first heating and then cooling during the evolution of the flare. Moreover, the high energy flux appears to precede the low energy flux by a few to several minutes, as also typically observed in solar flares. Within our limited sample there is no obvious correlation of flare temperatures and time scales with other stellar parameters. We found instead a correlation between quiescent X-ray emission and flare energy, the largest flares occurring only on the most active stars.

The average rate of detection of X-ray flares was typically ≈ 1 flare every ≈ 10 hours for the stars observed by EXOSAT. However, flare detectability strongly depends on the quiescent X-ray level of the star and decreases sharply as the latter increases. The available data do not allow the determination of the flare frequency distribution vs. energy for each individual star in our sample. However, by assuming that all stars in the sample obey the same relationship, we found that the flare frequency distribution can be approximated at high flare energies as a power law (see Fig. 3), where N(>E) is the number of flares with total X-ray energy larger than E and α is the power-law index which turns out to be 0.7±0.1. At lower flare energies the distribution flattens owing to detection threshold effects. The shape of the derived flare frequency distribution is consistent with that found for optical flares (for which α varies from 0.4 to 1.2 depending on the star).

References

Agrawal, P.C., Rao, A.R., and Sreekantan, B.V.: 1986, *Montly Not. Roy. Astron. Soc.* **219**, 225

Ambruster, C.W., Sciortino, S., and Golub, L.: 1987, *Astrophys. J. Suppl.* **65**, 273

Butler, C.J., Rodonò, M., Foing, B.H., and Haisch, B.M.: 1986, *Nature* **321**, 679.

Collura, A., Pasquini, L., and Schmitt, J.H.M.M.: 1988, *Astron. Astrophys.* **205**, 197

Haisch, B.M.: 1983, in *Activity in Red-dwarf Stars*, eds. P.B. Byrne and M. Rodonò, Dordrecht: Reidel Publ. Co., p. 255

Pettersen, B.R.: 1976, *Catalogue of Flare Star Data*, Inst. Theor. Astrophys., Blindern, Report No. 46

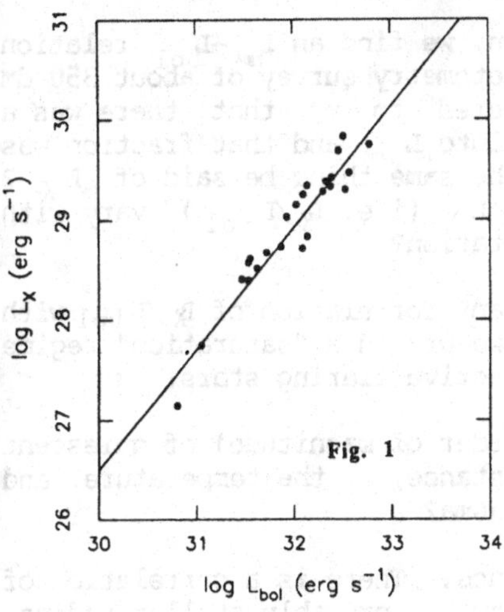

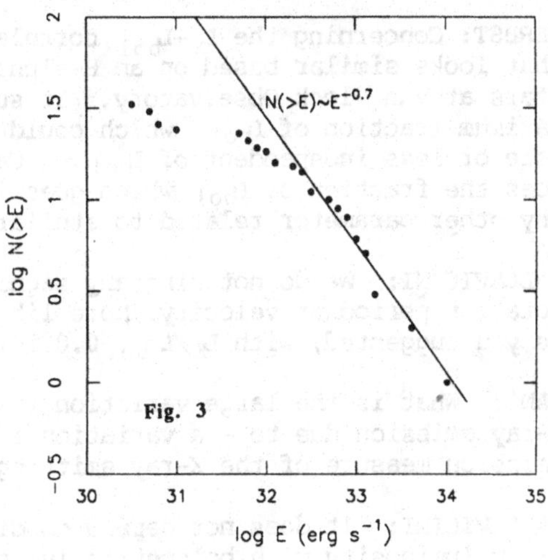

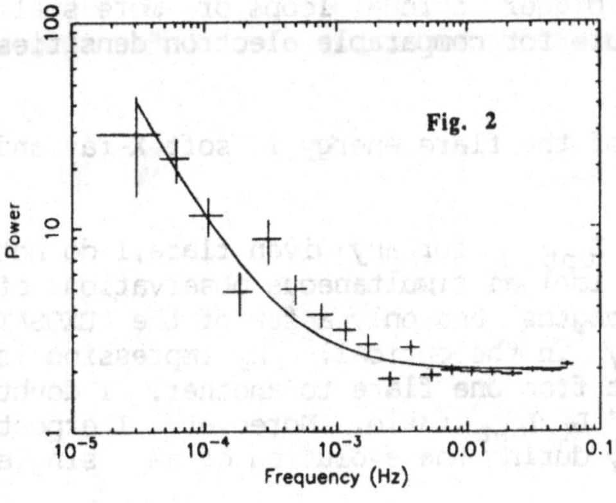

Fig. 1: X-ray quiescent luminosity versus bolometric luminosity

Fig. 2: Average power spectrum of UV Ceti

Fig. 3: X-ray flare frequency distribution

GIAMPAPA: Since L_X is correlated with L_{bol} but not with rotation period does this really mean that L_X is correlated with radius (surface area), i.e.the surface is "saturated" with magnetic flux tubes at high rotation rates?

PALLAVICINI: A correlation of L_X with radius for very active flare stars is probably the most plausible interpretation at present. If this occurs, certainly we must be in a "saturated" stage, since no such correlation between L_X and L_{bol} is observed for non-flaring dM stars.

GIAMPAPA: Do you see periodic structure in the quiescent or flare X-ray emission?

PALLAVICINI: A detailed variability study carried out over our EXOSAT observations did not show any periodic structure in the quiescent or flare emission.

HERBST: Concerning the L_X-L_{bol} correlation, we find an $L_{H\alpha}$-L_{bol} relation that looks similar based on an H-alpha photometry survey of about 350 dM stars at Van Vleck Observatory. It suggested to us that there was a maximum fraction of L_{bol} which could go into $L_{H\alpha}$ and that fraction was more or less independent of L_{bol}. Can the same thing be said of L_X ? Does the fraction of L_{bol} which goes into L_X (i.e. L_X/L_{bol}) vary with any other parameter related to stellar rotation?

PALLAVICINI: We do not find any significant correlation of L_X/L_{bol} with rotation period or velocity. More likely we are in a "saturation" regime as you suggested, with $L_X/L_{bol} \simeq 0.001$ for active flaring stars.

LANG: What is the large variation (two order of magnitude) of quiescent X-ray emission due to - a variation in distance, or the temperature and emission measure of the X-ray emitting plasma?

PALLAVICINI: It does not depend on distance. There is a correlation of X-ray luminosity with bolometric luminosity, or probably stellar radius, and also emission measure variations should play a role.

LANG: Bigger stars probably have bigger coronal loops or more small ones, hence a greater emission measure for comparable electron densities and greater X-ray luminosity.

GRININ: What is the typical ratio of the flare energy in soft X-ray and optical?

PALLAVICINI: If you mean the ratio L_{opt}/L_X for any given flare, I do not have an easy answer at hand. This implies simultaneous observations of flares at X-ray and optical wavelengths and only a few of the EXOSAT flares were observed simultaneously in the optical. My impression is that this ratio can vary quite a bit from one flare to another. I doubt that we can even talk of a "typical" L_X/L_{opt} ratio. Moreover, I expect that this ratio changes considerably during the evolution of any single flare.

BENZ: Is there a correlation between soft X-ray emission of M dwarfs with their quiescent radio emission?

PALLAVICINI: There are only a handful of flare stars for which real "quiescent" radio emission has been detected (at levels of about 1 mJy or less). There is no obvious correlation between X-ray luminosity and quiescent radio luminosity for the limited sample available at present. Detailed comparison of X-ray and radio emission for 4 stars observed by us with EXOSAT and VLA (Kundu et al. 1988) also indicates poor correlation. More observations in the radio would be useful to address this question.

Balmer and Soft X-ray Emission from Solar and Stellar Flares

C.J. Butler

Armagh Observatory, N. Ireland

Summary: Integrated soft X-ray (8-12A) fluxes for solar flares have been scaled to the equivalent EXOSAT fluxes using spectra obtained from a variety of rocket-based experiments. The data show good agreement with the soft X-ray - $H\gamma$ correlation established by Butler et al. (1988) for stellar flares and confirm the basic similarity, in this respect, of flares on the Sun to those on dMe stars.

1 Introduction

It was recently shown by Butler, Rodono and Foing (1988), (henceforth Paper I), that there exists a well defined linear correlation between the integrated energy emitted in H γ and soft X-rays in stellar flares. Their data also included a single solar flare for which the equivalent soft X-ray flux in the EXOSAT CMA passband (0.04-2 KeV) had been computed by Doyle et al. (1988). This single point appeared to confirm that the relationship established for stellar flares was also appropriate to flares on the Sun.

Nearly two decades ago a similar correlation had been proposed by Thomas and Teske (1971), based, not on the direct, simultaneous, observations of individual flares in Balmer lines and soft X-rays as used in Paper I, but by observations of solar flares for which the H-alpha class was known. In their work, the H α and X-ray observations were made at different times, but could be related by a common, photographically determined, H α class. It is difficult to compare their plot for solar flares directly with those for stellar flares due to the fact that their soft X-ray fluxes were measured by OSO III over a wavelength range 8-12 A which only partly overlaps the much broader bandwidths of the EXOSAT CMA (6-280 A) and the EINSTEIN IPC (3-62 A) with which the stellar flares were observed.

2 Scaling Factors for Solar Flares

The question arises as to whether the soft X-rays over the 8-12 A region can be scaled to determine an equivalent flux over the 6-280 A wavelength region used for stellar flares. Regrettably, there is a relative scarcity of data in the soft X-ray - EUV region for solar flares, and much that exists is from the early days of rocket experiments. We have, in fact, not been able to find a single spectrum that covers the entire region, either for a solar flare or for the quiet Sun. The problem is further complicated by the large number of emission

L. V. Mirzoyan et al. (eds.), Flare Stars in Star Clusters, Associations and the Solar Vicinity, 153–157.

lines present in the soft X-ray/EUV region. Nevertheless, fluxes have been published for several spectra, which could allow us to build up scaling factors from one wavelength region to another, and thereby determine an approximate overall scaling factor to relate F_{8-12} to F_{6-280}. The following spectra were used in the determination of these scaling factors:

(i) A flare spectrum from 6-25 A by Neupert et al. (1973)

(ii) A flare spectrum obtained simultaneously in two wavelength ranges: 15-50 A and 139-430 A by Freeman and Jones (1970)

(iii) A spectrum of the quiet Sun from 15-90 A by Freeman and Jones (1970)

(iv) A spectrum of the quiet Sun from 30-130 A by Manson (1967).

The wavelength regions where there is known to be strong emission, e.g. 5-50 A and 139-280 A, are covered adequately by the three flare spectra (i) and (ii), however for the region from 50-139 A we have not been able to find a suitable calibrated flare spectrum. Fortunately the flux in this region is of the order of a factor 3 or so less than the combined flux in regions 6-50 A and 139-280 A where the strongest lines occur. Whilst there are undoubted differences in the fluxes of emission lines during flare as compared to quiet periods we have, in the absence of flare data for the spectral region 50-139 A, been forced to assume that the *ratio* of fluxes in the appropriate wavelength bands will be approximately the same during flare and quiet periods. Also the fluxes listed in the above publications are often only for emission lines and not for continua. Whereas the emission line fluxes, in general, dominate the energy in the 6-280 A range, the continuum can sometimes represent a reasonable fraction of the total flux. Here we have assumed that the continuum will scale roughly as the emission lines flux, an assumption which is probably reasonable given that much of the continuum appears to be due to unresolved lines. In this case the relative flux over different wavelength ranges will be unaffected. With these provisos in mind we give in Table 1 values for the flux in units of F_{8-12}.

Table 1: Soft X-ray flux in units of F_{8-12}

F_{6-25}	F_{25-90}	F_{90-130}	$F_{139-280}$
4.0	4.8	0.9	5.5

It will be noticed that the region 130-139 A is not included; unfortunately we have not been able to find a calibrated spectrum for this range. Assuming that the contribution from this range is negligible, the total flux in the 6-280 A, or 0.04-2 KeV, range is found to be approximately 15 times the flux in the 8-12 A range used by Thomas and Teske (1971).

Applying the factor of 15, the sum of the fluxes in Table 1, to convert the 8-12 A flux to the equivalent EXOSAT CMA flux, and the factor of 1/3, as was used by Paper I, to convert the Hα flux to $H\gamma$ flux we may plot the solar flare data of Thomas and Teske (1971) on the same diagram as the stellar data from Paper I. In figure 1 we see that the data from solar and stellar flares are, to within the errors of measurement, in complete agreement, and that a single line will satisfy both sets of data. Thus we confirm that the solar flares obey the same, apparently linear, relationship between the Balmer and soft X-ray flux as do their stellar counterparts.

3 Interpretation

As discussed in Paper I and by Haisch (1988), the high degree of correlation between the Balmer and soft X-ray emission, is unexpected for the following reasons:

(i) The temperature regimes, in which the soft X-rays and Balmer emission arise, are widely different; the former at $10^6 - 10^7 K$ and the latter at 10^4K.

(ii) The time-profiles (light curves) of some flares show an appreciable impulsive component in the Balmer flux, (see the flare on AD Leo by Rodono et al. 1989), whereas the soft X-ray flux is commonly supposed to arise from the *thermal* or *gradual* phase of a flare.

(iii) The lower Balmer lines Hα, Hβ and Hγ etc are expected to be optically thick and therefore their integrated flux should depend on the shape of the plasma.

The relationship between the Balmer and soft X-ray emission should be taken into consideration in forming models of flares and would seem to mitigate against some that have been proposed. For instance one model proposes that the Balmer emission arises from cooling loops that lie beneath the arcade of hot loops containing the X-ray emitting plasma. Such a picture would suggest that the Balmer emission should always reach its peak emission after the X-ray peak. As Haisch (1988) as pointed out, this simple progression whereby the Balmer flux always follows the soft X-ray flux does not correspond to reality; in some cases the X-ray event precedes the Balmer emission and in some cases, vice versa. The stellar flare observed by Kahler et al. (1982) on YZ CMi is a clear example of a flare where the Balmer emission peaked earlier than the soft X-rays.

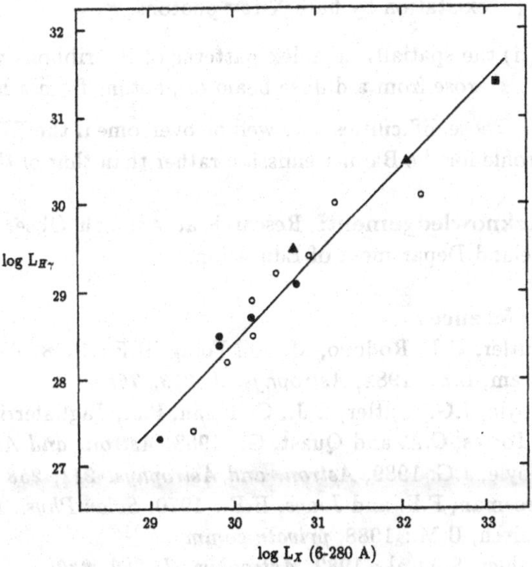

Figure 1. The equivalent Hγ and EXOSAT CMA flux for the solar flare data of Thomas and Teske (1971) - open cicles, plotted with the data for stellar flares in Paper I - filled symbols.

Paper I, from energy budget considerations, and assuming reasonable values for the Balmer decrement and the ratio of Balmer to Lyman flux, concluded that the total hydrogen emission was approximately two thirds of the downward-directed soft X-ray flux. However

the EUV flux, which is also capable of exciting hydrogen, was estimated to be only of the order of half the soft X-ray flux; an estimate we now believe to be eroneous. From a comparison of the two spectra of the quiet Sun by Freeman and Jones (1970); one of the 465-794 A region, and the other of the 15-90 A region, and the various scaling factors that have already been discussed, we find that the flux in the 465-794 A region is approximately 2.6 times greater than the flux in the 6-280 A region. Thus the total EUV flux capable of exciting hydrogen may be of the order of five times greater than the soft X-ray (6-280 A) flux. In this connection we may also note the recent results of emission measure analysis by Doyle (1989) which give values of the ratio of EUV/soft X-ray flux for nine dMe stars. These ratios vary from 0.4 to 6.3 with an average value of 3.5 .

As computations by Mullan and Tarter (1977) and Cram (1982) indicated a possible efficiency of 10-20% for conversion of soft X-ray photons to hydrogen emission, it would appear that there is sufficient energy in the soft X-ray and EUV regions combined to produce the observed hydrogen flux. Nevertheless, as Haisch (1988) has pointed out, there are two fatal problems with this explanation for the correlation between Balmer and soft X-ray flux:

(i) the lack of a one-to-one relationship between soft X-ray and Balmer flux over all phases of flares - as would be predicted if the Balmer flux were to arise soley from the excitation by flare X-ray photons,

(ii) the spatially complex patterns of Hα ribbons would not be expected if the excitation arose from a diffuse beam of photons from a high altitude coronal source.

These difficulties may well be overcome if the EUV flux can be shown to follow the time profile for the Balmer emission rather than that of the soft X-rays in solar and stellar flares.

Acknowledgement: Research at Armagh Observatory is grant-aided by the Northern Ireland Department of Education.

References

Butler, C.J., Rodono, M. and Foing, B.F.: 1988, *Astron. and Astrophys. 206, L1*

Cram, L.E.: 1982, *Astrophys. J. 253, 768*

Doyle, J.G., Butler, C.J., Callanan, P.J., Tagliaferri, G., de la Reza, R., White, N.E., Torres, C.A. and Quast, G.: 1988, *Astron. and Astrophys. 191, 79*

Doyle, J.G: 1989, *Astron. and Astrophys. 214, 258*

Freeman, F.F. and Jones, B.B.: 1970, *Solar Phys. 15, 288*

Haisch, B.M.: 1988, *private comm.*

Kahler, S. et al.: 1982, *Astrophys. J. 252, 239*

Manson, J.E.: 1967, *Astrophpys. J. 147, 703*

Mullan, D.J. and Tarter, C.B.: 1977, *Astrophys. J. 212, 179*

Neupert, W.M., Swartz, M. and Kastner, S.O.: 1973, *Solar Phys. 31, 171*

Rodono, M., Houdebine, E.R., Catalano, S., Foing, B.H., Butler, C.J., Scaltriti, F., Cutispoto, G., Gary, D.E., Gibson, D.M. and Haisch, B.M.: 1989, *Poster Paper, IAU Colloquium No 104, Stanford, Aug. 1988 - Catania Astrophys. Obs., p 53*

Thomas, R.J. and Teske,R.G.: *Solar Phys. 16, 431*

APPENZELLER: Can you tell from the observed Balmer decrement whether the Balmer lines are due to recombination or due to collisional excitation?

BUTLER: I think the observed Balmer decrement can be explained by collisional excitation. Some calculations on this have been made by Houdebine and by Katsova et al. However, the observations are not straight forward due to the coalescence of the higher Balmer lines and the difficulty of determining the continuum level. Also our observations only cover the blue region of the spectrum, not H-alpha or H-beta.

X–RAY VARIABILITY IN THE ORION NEBULA

JEAN–PIERRE CAILLAULT
Department of Physics and Astronomy
University of Georgia
Athens, Georgia 30602 USA

and

SAEID ZOONEMATKERMANI
Columbia Astrophysics Laboratory
Columbia University
New York, New York 10027 USA

ABSTRACT. We have examined each of the 172 EINSTEIN X–ray sources in Orion for both short (100–10,000 seconds) and long (1 day to 2 years) timescale variability. No strong flares were seen in 75,000 total seconds of observations. We compare the Orion variability results with those from the ρ Oph star–forming region, solar X–ray flares and dMe U–band flares.

1. Introduction

The Orion Nebula is the best known and best studied site of recent star formation in our galaxy. A rich history of optical, radio, and infrared work provides an invaluable background for observations in the X–ray regime, while its age and proximity mark it as an essential target in studies of the evolution of stellar properties.

Orion was observed by the EINSTEIN Observatory for a total of ~ 140 ksec over a 23–month span with 13 IPC (75 ksec) and 6 HRI (65 ksec) pointings within the central 2° x 2° region centered on the Trapezium. Some work on these observations has already been presented (Ku and Chanan 1979; Pravdo and Marshall 1981; Ku, Righini–Cohen, and Simon 1982; Smith, Pravdo, and Ku 1983), but these focussed on only 1 HRI and 2 IPC pointings (totalling ~ 25 ksec). More recently, Caillault and Zoonematkermani (1989) have discussed the unexpected X–ray emission from the B6–A3 main sequence stars in the Nebula. We discuss herein our analysis of the variability characteristics of the X–ray sources.

In particular, we compare our results with those obtained by Montmerle, et al. (1983) for the ρ Oph star–forming region. They concluded that the X–ray emission from the young stars in that cluster was the result of continual flaring. This was based, in part, on the fact that the amplitude distribution of flux variations was similar to those for solar X–ray flares and dMe U–band flares.

L. V. Mirzoyan et al. (eds.), Flare Stars in Star Clusters, Associations and the Solar Vicinity, 159–162.

2. The Observations

The X–ray data were obtained using the EINSTEIN Observatory imaging proportional counter (IPC) and high–resolution imager (HRI). The IPC provided a $1^{o} \times 1^{o}$ field of view with 1' resolution in the 0.15–4.5 keV band. The HRI provided a ~ 25' diameter field of view with ~ 4" resolution in the slightly softer 0.1–3.0 keV band; further instrumental details can be found in Giacconi, et al. (1979). The primary survey consists of 13 overlapping IPC pointings; effective exposure times ranged from 100 to 30,000 seconds, yielding minimum detectable soft X–ray fluxes of ~ 10^{-13}ergs cm^{-2}s^{-1} in individual fields (but an order of magnitude fainter upon merging). This corresponds to an X–ray luminosity threshold of ~ 2.5×10^{30}ergs s^{-1} assuming a distance to Orion of 450 pc. Six HRI images were also taken, with exposure times ranging from 3,000 to 22,000 seconds.

Each of the 13 IPC fields was analyzed using a source detection algorithm initially developed for study of the LMC (Wu, Hamilton, and Helfand 1988). This method incorporates, in addition to the normal telescope vignetting, mirror scattering and instrument dead time corrections, a flat–field correction, a cosmic–ray particle count rate correction, and allows for the subtraction of sources which would otherwise contaminate the calculation of the local background for source detection. We have used only those sources with significance $> 3.5\sigma$ in the subsequent discussion.

Most of the X–ray sources were analyzed for X–ray variability on a variety of timescales ranging from 100 sec (limited by counting statistics) to the 23–months separating the first and last observations. A χ^2 – distribution test was performed for each light curve, and although many sources varied over the long–term interval, none showed significant (> 3x) flaring activity during any one pointing ($< 3 \times 10^4$ s).

3. The X–Ray Variability

Although no single flares were seen, we have detected many sources which display large variability (> 10x); an example of this is the source $\alpha = 5^h32^m02.16^s$, $\delta = -5^o02'36"$ (tentatively identified as V652 Ori = P1496). The average count rate for the source on 1979 day 65 was 0.1270 counts s^{-1}, while on 1980 day 265 it was 0.0097 counts s^{-1}, a factor of 13 difference. Of course, this type of analysis may only be performed on sources which have been detected in more than one image.

There are a number of sources which were detected only once and then not seen (and with upper limits lower than the flux when detected) in subsequent observations of the same region; these are probably "flare" sources, too. However, we ignore these sources for the time being other than to note their existence and prevalence (there are 26 of them).

For those sources which have been detected in multiple fields, we follow an analysis similar to that which Montmerle, et al. (1983) conducted for the variable X–ray sources in the ρ Oph star–forming region. They assumed that all of the single X–ray sources belong to one type of X–ray object, seen at different levels of activity. They defined a quantity F_x as the ratio of the flux of an X–ray source in any observation to the flux of that source in its minimum, or "ground", state of activity. A histogram $N(F_x)$ of the normalized amplitude distribution of the fluxes of the Orion X–ray sources is shown in Figure 1. This histogram is derived from

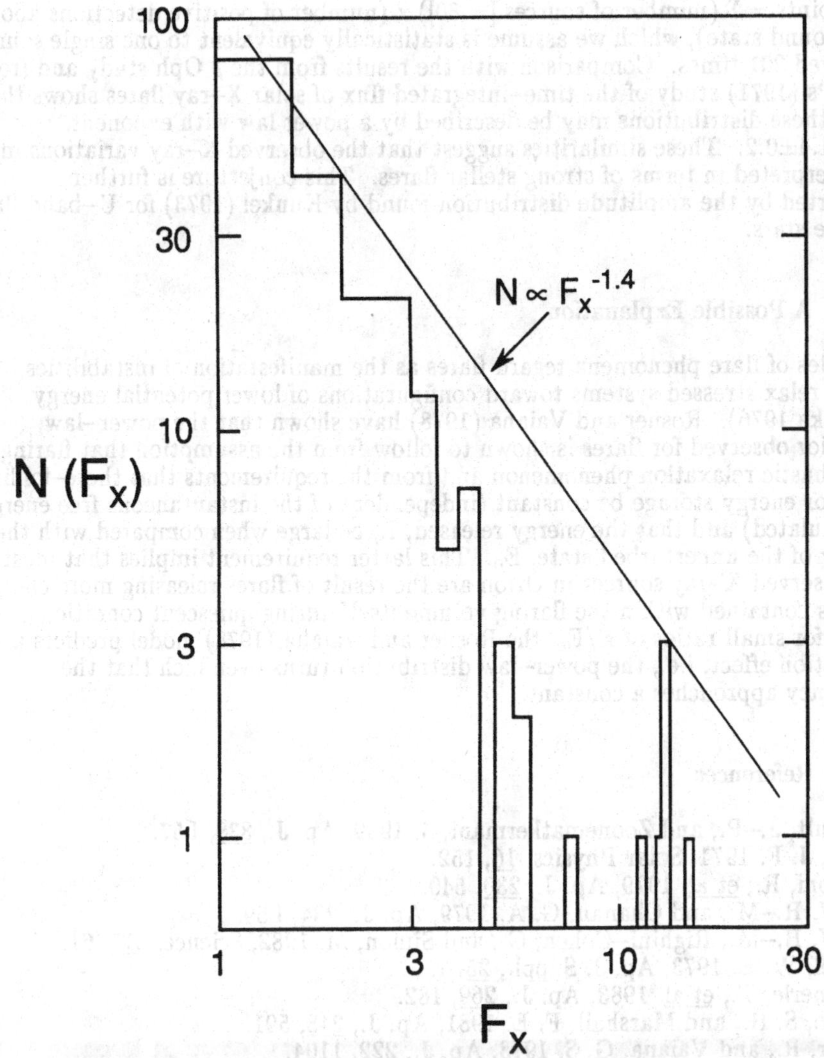

Figure 1. A histogram $N(F_x)$ of the normalized amplitude distribution of the fluxes of the Orion X–ray sources.

201 points = Σ (number of sources [= 80]) x (number of positive detections above the ground state), which we assume is statistically equivalent to one single source observed 201 times. Comparison with the results from the ρ Oph study and from Drake's (1971) study of the time–integrated flux of solar X–ray flares shows that all of these distributions may be described by a power law with exponent $\alpha \sim -1.4 \pm 0.2$. These similarities suggest that the observed X–ray variations may be interpreted in terms of strong stellar flares. This conjecture is further supported by the amplitude distribution found by Kunkel (1973) for U–band flares in dMe stars.

4. A Possible Explanation

Theories of flare phenomena regard flares as the manifestation of instabilities which relax stressed systems toward configurations of lower potential energy (Svestka 1976). Rosner and Vaiana (1978) have shown that the power–law behavior observed for flares is shown to follow from the assumption that flaring is a stochastic relaxation phenomenon and from the requirements that the *e*–folding time for energy storage be constant (independent of the instantaneous free energy accumulated) and that the energy released, E, be large when compared with the energy of the unperturbed state, E_0. This latter requirement implies that most of the observed X–ray sources in Orion are the result of flares releasing more energy than is contained within the flaring volume itself during quiescent conditions, since, for small ratios of E/E_0, the Rosner and Vaiana (1978) model predicts a saturation effect, i.e., the power–law distribution turns over such that the frequency approaches a constant.

5. References

Caillault, J.–P., and Zoonematkermani, S. 1989, Ap. J., 338, L57.
Drake, J. F. 1971, Solar Physics, 16, 152.
Giacconi, R., et al. 1979, Ap. J., 230, 540.
Ku, W. H.–M., and Chanan, G. A. 1979, Ap. J., 234, L59.
Ku, W. H.–M., Righini–Cohen, G., and Simon, M. 1982, Science, 215, 61.
Kunkel, W. E. 1973, Ap. J. Suppl., 25, 1.
Montmerle, T., et al. 1983, Ap. J., 269, 182.
Pravdo, S. H., and Marshall, F. E. 1981, Ap. J., 248, 591.
Rosner, R., and Vaiana, G. S. 1978, Ap. J., 222, 1104.
Smith, M. A., Pravdo, S. H., and Ku, W. H.–M. 1983, Ap. J., 272, 163.
Svestka, Z. 1976, Solar Flares (Dordrecht:Reidel).
Wu, X.–Y., Hamilton, T. T., and Helfand, D. J. 1988, B. A. A. S., 20, 974.

SOME RESOLUTIONS ON T TAURI STARS

V.A.AMBARTSUMIAN
Byurakan Astrophysical Observatory
Armenian Academy of Sciences, USSR

This symposium is dedicated to the results of studies of flare stars. The necessity of the detailed analysis of observational data on flare stars follows from the very simple fact, that until now the theory has not found for the answer to the question: why the young stars of low luminosity are flaring.

The inability to explain this fundamental property of young stars shows convincingly that theories of stellar structure and stellar evolution which have been proposed until now are in very primitive stage. Therefore in our quest for understanding the flare stars we shall try to find in some empirical way their place in the whole picture of phenomena related to the early stages of stellar evolution. This is the reason why we ask ourselves about the connections between flares and other forms of activity of young stars, on the significance of the presence of flare stars in young stellar groups and clusters and about the correlation between relative frequency of flare stars and the age of the given cluster. In this way many interesting data have been found and one can expect much more findings from the work in the same direction.

Already the first results have shown that proportion of flare stars in young open clusters is much larger than in old open clusters. No flare stars are present in globular clusters. These simple facts show us that the flare stars represent indeed the early stages of stellar evolution, they are relatively young. The observations show also that the flare stars in older open clusters are in the average much fainter (less luminous) than in younger clusters. From this we can conclude that the flare activity in the stars of lower luminosities lasts much longer than in stars of higher luminosities.

Thus these simple and very clear empirical correlations bring us to very important conclusions about the earlier stages of stellar evolution.

L. V. Mirzoyan et al. (eds.), Flare Stars in Star Clusters, Associations and the Solar Vicinity, 163–167.

More complicated and more interesting are data on flare stars in T-associations.

The reason for this is the fact that many of flare stars in such associations have been identified with T Tauri stars. It became clear that at some stage of the evolution a young star can show simultaneously both the flare activity and T Tauri activity. In the paper of Dr.Parsamian the correlation between these two forms of activity has been investigated in detail. It was shown that more than 40 percent of T Tauri variables with registered amplitudes of continuous varations $A>1^m$ are at the same time flare stars. This percentage is much higher when we consider T stars which are more active in continuous luminosity variations, reaching 64% in the case of stars having $A>2^m$.

Since in the paper of Parsamian all data refer to photographically observed flares, it is clear that if we take into account also the very small-amplitude flares, almost all T Tauri stars that have cosiderable amplitude of continuous variations will be recognized as flare stars. The inevitable conclusion is that during the evolution of new born stars the T Tauri phase and the flare phase overlap, but the flare phase is usually much longer. And the T Tauri phase almost completely represents the initial part of flare phase.

It is quite possible that the flare activity is more intense during the period when the slow brightness variations are stronger. However this question deserves further study. But the simplest assumption is that the flare activity streches over the whole length of T Tauri life.

But we know that during the T Tauri phase a star shows also other forms of activity which are expressed first of all in the variations of its spectrum.

First of all there are emission lines, that changed not only their intensity, but also their wave length and breadth. The presence of emission lines is an evidence of very extended atmosphere. Their variability suggests us that not only the extent of the chromosphere is changing, but also that the ionisation degree is variable.

Since the T-associations are as a rule embedded in the dark nebulae where we have different streams of gases one can try to assume that these changes of emission lines are caused by influence of such external streams. However the details of these changes speak in favour of internal causes, especially because they are often accompanied with variations of the brightness of the star.

Thus the spectral changes are evidences of very intensive internal underphotospheric processes.

But one of persistent properties of the spectra of majority of T Tauri stars - the presence of an absorption component on short wave length side of Hα emission line is a strong evidence of the continous ejection of gaseous matter from the atmospheres of these stars. Owing to this phenomenon of the mass outflow from many members of T- associations the total mass of the nebula is increasing and the total mass of stars within it decreases. I would like to emphasize the evolutionary significance of the outflow phenomenon which persists at the earliest stages of the evolution of low mass stars, similar to what happens with many massive stars.

However, according to the spectrophotometric data the mass ejected during such outflow is of the order of $10^{-8}\mathfrak{M}_{\odot}$ per year. Taking into consideration the existing estimates of the age of T- associations (of the order of 10^7 years) we conclude that the total mass ejected by a star during the existence of an association is of the order of one tenth of solar mass. But the total mass ejected in the nebula by all members of the association is of the almost same order of magnitude as the whole mass of the nebula itself.

In some cases we observe larger descrete formations ejected from T Tauri stars. For example the so called HH objects originate as results of ejection from T Tauri stars, sometimes symmetric ejections. Of course the observations cannot decide definitely whether the main masses of these objects are also ejected from the atmosphere or accelerated by jets from the stars.

However, in both cases the process must be accompanied with very intense loss of mass from outer layers of the T Tauri star and with enrichment of the interstellar medium.

We can assume that within one kiloparsec around us there are about 1.000 T Tauri stars brighter than M=+6. On the other hand in the same volume there are already registered more than 100 HH objects.

Since the age of the observed HH objects is of the order of 10^4 years or less, we can assume that not less than one HH object is formed in that volume of space per each century.

If we suppose that every T Tauri star can form HH objects we come to the conclusion that in the average every T Tauri star can produce one HH object per 10^5 years.

In addition the T Tauri stars can produce also the cometary nebulae. A good example is the variable star PV Cephei. We see that this form of the activity of T Tauri stars is also connected with the ejection of comparatively large mass of material.

But we need to be here more coutious in generalizations since some of the stars generating certain cometary nebulae (for example R Monocerotis) don't correspond exactly to the description of the T Tauri stars.

Finally, the most fascinating kind of the activity of T Tauri stars is their transformation in to Fuors. We can formulate this saying that T Tauri stars are sometimes fuoring. In his recent paper Herbig has shown by means of rough estimates similar to those given above about the frequency of formation of HH objects that in the average each T star during every 5 thousand years is transforming itself into a Fuor - a comparatively luminous star.

Regretably enough we don't know the average duration of Fuor-flare. We can state only that FU Orionis - the first discovered Fuor after reaching maximum and then slightly fading, remains bright more than half of century. Other Fuors remain bright during several decades. But if we will not found something extravagant about Fuors we will apparently conclude that their mean lifetime must be not much longer than a century. Otherwise we should have find some stars that have fuored during the last century or even earlier. But of course this is true only when we assume that we can recognize Fuors even during hundreds years after the brightening.

The most interesting thing which we know about Fuors is that the emission line Hα in their spectra has the P Cygni like absorption component. This is an evidence of very intense outflow from the star.

But if we discuss Fuors and the fact that their spectra contain lines of P Cygni structure let us remember that P Cygni itself has sometimes (in the beginning of 17-th century) brightened then slightly faded and after this remains almost constant, thus reminding in some degree the process of fuoring. As I mentioned above the FU Orionis also has somewhat diminished in brightness after the maximum and then remains almost constant in brightness.

Of course I don't suppose that before the brightening the star P Cygni and the similar supergiants in the Galaxy were sometimes low luminosity dwarfs. However it is not excluded that they had some median luminosity. But this is a special question which we must refer to people who study the O-associations and the evolution of early type supergiants.

MITSKEVICH: Can we say that ejection of matter is the main process for T Tauri stars or is this only the result of accretion?

AMBARTSUMIAN: I believe that the ejection is the main process for T Tau stars because the observations show so. But of course there is a small accretion also. Sometimes we can see it. Never the less, as I believe, the main process is ejection.

RODONO: You have called attention to the important role of mass ejection in stellar evolution, in addition to mass and chemical composition. Now we have indirect and direct evidence of the important role of magnetic fields in active stars and in flares. Can you comment on the role of magnetic fields in the evolutionary scenario you have presented?

BLAAUW: You referred to the occurrence of flare stars in groups of a great variety of ages and concluded that stars of all masses go through a phase of flaring in their youth. I know that as reference groups we have the group around the Trapezium(about 1 million years), the Pleiades (about 100 million years) and the field stars (still older on the average). Do we have any group with age around 10 million years of which we can say that it contains flare stars of that age?

MIRZOYAN: The observations of flare stars are very selective. The observers try to observe mainly the clusters and associations which are not far away. This gives a possibility to observe many flares. Probably this is the reason why in practice we have no observations of flares in groups that you have mentioned.

THE VARIABILITY OF T TAURI STARS

W. HERBST
Van Vleck Observatory
Wesleyan University
Middletown, CT 06457
U.S.A.

For many years we have monitored the variability of some bright T Tauri stars with the 0.6 m "Perkin telescope" at Van Vleck Observatory. Collaborators in this endeavor have included several astronomers at the Crimea Astrophysical Observatory, at Capidomonte Observatory, at the U.S. Naval Observatory, and even some sophisticated amateurs.

As has long been known, the typical variation exhibited by T Tauri stars is irregular in nature. While several studies have shown periodic components (or even purely cyclic behavior) in the light variations of some T Tauri stars (especially the "weak" or "naked" T Tauri stars; e.g. Rydgren and Vrba, 1983; Herbst et al., 1986; Bouvier and Bertout, 1989), these remain the exceptions. Exhaustive studies of many stars fail to reveal distinctive periodicities (Herbst et al., 1987). Periods found at one epoch may also be different from those at other epochs, an example being TW Hya (Herbst and Koret, 1988). The periodic variations are attributed to spots - either cool or hot - on or close to the stellar surfaces. The cause of the irregular variations remains unknown. In this paper we discuss the best studied example of a spotted T Tauri star, V410 Tau, and then mention a newly discovered constraint on the mechanism for the irregular variation.

V410 Tau is a K3 star with weak Hα emission and little or no infrared or ultraviolet excess. It is a variable, non-thermal radio source and lies quite far from the main sequence in the HR diagram. Conventional pre-main sequence tracks imply that it is about a one million year old, one solar mass star. Its light variation was discovered by Mosidze (1968) and was believed to be irregular. Rydgren and Vrba (1983) showed, however, that it was purely cyclic in 1981, and Vrba, Herbst and Booth (1988; Paper I) and Herbst (1989; Paper II) have confirmed that this is so at all epochs. The period is 1.871 days, and is interpreted as the rotation period of the star. Color variations show that the spots on this star must be cool - about 1300 K cooler than the photosphere.

The amplitude and shape of the light curve (but not its period) has changed substantially over the years as

L. V. Mirzoyan et al. (eds.), Flare Stars in Star Clusters, Associations and the Solar Vicinity, 169–172.

shown in Papers I and II. At least two spots must be present to account for the double-peaked nature of the variation at some epochs. The spots must be enormous to explain the very large amplitude of variation (up to 0.55 mag in V). The inclination angle of the rotation axis to the line of sight must also be close to 90° to account for the large v sin i and the shape of the light curve. Finally, we can infer from the light curve that the spots must be centered at latitudes of 50° or more from the equator.

In Paper II detailed modeling of the spot variations has been carried out which substantiates the claims of the previous paragraph. The results of this modeling are summarized in Table 1.

Table 1. Results of Spot Modeling for V410 Tau (Paper II)

Rotation Period: 1.871 days
Photospheric Temperature: 4400 K
Spot Temperature: 3100 K
Unspotted apparent V mag.: 10.6
inclination angle: 75°-85°
Number of spots: at least two
spot latitudes: +55° to +75° and -50° to -65°
spot radii: 40° to 60°
Spot lifetimes: The principal spot has been present since at least 1983 and perhaps much longer

The observed changes in amplitude and shape of the light curve in V410 Tau can be understood as changes in the relative longitudes of the two spots. The two spots approached each other from 1983 until 1986 reaching a minimum separation of about 90° and are now receding. It is expected that the light curve amplitude will diminish for a couple of years and become double-peaked again as the spots move to opposite sides of the star. Continued observation of this very important star is highly desirable.

It has often been suggested that there is a connection between the periodic and aperiodic variation of T Tauri stars, the latter being attributed to spots which change their extent on time scales comparable to or shorter than a rotation period (e.g. Holtzman, Herbst and Booth, 1986; Bouvier and Bertout, 1989). In order to test the hypothesis that cool spots are responsible for the irregular variations R. Levreault and I undertook a program of Wing system photometry of six T Tauri stars. The bluest filters of this system are designed for study of the TiO feature which is expected to be enhanced in cool spot regions. We observed T Tau, RY Tau, CO Ori, RW Aur, SU Aur, and V410 Tau at Van Vleck Observatory during the 1988/89 Milky Way season. The results have been submitted for publication to the Astronomical Journal but a brief summary is given here.

The principal result is that we detected variation in the TiO feature in only one star - V410 Tau. This positive result is important because it confirms, both qualitatively and (to within the errors) quantitatively the general picture of spot induced periodic variations. In other words, it gives us direct evidence of the presence of cooler regions associated with the star V410 Tau and demonstrates that when there is an increase in the fraction of light coming from these cooler regions, the total light of the star is diminished. We also find that, for its color, V410 Tau is unusually enhanced in TiO absorption even near maximum light. This demonstrates that we do not see the spot-free star, even then. In principle this should eventually help us construct better spot models by providing a constraint on the apparent magnitude of the unspotted star.

The absence of any change in TiO strength even during large excursions in brightness is, at least in the case of RY Tau, rather strong evidence that its irregular variations are not caused by changes in the cool spot coverage. Furthermore, the fact that, for its color, it does not appear unusual in its TiO index, at any light level, suggests that it simply is not extensively covered with spots at any brightness level. It appears that we will have to search elsewhere for the cause of the irregular light variations in this and, presumably, other T Tauri stars. Hot spots are one possibility, generated either by solar flare-like phenomena or by accretion. Another, possibly related possibility is changes in contributions to the total light from disks and/or boundary layers.

It is a pleasure to acknowledge the assistance of the Perkin Fund and the NSF in supporting astronomy at Wesleyan University. I also thank the large number of undergraduates who have contributed to these studies over many years and my collaborators in the Soviet Union and elsewhere.

References

Bouvier,J. and Bertout,C.(1989), A&A **211**, 99.
Herbst,W., Booth,J.F., Chugainov,P.F., Zajtseva,G.V., Barksdale,W., Covino,E., Terranegra,L., Vittone,A. and Vrba, F. J. (1986) Ap.J. (Letters) **310**, L71.
Herbst,W., Booth,J.F., Koret,D.L., Zajtseva,G.V., Shakhovskaya,N.I., Vrba,F.J., Covino,E., Terranegra,L. Vittone,A., Hoff,D., Kelsey,L., Lines,R. and Barksdale,W. (1987), A.J. **94**, 137.
Herbst, W. and Koret, D. L. (1988), A.J. **96**, 1949.
Herbst, W. (1989), A.J. (in press).
Holtzman,J.A., Herbst,W. and Booth,J.F. (1986), A.J. **92**, 1387.
Mosidze, L. N. (1968), Astr. Tsirk. (Kazan) No. 474, p. 6.
Rydgren, A. E. and Vrba, F. J. (1983), Ap.J. **267**, 191.
Vrba,F.J., Herbst,W., and Booth,J.F. (1988), A.J. **96**, 1032.

PETROV: The large spots on V410 Tau could be seen by the Doppler imaging technique. Has anyone observed this star spectroscopically with high resolution?

HERBST: A few high resolution spectra exist, taken by Strom and by Basri. An attempt to use these for Doppler imaging is underway. It would be important to abtain more.

GAHM: V410 Tau is a very rapid rotator. Is it the record among T Tauri stars or are there stars spinning even faster?

HERBST: I believe there are stars with larger v sin i's, although this one is very fast.

PERIODICAL VARIATIONS OF THE BRIGHTNESS OF T TAURI

G.V. ZAJTSEVA
State Sternberg Astronomical Institute, USSR

1. INTRODUCTION

T Tauri-type stars are characterized by the fast irregular variability. It is possible to select several components in brightness variations: slow components with timescales of years and fast ones with timescales of days and tens of days. Besides, up to 1 mag flares in the ultraviolet part of the spectrum occur during some nights. The overall change of the brightness may reach 2-3 mag.

Periodical light variations have been discovered in a number of T Tauri stars with amplitudes of several hundredths of the stellar magnitude and periods of several days. Periods are determined with different degrees of accuracy using observations concentrated on restricted time intervals. Only for V410 Tau the period of 1.8 days was confirmed by Vrba et al.(1988) from observations at different epochs. Periods are thought to be axial rotation periods and observations are interpreted taking the hypothesis of the presence of temperature inhomogeneities (spots) on star surface in analogy with BY Dra-type stars.

2. OBSERVATIONS

The photoelectric observations of T Tau obtained at the Crimean Station of the State Sternberg Institute from 1971 are plotted in Figure 1. One can note that the light variability is more quiet than for other stars of this type. Nevertheless three components of variations are quite noticeable: two 2000 days waves of the slow component, variations from night to night of the order of 0.1-0.2 mag and ultraviolet flares reaching sometimes 0.8 mag.

The period of 2.8 days was discovered by Herbst et al. (1986) from cooperative observations obtained in the season of 1985-86. This period was confirmed by Zajtseva (1989) using numerous observations obtained in Crimea. Also, observations by Zajtseva et al. (1988) in 1987 and Herbst and Coret (1988) in 1988 have shown the presence of the periodical component in light variations of T Tauri.

Amplitudes A_V and phases ϕ_{max} derived from 4 series of observations of T Tau are compared in Table 1. For each series the number of nights n and the number of observations N are given. Phases are computed with a period of 2.80 days and the initial epoch is JD 2440000.0. Corresponding phase diagrams are presented in Figure 2.

L. V. Mirzoyan et al. (eds.), Flare Stars in Star Clusters, Associations and the Solar Vicinity, 173–176.

Table 1. A comparison of amplitudes and phases of the periodical comp.

Season		J.D.	n	N	A_V	φ_{max}
I	1976-77	2443049-43232	41	88	0.038	0.10
II	1985-86	2446323-46526	203	206	0.037	0.60
III	1987	2447099-47102	4	44	0.070	0.93
IV	1988	2447162-47174	13	13	0.048	0.53

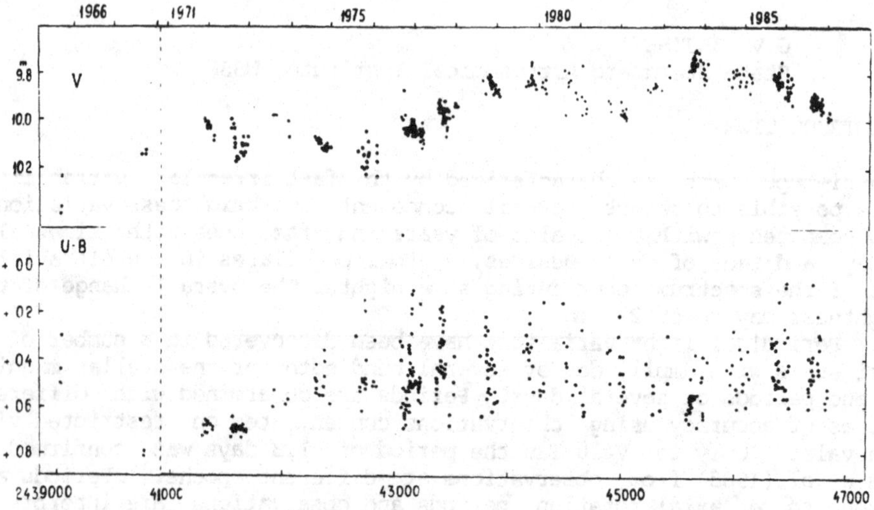

Figure 1. Variations of V and (U-B) for T Tauri.

One can see that the amplitude is the largest for series IV when 44 observations were distributed over 4 nights. The amplitude is smallest for series II with 206 observations over 203 nights. This result indicates that the effect of irregular variations is to decrease the amplitude of the periodical component. Noteworthy are the phase changes which are obviously irregular and could be hardly explained by variations of the period. Especially high is the phase difference between series III and IV, which are separated by two months. Probably the change is due to the longitude shift of an active area (spot).

3. DISCUSSION

Let us suppose that a spot exists on the surface of T Tau and the spot is characterized by the relative area f and the temperature T_s. The amplitude of the light variations due to the spot appearance can be evaluated as

$$\Delta m(\lambda)=-2.5 \log(1-f(B_\lambda(T_s)/B(T_*)))$$

where λ is the wavelength, B(T) is the Planck function, T_* is the temperature of the unspotted part of the star's surface. The temperature adopted for T Tau is T_*=5500 K. The observed amplitudes for the 1987 observations of T Tau are ΔU=0.123, ΔB=0.084, ΔV=0.070, ΔR=0.057. These correspond to the values of T_*, T_s and f given in Table 2. Thus dark and bright spots in equal measure obey observations.

Table 2. Computed temperatures and areas of the spot and amplitudes of the light variations

T_*(K)	T_S(K)	f	ΔU	ΔB	ΔV	ΔR
5500	6500	0.065	−0.107	−0.101	−0.075	−0.058
5500	5000	0.170	+0.100	+0.086	+0.073	+0.060

We have demonstrated the presence of the rotational modulation of the brightness using series of photoelectric observations of T Tau differing in length and number. From these the evolution of spots and other information can be derived. Assuming a constant rotation period of 2.8 days we have found phase changes which probably indicate spot migration.

References

Herbst,W., Booth,J.F., Chugainov, P. F., Zajtseva, G. V., Barksdale, W., Covino,E., Terranegra,L., Vittone,A., Vrba,F., 1986, Astrophys. J. 310, L71.
Herbst,W., Booth, J.F., Koret, D.L., Zajtseva, G.V., Shakhovskaya, N.I., Vrba,F.J., Covino,E., Terranegra,L., Vittone,A., Hoff,D., Kelsey, L., Lines,R., Barksdale,W., 1987, Astron. J. 94, 137.
Herbst,W., koret,D.L., 1988, Astron. J. 96, 1949.
Vrba,F.J., Herbst,W., Booth,J.F., 1988, Astron. J. 96, 1032.
Zajtseva,G.V., Tarasov,K.V., Chernova,G.P., 1988, Astr. J. USSR 14, 610.
Zajtseva,G.V., 1989, Astrofizika (in press).

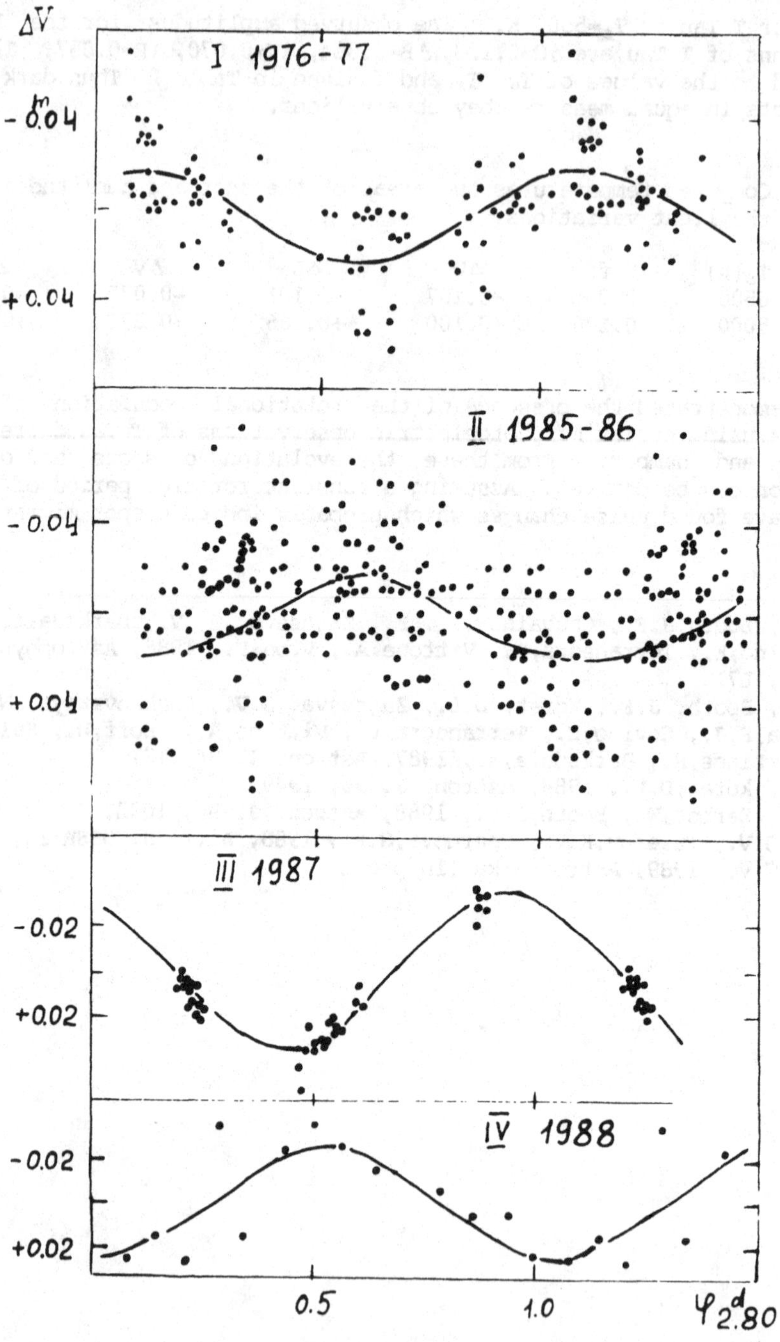

Figure 2. Phase diagrams for 4 seasons of observations of T Tauri.

THE POLARIZATION DISCOVERY AND INVESTIGATION OF T TAU STARS IN BYURAKAN

R.A.VARDANIAN
Byurakan Astrophysical Observatory
Armenian Academy of Sciences, USSR

The results of the polarization observations of T Tau, RY Tau and of field stars are presented (Fig.1). The ratio p/A_v for RY Tau is equal to 0.066 and is larger than the same ratio for surrounding field stars. The observed variations of the polarization parameters of T Tau and RY Tau as well as the maximum value of the ratio p/A_v for RY Tau have permited already in 1964 to suggest that these stars have intrinsic polarization.

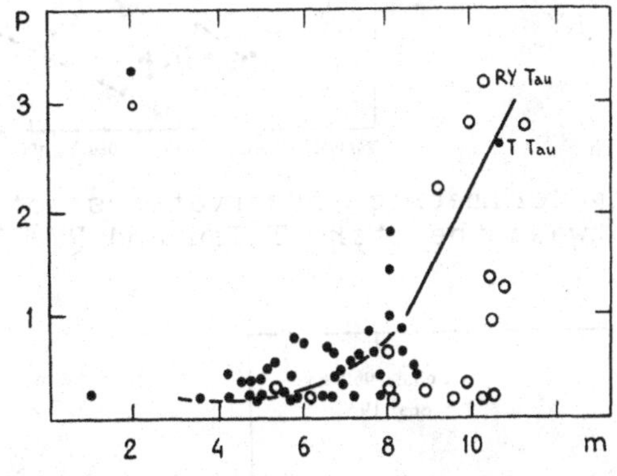

Figure 1. The dependence of the polarization (P) from the brightness (m) for the stars in the regions of T Tau and RY Tau stars.

Then it has been shown, that the directions of axies of cometary nebulae coincides nearly with the plane of polarization of the surrounding stars. This means that these directions for the cometary nebulae are conditioned by general or local galactic magnetic fields.

The electropolarimetric, photometric and spectroscopic observations of stars T Tau, RY Tau, NU Ori are given (Fig. 2), which confirm the existence of the intrinsic polarization and its changes in time.

L. V. Mirzoyan et al. (eds.), Flare Stars in Star Clusters, Associations and the Solar Vicinity, 177–178.

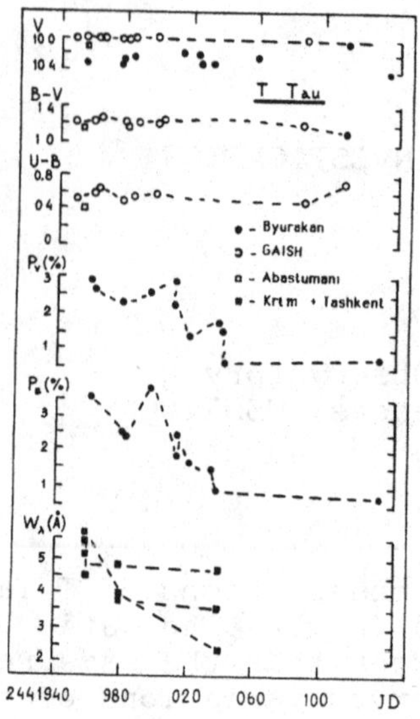

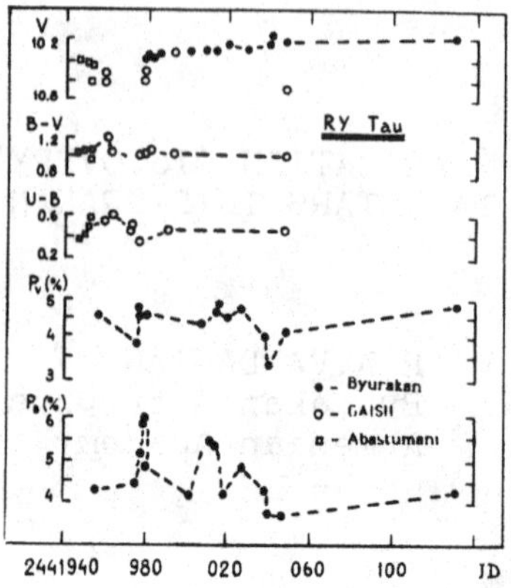

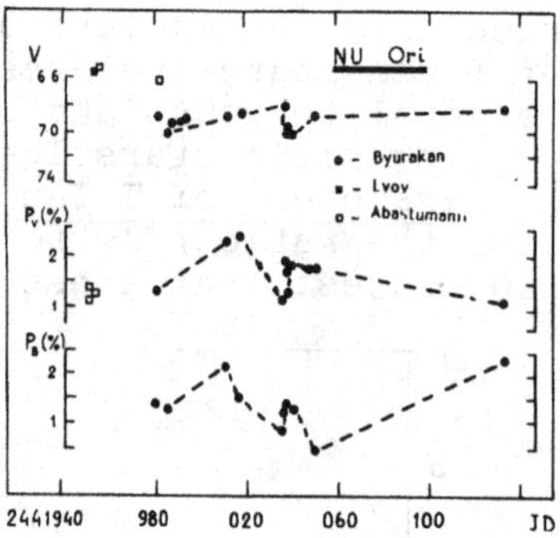

Figure 2. Photometric, polarimetric and spectral observations of T Tau, RY Tau and NU Ori.

At last electropolarimetric observations of stars located in the area between the stars T Tau and RY Tau are carried out (Fig. 3).

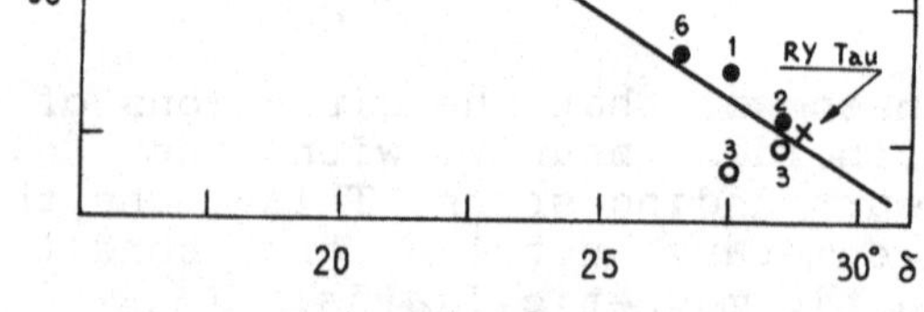

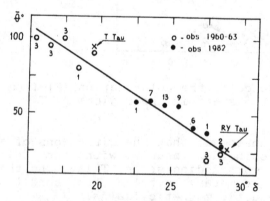

Figure 3. Correlation between the position angle of polarization (θ) and the declination of stars in the regions of T Tau and RY Tau stars.

RECENT RESULTS ON POLARIZATION OF T TAURI STARS AND OTHER YOUNG STELLAR OBJECTS

PIERRE BASTIEN and FRANÇOIS MÉNARD
Observatoire du Mont Mégantic, and Département de Physique,
Université de Montréal
B. P. 6128, Succ. A, Montréal, Québec, H3C 3J7
Canada

ABSTRACT. Recent results on observations and models of polarization of T Tauri stars (TTS) and other young stellar objects (YSO's) are presented. In particular, the difference in polarization properties between the classical T Tauri stars (CTTS) and the weak-line T Tauri stars (WTTS) is made. A correlation between polarization and rotation period is searched for but not found. Observations of polarization maps for many young stars are considered. Two types of models have been proposed to explain these observations. Current evidence favors single and multiple scattering in flattened, optically thick, structures, i.e. disks, around many TTS. In particular, the size of the optically thick part of the disks, and their inclination to our line of sight can determined for published polarization maps.

1. Introduction

The observations for the first important paper on linear polarization observations of TTS were carried out here, at the Byurakan Observatory, more than 25 years ago (Vardanian 1964). The observations of RY Tau are plotted in Fig. 1 as a function of time. The star shows strong variations in both P (from 2% to 4.5%) and θ (from 0° to 80°). The main conclusions of the paper are best summarized by this sentence from the English abstract to the paper: "The observed variations of the polarization parameters of T and RY Tau as well as the value of the ratio P/A_V for RY Tau permit to suggest that these stars have intrinsic polarization". This suggestion has been confirmed by subsequent observations for these two and many other TTS.

A discussion of subsequent polarization observations of TTS with a detailed review of their polarization properties and models has been given recently by Bastien (1988). In this paper, we point out a few new results on this subject. General review papers on TTS have been written by Bertout (1989) and by Appenzeller and Mundt (1990).

2. New Results

X-ray surveys of dark clouds have led recently to the discovery of another group of low mass TTS, the weak-line T Tauri stars (WTTS)(Walter *et al.* 1988). WTTS are believed to

L. V. Mirzoyan et al. (eds.), Flare Stars in Star Clusters, Associations and the Solar Vicinity, 179–184.

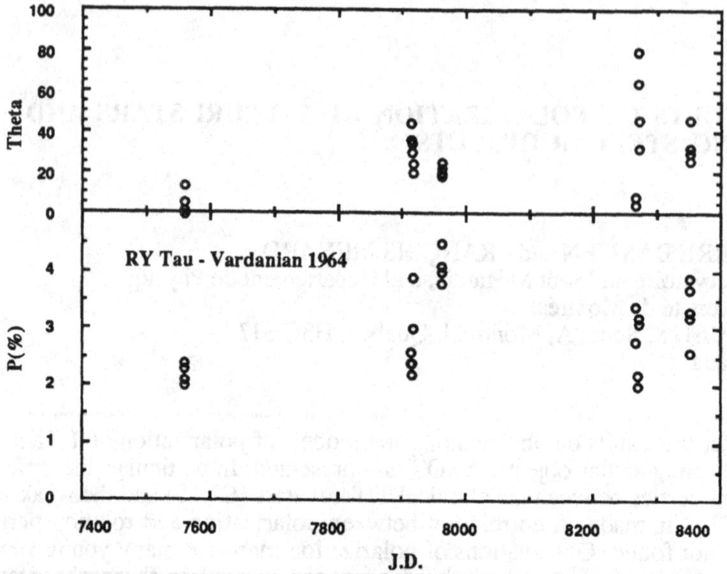

Figure 1. Linear polarization of RY Tau measured by Vardanian (1964) The three observing seasons, 1961, 1962 and 1963 (August to November) can be found easily. The error bars are about 0.3%. The data shows clearly the polarization variability of the star.

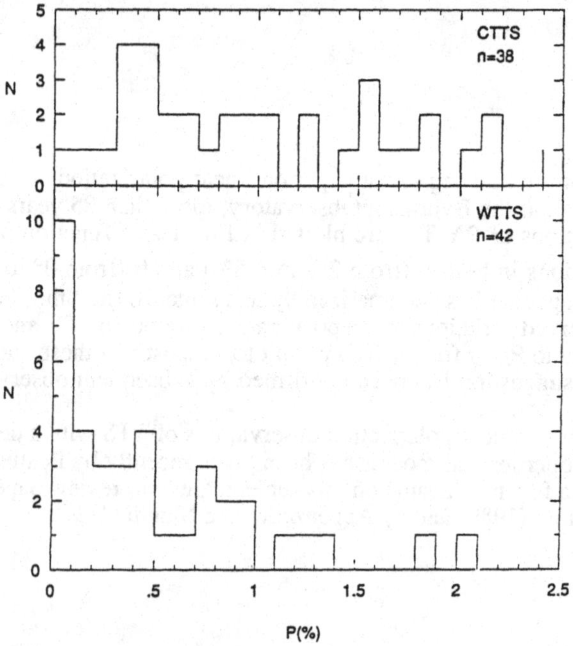

Figure 2. Histograms of linear polarization for CTTS and WTTS in the Taurus-Auriga dark cloud complex.

have no or very weak circumstellar disks, which distinguishes them from the classical TTS (CTTS) whose main properties are believed to be due to the presence of a disk. Infrared observations provide support for this interpretation, as the WTTS appear to have much weaker infrared excesses (Strom *et al.* 1989). Another check is to measure and compare the linear polarization of the two groups of stars since the polarization is due to circumstellar dust. In Fig. 2, we present two histograms of polarization data for the CTTS and WTTS in the Taurus-Auriga T associations. CTTS have larger values of polarization which are clearly different from zero, whereas the WTTS show a strong peak at zero polarization. This implies that most (but not all) WTTS have no or very little circumstellar material.

The results of polarization surveys of TTS (Bastien 1982, 1985) have been used to search for correlations between polarization and other properties of these stars. In particular, there is a very strong correlation between polarization and the infrared excess. Since then, rotation periods have been determined for many TTS (see Herbst 1990 and references therein). We present in Fig. 3 a plot of polarization against rotation period. As can be seen, there seems to be no correlation.

Linear polarization maps of nearly 60 YSO's have been published in the past few years. In these maps, the polarization vectors are usually perpendicular to the radius vector from the central illuminating source. Such a centrosymmetric pattern is typical of reflection nebulae where single scattering is responsible for the polarization since the polarization vectors are usually perpendicular to the scattering plane. However, about 60% of these maps show a pattern of aligned polarization vectors close to the central source (Bastien and Ménard 1988, 1990, hereafter BM1 and BM2 respectively). The polarization reaches typically 10%-15% in this region, the highest value noted so far being 30%.

Two interpretations have been proposed for the patterns of aligned polarization vectors. (1) Dichroic extinction by aligned nonspherical grains in a circumstellar disk has been suggested by many authors (e. g. Warren-Smith 1987). Various alignment mechanisms are possible. Alignment by a toroidal magnetic field is usually preferred. However, no model calculations for this interpretation have been performed. (2) Multiple scattering in and around an optically thick circumstellar disk was proposed by BM1, where they presented calculations done under the assumption that only two scatterings occur in the disk, and showed that it is possible to obtain patterns of aligned vectors with multiple scattering, provided one has an optically thick surface.

There are many objections which can be raised against dichroic extinction (BM1, BM2). No polarization observations support the presence of aligned grains around TTS and YSO's. This means that magnetic fields, if present, are not efficient at aligning grains in the circumstellar environment. However, all of the arguments against dichroic extinction support the idea of scattering on dust grains. In particular, circular polarization has been predicted and detected in the region (disk) where multiple scattering is occurring (Ménard, Bastien, and Robert 1988). When the observed maps are compared with computed maps, one can obtain the inclination of the disk and also the size of the disk for which the optical depth is ≥ 1 (BM2). The BM1 model has now been confirmed with a Monte Carlo code which includes as many scatterings as there should be for given geometry and dust properties (Ménard 1989, Ménard and Bastien 1990). In Fig. 4 we show a computed map which corresponds to an average inclination angle of 93.2°. The density law used decreases linearly with distance r from the star in the equatorial plane, and exponentially with distance z from that plane. Graphite grains with a radius of 0.2 μm have been used, giving an optical depth in the plane of the disk of 36.

We are grateful to the Natural Sciences and Engineering Research Council (NSERC) of Canada for financial assistance.

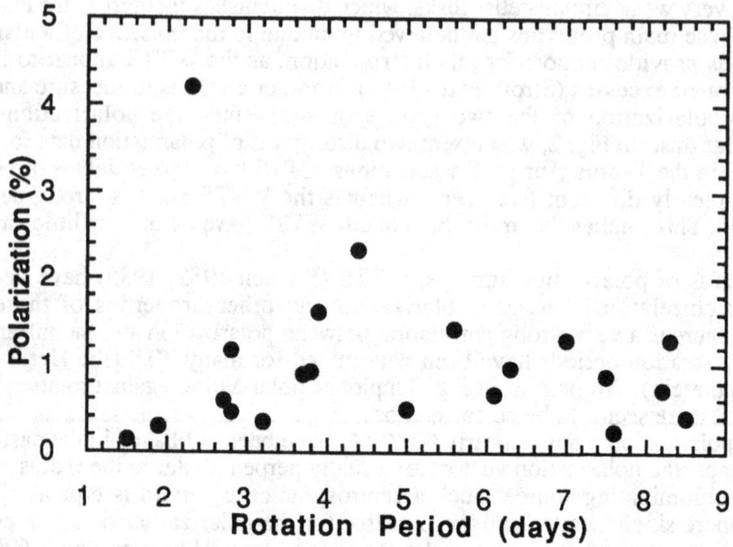

Figure 3. Polarization in a red passband of TTS plotted against their rotation periods. The star with a large polarization and a small rotation period is LHα 332-20.

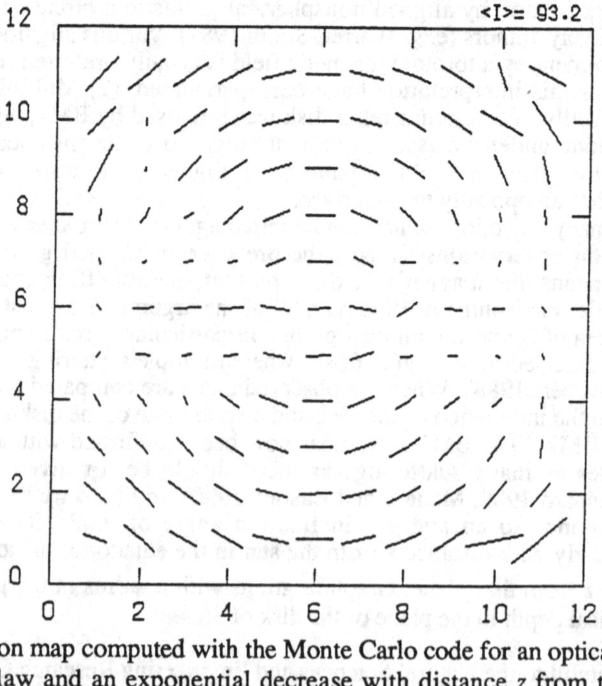

Figure 4. Polarization map computed with the Monte Carlo code for an optically thick disk with a r^{-1} density law and an exponential decrease with distance z from the equatorial plane. The region shown corresponds to 10 AU on the side.

REFERENCES

Appenzeller, I., and Mundt, R. (1990) 'Observational Properties of T Tauri Stars', *Astron. Astrophys. Rev.*, in press.

Bastien, P. (1982) 'A Linear Polarization Survey of T Tauri Stars', *Astron. Astrophys. Suppl.*, **48**, 153-164, and **48**, 513-518.

Bastien, P. (1985) 'A Linear Polarization Survey of Southern T Tauri Stars', *Astrophys. J. Suppl.*, **59**, 277-291.

Bastien, P. (1988) 'Polarization Properties of T Tauri Stars and Other Pre-Main Sequence Objects', in G. V. Coyne *et al.* (eds.), *Polarized Radiation of Circumstellar Origin*, Vatican Press, Vatican City, 541-582.

Bastien, P., and Ménard, F. (1988) 'On the Interpretation of Polarization Maps of Young Stellar Objects', *Astrophys. J.*, **326**, 334-338 (BM1).

Bastien, P., and Ménard, F. (1990) 'Parameters of Disks Around Young Stellar Objects From Polarization Observations', submitted (BM2).

Bertout, C. (1989) 'T Tauri Stars: Wild as Dust', Ann. Rev. Astron. Astrophys., **27**, 351-395.

Herbst, W. (1990) 'The Variability of T Tauri Stars', this volume.

Ménard, F. (1989) 'Étude de la polarisation causée par des grains dans les enveloppes circumstellaires denses', Ph. D. Thesis, Univ. de Montréal.

Ménard, F., and Bastien, P. (1990), in preparation.

Ménard, F., Bastien, P., and Robert, C. (1988) 'Detection of Circular Polarization in R Monocerotis and NGC 2261: Implications for the Polarization Mechanism.', *Astrophys. J.*, **335**, 290-294.

Strom, K. M., Strom, S. E., Edwards, S., Cabrit, S., Strutskie, M. F. (1989) 'Circumstellar Material Associated With Solar-type Pre-Main-Sequence Stars: A Possible Constraint on the Time Scale for Planet Building', *Astron. J.*, **97**, 1451-1470.

Vardanian, R. A. (1964) 'The Polarization of T and RY Tau', *Contrib. Byurakan Obs.*, **35**, 3-23 (in Russian).

Walter, F. M., Brown, A., Mathieu, R. D., Myers, P. C., Vrba, F. J. (1988) 'X-Ray Sources in Regions of Star Formation. III. Naked T Tauri Stars Associated With the Taurus-Auriga Complex', *Astron. J.*, **96**, 297-325.

Warren-Smith, R. F. (1987) 'Magnetic Fields and the Optical Polarization of Star Formation Regions', *Quart. J. R. A. S.*, **28**, 298.

LADA: Can you apply your modelling technique to T Tauri stars without such extended reflection nebulae to obtain disk inclinations? Such observations would be very important for understanding the energy budget of individual YSO's. Certainly with knowledge of disk inclination we would be able to obtain more accurate estimates of the bolometric luminosities of T Tauri stars and perhaps estimates of their accretion luminosity as well.

BASTIEN: Yes, provided one has high spatial resolution observations of these stars. For example, if polarization measurements can be made with the Space Telescope one should be able to derive inclination angles for these stars which have less extended nebulosities around them.

LANG: IRAS infrared observations show evidence for circumstellar disks, and in one case (beta Pictoris) the disk has been imaged optically. Is there linear polarization for these objects and does it confirm your theory of scattering from a circumstellar disk?

BASTIEN: Yes, the linear polarization of the disk associated with beta Pictoris has been measured. The polarization is perpendicular to the disk, which means that the optical depth is so small that single scattering only is taking place in the disk.

LANG: Are there similar infrared observations of T Tauri stars that show strong linear polarization for a circumstellar disk?

BASTIEN: A few young stars have been imaged in the infrared from the ground and do show evidence for a circumstellar disk. Some polarization measurements have also been made at 2 micron and show evidence for circumstellar disks.

VARDANIAN: Was the role of interstellar polarization taken into account during the modelling?

BASTIEN: No, it was not taken into account because it does not have to. The interstellar polarization is usually small (about 0.1 % in Taurus to < 2 % in Cygnus, for example) whereas the polarization in the models is large (> 10 %) and would dominate interstellar polarization.

MITSKEVICH: Did you try to find any correlation between the polarization of T Tauri stars and displacement of their molecular lines, for example CO-lines?

BASTIEN: No, I did not.

THE UBVRI PHOTOMETRIC AND POLARIMETRIC OBSERVATIONS OF THE T TAU STAR HDE 283572

N.I.SHAKHOVSKAYA
Crimean Astrophysical Obsevatory
Nauchny
Crimea 334413
USSR

ABSTRACT.The preliminary analysis of UBVRI photometric and polarimetric observations of the naked T Tau star HDE 283572 indicates that:
(i)there is intrinsic and variable component of polarisation;
(ii)the distribution and the area of the active regions on stellar surface are variable on time scale of a year.

1. Introduction

HDE 283572 is ninth-magnitude G-star south of RY Tau. As suggested by Walter et. al.(1987),HDE 283572 is a good example of a "naked T Tauri" (NTT) stars,i.e.,stars indentical to the low mass T Tauri stars save for the lack of significant circumstellar envelope. It was proposed, that the periodic ($\gtrsim 1^{d}.5$) modulation of magnitude in V ($\gtrsim 0^{m}.2$) is caused by asymmetric distribution of the active regions on the surface. This star may be very useful for studies of how the stellar (not circumstellar) activity levels behave in very young convective stars. Presented here are the preliminary results of my photometric and polarimetric observations.

2. Observations and results

The obsevations were made with the 125cm telescope of Crimean Astrophysical Observatory using the simultaneous five color (UBVRI) photometer-polarimeter of Helsinki University.

2.1 PHOTOMETRY

In Fig.1 V magnitudes folded on the ephemeris:J.D.=2445600.173+1.548E by Walter et al. (1987) for the different seasons are given. The observations in 1983 were taking from Walter et al (1987). As follows from Fig 1, Walter's period does not contradict to my observations in the individual seasons. There are, however, a phase shift between different seasons and I could not improve period like that to fit all the observations. The amplitudes ,maximal and minimal magnitudes of the light curve are variable.

L. V. Mirzoyan et al. (eds.), Flare Stars in Star Clusters, Associations and the Solar Vicinity, 185–188.

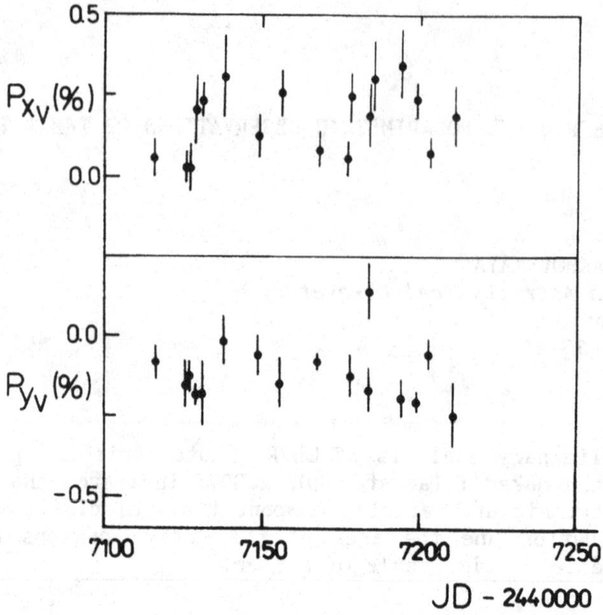

Figure 2. The components of polarization vector Px(%) and Py(%) in V band in 1987-88, plotted as function of Julian Days.

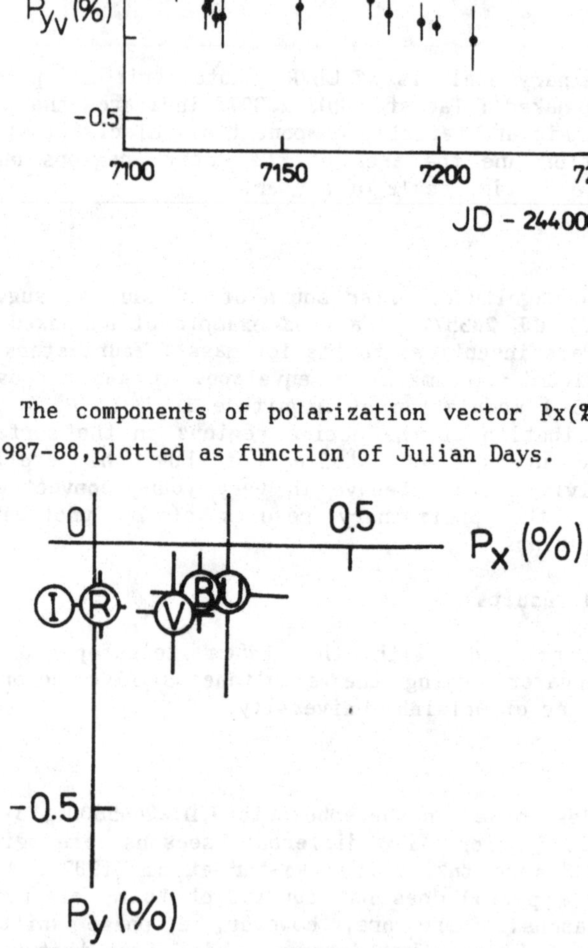

Figure 3. The polar diagram (Px,Py)(%) for averaged data in UBVRI.

The variability of the light curves and the phase shift could be explained by the change of the distributions and the area of the active regions on the stellar surface on the time scale of a year.

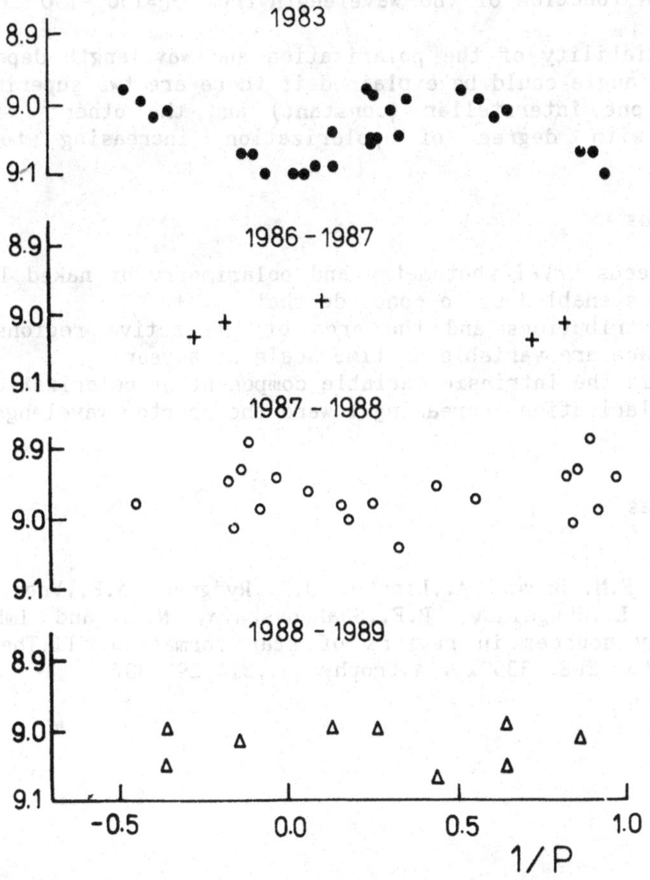

Figure 1. The light curves in V band on the ephemeris by Walter et al.

There are strong linear correlations between magnitudes in the different bands, so: $\Delta U/\Delta V=1.17\pm0.10$; $\Delta B/\Delta V=1.10\pm0.06$; $\Delta R/\Delta V=0.82\pm0.04$; $\Delta I/\Delta V=0.73\pm0.03$. Thus, the brighter the star the bluer it is.

2.2. POLARIMETRY

The polarization in the R and I bands remained constant. But in the UBV the variation of polarization with amplitude 0.5 % was observed. As an example in Fig.2 the parameters Px and Py in V band in 1987-88 are given ($P_x=P\cos2\vartheta$ and $P_y=P\sin2\vartheta$ are the components of the polarization vector P in the $(P,2\vartheta)$ plane). But I could not find any correlation neither between Px,Py in different bands, nor between the polarization and photometric variability.

In Fig.3 the Px and Py for UBVRI bands averaged for all data are given. As follows from Fig.3, there is strong rotation of the position angle ϑ as a function of the wavelength, from $\vartheta_U=150^0-190^0$ to $\vartheta_I=110^0-120^0$.

The variability of the polarization and wavelength dependence of the position angle could be explained if there are two superimposing components: one interstellar (constant) and the other - intrinsic (variable) with degree of polarization increasing toward the ultraviolet.

3. Conclusions

The simultaneous UBVRI photometry and polarimetry of naked T Tau star HDE 283572 has enabled us to conclude that:

i) The distributions and the area of the active regions on the stellar surface are variable on time scale of a year.

ii) There is the intrinsic variable component of polarization with degree of polarization increasing toward the shorter wavelenght.

References

Walter, F.M.,Brown, A.,Linsky, J.L.,Rydgren, A.E.,Vrba, F.,Roth, M.,Carrasco, L.,Chugainov, P.F.,Shakhovskaya, N.I. and Imhoff C.L. (1987) 'X-ray sources in regions of star formation. II.The pre-main sequence G star HDE 283572', Astrophys.J.,**314**,297-307.

AXIAL ROTATION OF BY DRA-TYPE STARS AND RELATED OBJECTS

P. F. CHUGAINOV
Crimean Astrophysical Observatory
Nauchny, Crimea 334413, USSR.

ABSTRACT. Periods of the rotational modulation of the brightness of BY Dra-type stars, Pleiades spotted stars, naked T Tau stars and T Tau-type stars are compared with absolute bolometric stellar magnitudes. Arguments are given that the majority of BY Dra-type stars have the age of 10**8 years and relatively fast as well as slowly rotating objects are met between them.

1. INTRODUCTION

In this paper the recent works on the search for rotational modulation of the brightness of young dwarf late-type stars are summarized. The aim is to study the difference in periods of the axial rotation between stars pf different ages with masses nearly equal to the mass of the Sun. Four groups of stars are selected. The names of stars are given in Table 1.

Table 1. Names of stars

BY Dra type: HD 1835, FF And, HD 8358, CC Eri, VY Ari, HD 22403, EI Eri, V833 Tau, V1005 Ori, AB Dor, OU Gem, YY Gem, YZ CMi, HD 82558, DH Leo, DK Leo, HD 91816, BF CVn, DF UMa, EQ Vir, ξ Boo A, HD 143313, TZ CrB, V722 Her, BY Dra, V815 Her, V775 Her, V478 Lyr, AU Mic, FK Aqr, EV Lac, BD-16°6218, KZ And, II Peg
Pleiades spotted: HII 34, 152, 296, 324, 335, 625, 686, 727, 739, 879, 882, 1124, 1332, 1531, 1883, 2034, 2244, 2927, 3163
Naked T Tau: HP Tau/G2, V410 Tau, V819 Tau, V826 Tau, V827 Tau, V830 Tau, V836 Tau, HD 283447, HD 283572
T Tau-type: RY Tau, T Tau, UX Tau A, BP Tau, DN Tau, AA Tau, DH Tau, DI Tau, SY Cha, LH_α 332-20, RY Lup, ROX 21, ROX 29

2. THE STARS AND THEIR PROPERTIES

2.1. BY Dra-type stars

In this group are included according to [1-10] 11 single stars and 23 spectroscopic binaries of spectral types from G to M which are believed to differ not significantly in luminosity from those of main sequence

L. V. Mirzoyan et al. (eds.), Flare Stars in Star Clusters, Associations and the Solar Vicinity, 189–192.

stars, although for several stars only spectroscopic luminosities are known. For a few exceptions the masses are thought to be in the range of 0.5–1 M⊙. It should be noted that 8 spectroscopic binaries have been classified as RS CVn stars in [6].

In contrast to the other stars considered here the BY Dra-type stars do not constitute a spatial group of common origin. Therefore the age should be estimated for each star separately. The unusually high eccentricity of the orbit of BY Dra itself, the rather large radius of its main component and the non-synchronism of axial and orbital rotation have been considered as evidence of an age of about 10**8 years [11-13]. Indications of youth are also the presence of H,K Ca II and H emission lines in the spectra and the high lithium abundance which has been found in the following 7 stars: HD 1835, VY Ari, V1005 Ori, AB Dor, HD 82558, Boo A, V815 Her, V478 Lyr [2, 7, 13-19]. According to the information contained in [20],21 of 22 BY Dra-type stars with known space velocities belong to young-disk population stars with ages not exceeding 5 10**8 years. Thus it is highly likely that for the majority if not all BY Dra type stars the age is of the order of 10**8 years.

2.2. Pleiades spotted stars

The rotational modulation of brightness has been discovered for 18 stars belonging to the Pleiades cluster [21-23]. Spectral types and masses are approximately the same as for BY Dra-type stars, the mass is about 0.8 M⊙. The age of the Pleiades cluster is about 10**8 years.

2.3. Naked T Tau stars

The complete list of these stars and evaluation of their ages (1-40) 10**6 years are given in [24]. Here are considered 9 stars showing the rotational modulation [25-29]. Their spectral types are G1-M0, masses may be approximately equal to 1 M⊙.

2.4. T Tau-type stars

The information on 13 T Tau-type stars which show the rotational modulation of the brightness may be found in [27, 28, 30-32]. According to [33] masses and ages of these 13 stars are less than 1.2 M⊙ and 5 10**6 years.

3. DISCUSSION

One can see from Figure 1 that the photometric periods of these stars do not depend on the absolute bolometric magnitude. Differences between the four groups of stars will be discussed further. Note that for each of the groups the largest value of the period does not exceed 8-10 days.

The mean values of the bolometric magnitude are 6.4 for BY Dra-type stars, 6.7 for Pleiades spotted stars, 4.1 for naked T Tauri stars and 3.8 for T Tau-type stars. The range of these values correspond to the

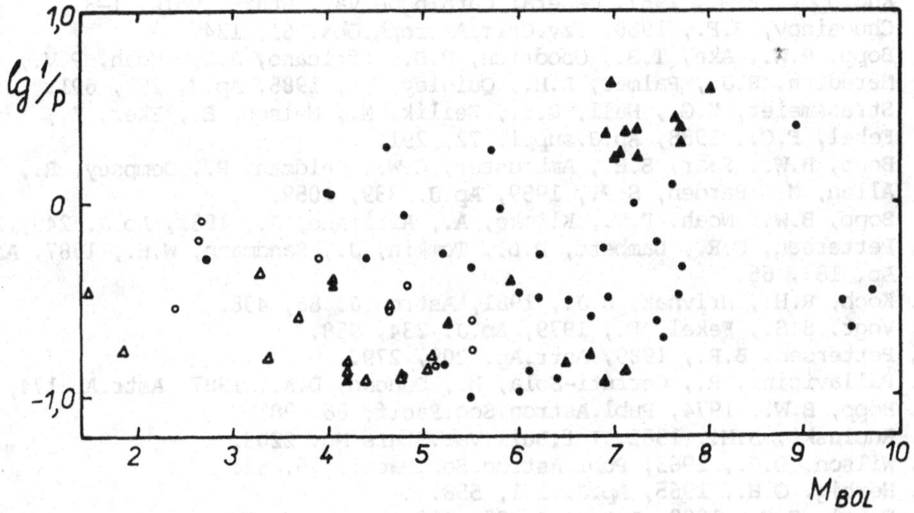

Figure 1. The relation between bolometric magnitudes and periods of stars. Filled circles=BY Dra-type stars, filled triangles= Pleiades spotted stars, open circles=naked T Tau stars, open triangles=T Tau-type stars.

evolutionary track of the 0.8 M⊙ star for 10**6 to 10**8 years. The agreement with estimated masses and ages of groups is rather good.

The differences of mean periods between the four groups are not large. The mean values of periods are 3.3 days for BY Dra-type stars, 2.65 days for Pleiades spotted stars, 3.4 days for naked T Tau stars, and 5.4 days for T Tau stars. However, within two groups the differences of extreme periods are striking: 0.5 days and 10 days for BY Dra-type stars, 0.24 days and 8 days for Pleiades spotted stars. One may suppose that the axial rotation of a star of 0.8 Mo changes drastically when its age is about 10**8 years. Such conclusion regarding the Pleiades spotted stars was drawn in [34]. On the other hand it is shown in this paper that for majority of BY Dra-type stars masses and ages are about 0.8 Mo, 10**8 years. Thus BY Dra-type stars as well as Pleiades spotted stars may be in the stage of evolution which is characterized by a strong change of the axial rotation. Partly the period differences within groups may be explained as an effect connected with the mass difference of stars entering the group. Probably the age of the star which exhibits the change of its period depends on the mass. The extreme periods are 2.3 days and 8.2 days for T Tau-type stars, 1.2 days and 6.99 days for naked T Tau stars.

References

1. Chugainov, P.F., 1976, Izv.Crim.Astroph.Obs. 54, 89.
2. Fekel, F.C., Moffett, T.J., Henry, G.W., 1986, Ap.J. suppl. 60, 551.

3. Kholopov, P.N., 1987, General Catalogue Var. Stars, vols. 1-3.
4. Chugainov, P.F., 1980, Izv.Crim.Astroph.Obs. 61, 124.
5. Bopp, B.W., Ake, T.B., Goodrich, B.D., Africano, J.L., Noah, P.V., Meredith, R.J., Palmer, L.H., Quigley, R., 1985, Ap.J. 297, 691.
6. Strassmeier, K.G., Hall, D.S., Zeilik, M., Nelson, E., Eker, Z., Fekel, F.C., 1988, Ap.J.suppl. 72, 291.
7. Bopp, B.W., Saar, S.H., Ambruster, C.W., Feldman, P., Dempsey, R., Allen, M., Barden, S.P., 1989, Ap.J. 339, 1059.
8. Bopp, B.W., Noah, P.V., Klimke, A., Africano, J., 1981, Ap.J. 249, 21
10. Pettersen, B.R., Lambert, D.L., Tomkin, J., Sandmann, W.H., 1987, Ast Ap. 183, 66.
11. Koch, R.H., Hrivnak, B.J., 1981, Astron.J. 86, 438.
12. Vogt, S.S., Fekel. F., 1979, Ap.J. 234, 958.
13. Pettersen, B.R., 1989, Astr.Ap. 209, 279.
14. Pallavicini, R., Cerruti-Sola, M., Duncan, D.K., 1987, Astr.Ap.174, 1
15. Bopp, B.W., 1974, Publ.Astron.Soc.Pacif. 86, 281.
16. Rucinski, S.M., 1982, Inf.Bull.Var.Stars No. 2203.
17. Wilson, O.C., 1963, Publ.Astron.Soc.Pacif. 75, 62.
18. Herbig, G.H., 1965, Ap.J. 141, 558.
19. Fekel, F.C., 1988, Astron.J. 95, 215.
20. Gliese, W., 1969, Veroff. Astr. Rechen-Inst. Heidelberg No. 22.
21. van Leeuwen,F., Alphenaar, P., Meys, J.J.M., 1987, Astr.Ap.suppl.67,4
22. Stauffer, J.R., Schild, R.A., Baliunas, S.L., Africano, J.L., 1988, Publ.Astron.Soc.Pacif. 99, 471.
23. Magnitski, A.K., 1987, Dissertation Moscow.
24. Walter, F.M., Brown, A., Mathieu, R.D., Myers, P.C., Vrba, F.J., 1988 Astron.J. 96, 297.
25. Rydgren, A.E., Vrba, F.J., 1983, Ap.J. 267, 191.
26. Rydgren, A.E., Vrba, F.J., Chugainov, P.F., Zajtseva, G.V., 1984, Astron.J. 89, 1015.
27. Vrba, F.J., Rudgren, A.E., Chugainov, P.F., Shakhovskaya, N.I., Zak, D.S., 1986, Ap.J. 306, 199.
28. Vrba, F.J., Rydgren, A.E., Chugainov, P.F., Shakhovskaya, N.I., Weave W.B., 1989, Astron.J. 97, 483.
29. Walter, F.M., Brown, A., Linsky, J.L., Rydgren, A.E., Vrba, F.J., Rot M., Carrasco, L., Chugainov, P.F., Shakhovskaya, N.I.,Imhoff, C.L., 1987, Ap.J. 314, 297.
30. Bouvier, J., Bertout, C., Benz, W., Mayor, M., 1986, Astr.Ap. 165, 11
31. Herbst, W., Booth, J.F., Chugainov, P.F., Zajtseva, G.V., Barksdale, Covino, E., Terranegra, L., Vittone, A., Vrba, F.J., 1986, Ap.J.310,L
32. Herbst, W., Booth, J.F., Koret, D.L., Zajtseva, G.V., Shakhovskaya, N Vrba, F.J., Covino, E., Terranegra, L., Vittone, A., Hoff, D., Kesele L., Lines, R., Barksdale, W., 1987, Astron. J. 94, 137.
33. Cohen, M., Kuhi, L.V., 1979, Ap.J.suppl. 41, 743.
34. van Leeuwen, F., 1986, in Flare Stars and Related Objects (ed. L.V. Mirzoyan), Yerevan: Publ. House Acad. of Science, p. 289.

FLARES ON T TAURI STARS

G. F. GAHM
Stockholm Observatory
S-133 36 Saltsjöbaden
Sweden

ABSTRACT. An overview of the characteristics of short-term light variability on T Tauri stars is given. The evidence of the occurrence of flares comes from observations mainly at X-ray energies and from patrole observations in the ultraviolet spectral region. From such observations some limits on the peak fluxes and total energies of the largest flare-like events can be set. In addition, the frequency of such events can be deduced for a number of stars. It is demonstrated that there appears to be a qualitative difference between powerful flare-like events on the weak-line T Tauri stars (NTTs) and those on strong-line stars (CTTs). While it appears that the concept of surface flares occurring on NTTs may be correct there is the evidence that the disk-stars in addition produce flare-like events of a different nature. These events could be related to processes occurring not on the stars but in their circumstellar environment, for instance in a circumstellar disk. We also point at observations that could be of importance in clarifying the cause of flare-like activity on T Tauri stars and also comment on how this activity changes with stellar age.

1. Introduction

One characteristic property of T Tauri stars is that they vary in brightness with time. These variations occur on time-scales of minutes (or less) to years (or hundreds of years) and the pattern of variability may be drastically different from star to star, or even change with time for a given star. Regular as well as irregular fluctuations take place. Many of these properties were summerized by Herbig (1962) on the basis of existing observations in the optical spectral region. For reviews on more recent work concerning photometric variability of T Tauri stars the reader is referred to Bouvier (1986) and Gahm (1986, 1988), for studies of UV-variability to Lago (1988), for studies of X-ray variability to Feigelson (1987) and Montmerle (1989) and for studies of radio-variability to André et al. (1988). T Tauri flare models have been reviewed by Giampapa (1986).

In the present review we will be concerned mainly with flare-like phenomena as observed mainly at optical wavelengths, phenomena that have been studied by several scientists, not least at the Burakan Observatory. The appearance of short-lived bursts of light on these stars and the possible relation to similar phenomena on flare-stars was a major topic already long ago (see e.g. Ambartzumian, 1957; Haro, 1976). Rather than to strive for a complete overview of the large number of papers in the field I have instead compiled some of the published and

L. V. Mirzoyan et al. (eds.), Flare Stars in Star Clusters, Associations and the Solar Vicinity, 193–207.

unpublished photometric data, sought to bring them together to see what we can learn about the propertics of flare-likc phenomena on these stars.

2. The Time-Scales of Flare-Like Events

First we must agree on the definition of a flare-like event. In the present article I have taken the simple approach that a flare-like event occurs when the stellar light increases and decreases on time-scales similar to those of flares on flare stars and on the sun, that is over less than a few hours.The notion flare has often been attributed to light changes over time-scales of days and months, but such variations are likely to be of a very different nature than for ordinary stellar flares. As will be evident below it is not even clear that all flare-like events, defined in the present way, are similar to flares seen on for instance flare stars.

Regular fluctuations on time-scales of days occur on several stars and it has been convincingly demonstrated that for many objects this pattern is related to the rotational modulation of dark and bright areas on the stellar surface (Bouvier et al., 1986; Hartmann et al., 1986). This behaviour is particularly striking for T Tauri stars with very weak optical emission-line spectra, sometimes referred to as naked T Tauri stars (NTTs). Also for a number of stars with stronger emission lines, classical T Tauri stars (CTTs), periods have been reported but here superimposed irregular variations make the identification of periods difficult. Repeated observations over long base-lines in time do not always confirm the periods found (Fischerström and Gahm, 1989) and a warning in accepting published periods for CTTs is in place.

3. Observations of Flare-Like Events

Most photometric work on T Tauri stars has been made for purposes other than to record flare-like events. Consequently, there is a huge reservoir of data collected with typically two or three observations per star and night. Obviously, if changes occur on a star during one night very little can be said about how these changes evolve. On the other hand, such data are most useful in setting limits to what the typical or what the most extreme changes are that occur on different stars, on different time-scales and in different wavelength bands. In fact, we will make use of observations containing only 3 observations or more for statistical purposes in Chapter 4.

Even when several observations are collected per night for a given star, the probability to catch both the rise and fall of a flare-like event lasting a few hours is small. It is in many cases difficult to judge what has actually passed. As an example we find in Fig. 1 (left) that the star Thé 12 during one night declined in 3 of the 4 filterbands of the Strömgren system (u centered at 3500 Å, v at 4110 Å, b at 4670 Å and y at 5470 Å corresponding to the Johnson V-band). The slope resembles the declining part of a flare but whether there is a steep increase in flux just before the observations started or not can not be said. In Fig. 1 (right) is an example of an event taking place on the star RU Lup where possibly both the rise and the fall is observed. This is by our definition a flare-like event which is strong in all filterbands. It is, however, clear that the light curves have no resemblance to what is seen during flares on flare stars. While Thé 12 is an NTT, RU Lup is a CTT and it will be evident later that the probability to find an NTT in a phase of decline rather than rise, as in Fig. 1 a, is larger than for a CTT.

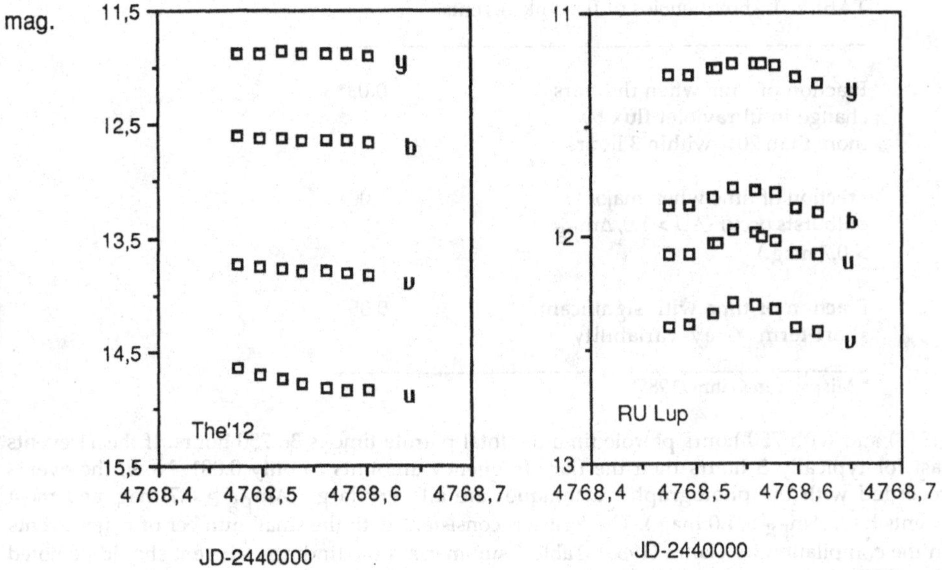

Figure 1. **left:** Lightcurves of Thé 12 obtained over 3 hours of observations in the Strömgren u, v, b and y filters when the star was decreasing in brightness in blue and violet light. **right:** A flare-like event observed on RU Lup.

3.1. FREQUENCIES OF FLARE-LIKE EVENTS

It has long been a rather wide-spread concept that T Tauri stars are very active and frequently undergo flare-like events. It is therefore rather disappointing to go through the literature and to find that major events, that is with rapid changes in the Johnson U-band of more than 20% in flux, are rather rare and that very few events for which both the rising and falling parts of the light-curve have been caught are at hand. A most striking example of such an event on T Tau itself has been found by Kilyachkov and Shevchenko (1976) but the authors state that during 4 sessions before and 15 sessions after the event no flares were found. On the basis of 750 hours of photographic patrole work in an area 5 x 5 degrees centered at R.A. 4^h 30^m, Dec. + 24° and covering numerous T Tauri stars 11 similar events, most of which reaching more than 1 magnitude above the quiescent level, were found by Hojaev (1987) on recognized T Tauri stars. Light-curves could be extracted for a few events but are not similar to those of stellar flares except for one event recorded on CI Tau (a CTT).

From the statistics made by Gahm (1986) on U-band variability on 22 stars over all together 886.5 hours of photometric patrole it was concluded that the fraction of time when the stars change in ultraviolet flux by more than 20% within 3 hours is 0.03. This value is similar to the derived fraction of time with significant short-term X-ray variability (0.05) but this does not mean that X-ray events and UV-events are related. Only very few events reach a total change in U-magnitude of > 1 mag.

We can now compare these frequencies with what can be derived from the work by Hojaev (1987). About 50 T Tauri stars reside in the area according to the catalogue by Herbig and Bell

TABLE 1. Frequencies of flare-like events.

Fraction of time when the stars change in ultraviolet flux by more than 20% within 3 hours	0.03*
Fraction of time when major outbursts occur ($\Delta U > 1.0$, $\Delta m_{pg} > 0.7$ mag.)	0.001
Fraction of time with significant short-term X-ray variability	0.05

* Misprint in Gahm (1988)

(1988) and with 750 hours patrole time the total patrole time is 36 750 hours. If the 11 events last for typically 3 hours then the flare frequency amounts to only 0.001. Now, the events recorded with the photographic techniques are all very large ($\Delta m_{pg} \geq 0.7$ mag. and most events have $\Delta m_{pg} > 1.0$ mag.). The figure is consistent with the small number of major events in the compilation by Gahm (1986). Table 1 summerizes the findings where it should be noted that CTTs with strong emission lines are more frequently changing than those stars with weak emission lines.

Quite naturally one would expect much higher frequencies of events for which $\Delta U < 20\%$ and this is also evident in the works by Kuan (1976), Schneeberger et al. (1979 a and b) included in Worden et al. (1981), Shevschenko and Shutiomova (1981) and Zaitseva et al. (1985). More observations of "micro-flaring" is required to establish the properties of these small and sometimes rapid fluctuations.

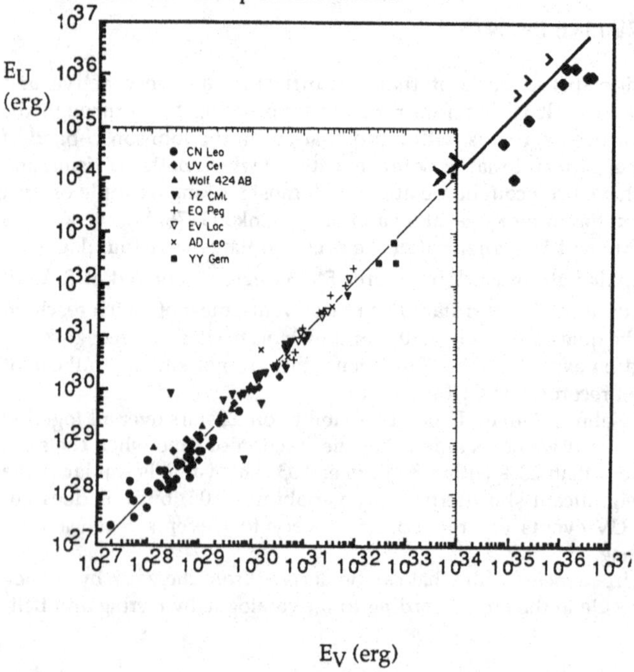

Figure 2. Total energies released in the Johnson U- and V-band (E_U versus E_V) for flares on flare stars according to Lacy et al. (1976), inserted frame, and on T Tauri stars (upper right, where > marks upper limits in E_V).

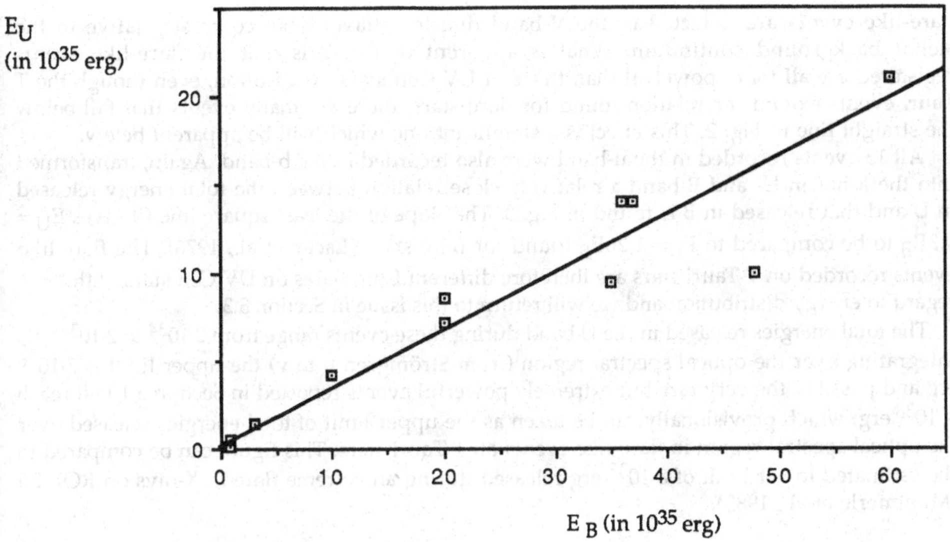

Figure 3. (above) Relation between energies E_U and E_B released during 13 flare-like events (4 events cluster at the lower left). The slope of the regression line is drastically different from that found for flare stars.

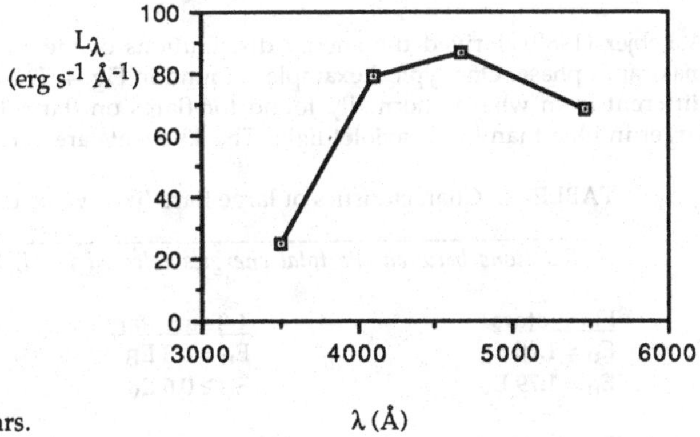

Figure 4. (right) Typical energy distribution at maximum phase of flare-like events on T Tauri stars. This particular case is for an event on RY Tau. Monochromatic luminosities (spherically symmetric case) are plotted versus wavelenghth after subtraction of background stellar flux.

3.2. TOTAL ENERGIES RELEASED IN MAJOR EVENTS

In the series of Strömgren photometric patrole made by C. Fischerström of the stars BP Tau, DI Tau, RY Tau, SU Aur, GW Ori, RU Lup and Thé 12 (= Sz 82) one can distinguish 13 flare-like events. It is possible to derive an estimate of the total energy released at the star in the 4 Strömgren filters (in several cases only as a upper limit). This analysis, which in part has been reported by Askebjer (1989) and in part is under preparation by Gahm (1989), leads to some interesting conclusions.

In Fig. 2 I have plotted the total energy released in the U-band (after proper correction for extinction and the difference in central wavelength and band-width relative to the u-band) against the total energy released in the V-band (with similar corrections) superimposed on the diagram showing the same quantities found for UV Ceti stars (Lacy et al., 1976). Not all flare-like events are detected in the V-band due to unfavourable contrast relative to the stellar background continuum. What is apparent in Fig. 2 is that the flare-like events registered are all more powerful than those of UV Ceti stars. In addition, even though the T Tauri events extend the relation found for flare stars, there are many events that fall below the straight line in Fig. 2. This effect is a significant one which will be apparent below.

All 13 events recorded in the u-band were also recorded in the b-band. Again, transformed into the Johnson U- and B-band a relatively close relation between the total energy released in U and that released in B is found in Fig. 3. The slope of the least-square line fit gives $E_U = 0.3E_B$ to be compared to $E_U = 1.20E_B$ found for flare stars (Lacey et al., 1976). The flare-like events recorded on T Tauri stars are therefore different from flares on UV Ceti stars with regard to energy distribution and we will return to this issue in Section 3.3.

The total energies released in the U band during these events range from $2 \cdot 10^{34}$ to $2 \cdot 10^{36}$ erg. Integrating over the optical spectral region (from Strömgren u to y) the upper limit is $7 \cdot 10^{36}$ erg and possibly the very rare but extremely powerful events reported in Section 3.1 will reach $\sim 10^{38}$ erg, which provisionally can be taken as the upper limit of total energies released over the optical spectral region in flare-like events on T Tauri stars. This figure can be compared to the estimated lower limit of $8 \cdot 10^{35}$ erg released during an extreme flare in X-rays on ROX 20 (Montmerle et al., 1983).

3.3. THE ENERGY DISTRIBUTIONS

Askebjer (1989) derived the energy distributions of the 13 events described above during maximum phase. One typical example is found in Fig. 4. The energy distribution is drastically different from what is normally found for flares on flare stars in that the fluxes are much larger in blue than in ultraviolet light. The 13 events are very similar in this respect. The

TABLE 2. Characteristics of large flare-like events on T Tauri stars

A. Relations between the total energies released in U, B and V

Flare stars	T Tauri stars
$E_u = 1.20\ E_B$	$E_u = 0.3\ E_B$
$E_u = 1.79\ E_V$	$E_u \geq 0.6\ E_V$

B. Maximum energies released:

$E_u = 2 \cdot 10^{36}$ erg (sample of 13 events)
($E_u = 2 \cdot 10^{34}$ erg for UV Ceti stars)
$E_{optical} = 7 \cdot 10^{36}$ erg (possibly $\sim 10^{38}$ erg $\sim 0.1\%$ of the time)
$E_X \sim 10^{36}$ erg

C. Maximum peak fluxes (sample of 13 events)

F_λ (Strömgren b) = $1.4 \cdot 10^{30}$ erg s^{-1} Å^{-1}
F_λ (Strömgren u) = $2.3 \cdot 10^{29}$ erg s^{-1} Å^{-1}

D. Black-body temperatures derived at maximum light

6200 K < T < 7300 K (13 events)

distribution is not similar to emission from a hydrogen plasma. Rather, it fits in general quite well to a black-body law. When the corresponding black-body temperatures are derived they fall in a narrow range 7300 K > T > 6200K. The indication is that optical flare-like events on T Tauri stars are cool, much cooler than for those on flare stars. We will return to a discussion on these specific properties in Chapter 4. The characteristics of large optical flare-like events on T Tauri stars, as derived in this chapter, are summerized in Table 2 where comparisons are made to flares on flare stars.

4. Characteristics of the Brightness Changes

We have seen in Chapters 2 and 3 that the light-curves of flare-like events on T Tauri stars are not always similar to what is found for flares on flare stars, that is with a rapid increase in flux followed by a slower decline. In order to further investigate this Askebjer (1989) and myself have made an inventory of published U-band measurements of T Tauri stars and included the unpublished Strömgren u-band measurements by C. Fischerström referred to above. A criterion has been that 3 or more observations were obtained per night and that an individual observer has observed the same star during several nights. Use is made also of corresponding measurements in V (=y). The data have been treated in the following way:

1. Both the U (or u) and V magnitudes are corrected for extinction, as a rule based on the values of A_V given by Cohen and Kuhi (1979) or more recent determinations if avaible and on an average interstellar extinction curve; 2. The relation between the so corrected U and V is plotted for each set of observations and as a rule the general slope can be described as linear with $\langle U \rangle = C \cdot \langle V \rangle$ + const, where C is determined. These "gross" variations, in many cases related to periodic fluctuations, are then eliminated from the U variations and only the residual changes in U are extracted; 3. For two consequtive observations during the same night the following quantities are determined: ΔU (or Δu) which is the change in magnitude after the above mentioned corrections and Δt which is the time difference between the two measurements (see Fig. 5).

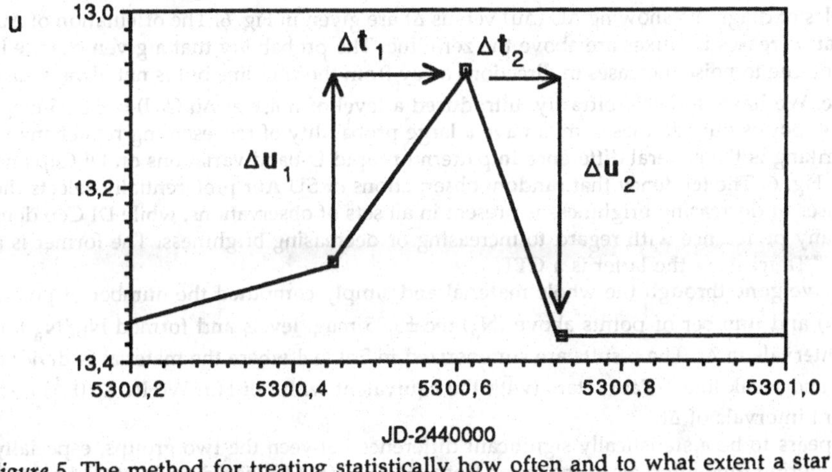

Figure 5. The method for treating statistically how often and to what extent a star increases or decreases in brightness is based on measurements of magnitude difference Δu and time difference Δt between consecutive observations made during one night. The figure shows observations in u of RY Tau with only 3 observations this night. These provide 2 points in the Δu versus Δt diagram (see Fig. 6).

TABLE 3. References to photometric data for the stars selected

Star	W(Hα)	Ref.	Star	W(Hα)	Ref.
CTTs:			NTTs:		
RU Lup	216	1, 2, 3	Thé 12	7	1
DG Tau	113	4, 5	SU Aur	4	1, 5, 11, 12, 13
DI Cep	95	1, 6, 7	WK 2	3	14, 15
TW Hya	86	8, 9	V 410 Tau	3	5, 14, 16, 17
RW Aur	84	1	FK 2	2	5
GW Ori	46	1, 5	WK 1	2	14, 15
BP Tau	40	5, 10	FK 1	2	5
RY Tau	20	1			

References 1: Gahm (1989); 2: Bastian and Mundt (1979b); 3: Whittet et al. (1985); 4: Shaimieva and Shutiomova (1985); 5: Bouvier et al. (1989); 6: Grinin et al. (1980); 7: Kardopolov and Filip'ev (1985); 8: Rucinski and Krautter (1983); 9: Rucinski (1988); 10: Vrba et al. (1986); Pugach (1975); 12: Herbst et al. (1983); 13: Herbst et al. (1987); 14: Rydgren and Vrba (1983); 15: Rydgren et al. (1984); 16: Mozidse (1970); 17: Calvet et al. (1985).

The stars were selected as to make a comparison possible between CTTs and weak-line T Tauri stars (NTTs). In Table 3 the stars are listed according to stregth of Hα emission with references to the works from which the photometric data were collected. A limit of 10 Å of the equivalent width of Hα was set to separate strong-line and weak-line stars (so defining the two groups CTTs and NTTs).

Examples of diagrams showing ΔU (Δu) versus Δt are given in Fig. 6. The orientation of ΔU is such that increases in fluxes are above the zero line. The probability that a given change is real and not due to noise increases in directions away from the zero line but is not always easy to estimate. We have, rather arbitrarily, introduced a level of noise at Δu (ΔU) = ± 0.05 mag. At any rate, points outside these limits have a large probability of representing real changes. What is striking is the general difference in pattern of rapid U-band variations on DI Cep and SU Aur in Fig. 6. The tendency that random observations of SU Aur preferentially selects the star in phases of decreasing brightness is present in all sets of observations, while DI Cep does not show any preference with regard to increasing or decreasing brightness. The former is a weak-line T Tauri star - the latter is a CTT.

We have gone through the whole material and simply computed the number of points below (N_b) and number of points above (N_a) the ±0.05 mag. levels and formed N_b/N_a for different intervals in Δt. The results are summerized in Table 4 where the material is divided into CTTs and weak-line T Tauri stars (with the equivalent width of Hα, W(Hα) ≤ 10 Å) and for different intervals of Δt.

It appears to be a statistically significant difference between the two groups, especially when considering time-intervals in the range from Δt = 0.05 to 0.1 Julian days (1.2 to 2.4 hours), which are rather typical time-scales of X-ray flares on T Tauri stars and of strong flares on flare stars. In this interval there is a much larger probability to find a weak-line T Tauri star in a phase of declining than increasing brightness while the CTT has more or less equal probability to increase or decrease in brightness (as illustrated by the light-curves in Fig. 1).

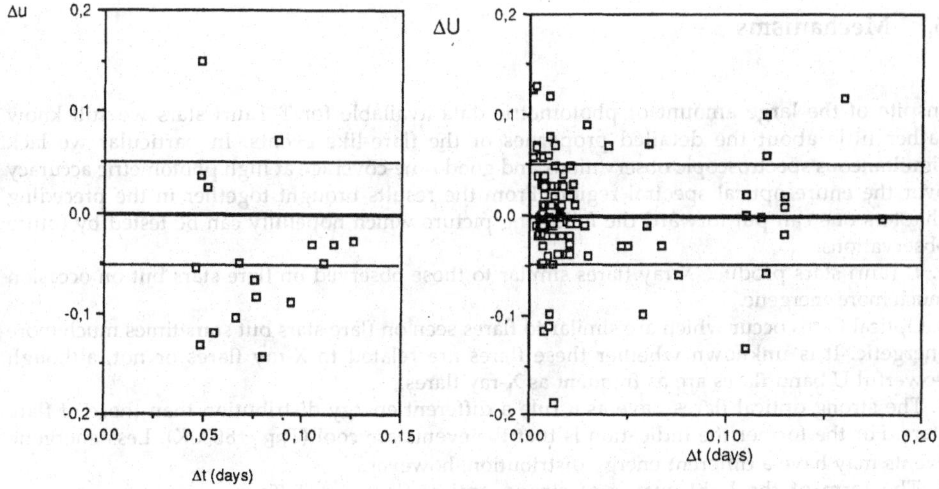

Figure 6. **left: Δu versus Δt for a subset of Strömgren photometric data of SU Aur showing that statistically this star was observed more frequently in a phase of decreasing rather than increasing brightness. A rise in ultraviolet flux results in positive values of Δu. right: ΔU versus Δt for DI Cep showing no preference in the direction of the changes of brightness above the noise levels at ± 0.05 mag.**

TABLE 4. Statistics on direction of brightness changes on the stars selected

Time interval in JD	N_b/N_a CTTs	NTTs
$0 < \Delta t < 0.05$	1.1	1.3
$0.05 < \Delta t < 0.1$	1.4	6.0
$0.1 < \Delta t < 0.2$	0.9	1.1

The general pattern of rapid U-fluctuations on NTTs is therefore consistent with the presence of flares which occur relatively frequently. For the CTTs, on the other hand, the rapid U-band fluctuations are statistically of a totally different character. There is no reason to doubt that flares do occur on the CTTs as well, but for these stars it appears that changes of a different nature dominates the variations in the ultraviolet. We will return to a discussion on these findings in Chapter 5. We close this chapter by noting that the difference found between the CTTs and NTTs is a statistical result including all stars. Some CTTs, notably RW Aur, have variability patterns more in line with the majority of the NTTs and some NTTs, notably V 410 Tau, behave similar to the CTTs.

5. Mechanisms

Inspite of the large amount of photometric data available for T Tauri stars we still know rather little about the detailed properties of the flare-like events. In particular we lack simultaneous spectroscopic observations and good time-coverage at high photometric accuracy over the entire optical spectral region. From the results brought together in the preceding chapters one can put forward the following picture which hopefully can be tested by future observations:

1. T Tauri stars produce X-ray flares similar to those observed on flare stars but on occasion much more energetic.
2. Optical flares occur which are similar to flares seen on flare stars but sometimes much more energetic. It is unknown whether these flares are related to X-ray flares or not, although powerful U-band flares are as frequent as X-ray flares.
3. The strong optical flares have as a rule a different energy distribution than those of flare stars. For the former the indication is that the events are cool ($T_{BB} \leq 8000K$). Less energetic events may have a different energy distribution, however.
4. The form of the light curves of strong optical flares on T Tauri stars is many times completely different than for typical flare star events, in particular for classical T Tauri stars where there is a slow rise followed by a slow decline in brightness. It is possible that on these CTTs, the strong events are of a completely different nature than stellar flares.

The present day data seem to indicate that ordinary flares over a large range of total energies occur frequently on T Tauri stars but that these events are masked by gross variations of a different kind on many CTTs (in particular). These stars are currently thought to be surrounded by circumstellar disks and it is natural to think that the flare-like events originate in the disk rather than on the stellar surface.

In the accretion model hypothesis developed by Shakura and Sunyaev (1973) and Lynden-Bell and Pringle (1974) and recently applied to T Tauri stars by e.g. Bertout et al. (1988) the rate of mass accretion, $\dot{M}$, will lead to a luminosity of the boundary layer between the star and the disk of:

$$L = GM_*\dot{M}/(2R_*)$$

where M_* and R_* are the stellar mass and radius, respectively. It is then seen that a temporal increase in $\dot{M}$ over the boundary layer will produce an increase in the flux from the boundary layer, which could be one mechanism behind the peculiar flare-like events. However, the velocity of the flow of matter drifting inwards in the disk radially to the star is:

$$v_r = -\dot{M}/(2\pi R\Sigma)$$

where R is the distance to an annulus of the disk and Σ the surface density (see Pringle, 1981). In order to achieve time-scales of the order of a few hours a very thin boundary layer is required with current guesses on the surface density.

As is apparent in other articles in this volume, there is the possibility that the boundary layer of the disk is at a large distance from the stellar surface. Clearly, the peculiar flare-like events may be related to stellar surface flares behind large columns of gas in the line-of-sight. The events may simply be secondary phenomena in a circumstellar plasma that is heated by underlying flares and then left to cool.

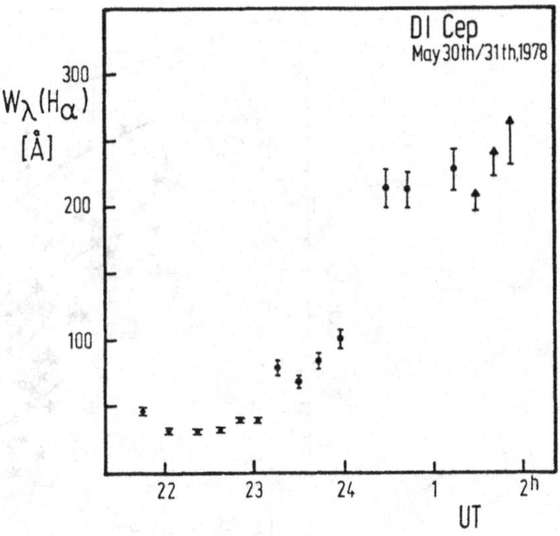

Figure 7. Rapid increase in Hα flux on DI Cep observed during one night by Bastian and Mundt (1979a). The equivalent width of Hα is plotted versus time.

For the events that are similar to flares on flare stars one is directed to extrapolations of flare models, such as investigated by van den Oord (1987). Models of stellar flares are extrapolations of models of solar flares and these are still uncertain. An extrapolation of a solar flare model including an electric double layer (see Carlqvist, 1986) was made by Askebjer (1989) to account for the energetic flares on T Tauri stars.

6. Further Aspects of Flares on Young Stars

In this chapter we first want to draw the attention to observations of rapid variability of T Tauri stars, observations that point at different processes than discussed so far. We will then end this overwiew by discussing how flare-like activity changes with age of the stars.

There are some reports on changes of profiles and/or intensities of emission lines on time-scales of < 2 hours. A dramatic example is shown in Fig. 7, where Bastian and Mundt (1979a) found the equivalent width of Hα of DI Cep to increase by a factor of ~4 over 1 hour. They stated that significant variations of the strength of the emission lines occurred on this star on 2 of 6 observing nights on time-scales down to ≤ 15 min.

Almost nothing is known about the relation between the broad-band events discussed above and the emission line changes. This is certainly an important area of future studies. For instance, if the major events are cool and not due to increased emission from a hydrogen plasma, will major changes nevertheless occur in the Balmer line fluxes? Can relations be found between U-band flux and Hγ flux, such as has been found for flare stars (Butler et al., 1988)? Most certainly new insights to the physics and dynamics of the processes involved will be gained when line profiles are monitored simultaneously with the broad-band fluxes.

Another intriguing phenomenon is the rapid flips in colour observed by Gahm et al. (1989) on RY Lup when the star was faint. This star may be seen through the outer "fluffy" regions of a circumstellar disk and an interpretation is that when the star becomes occulted by foreground dust we see rapid phenomena in the disk or in a boundary layer. Do other stars show similar phenomena?

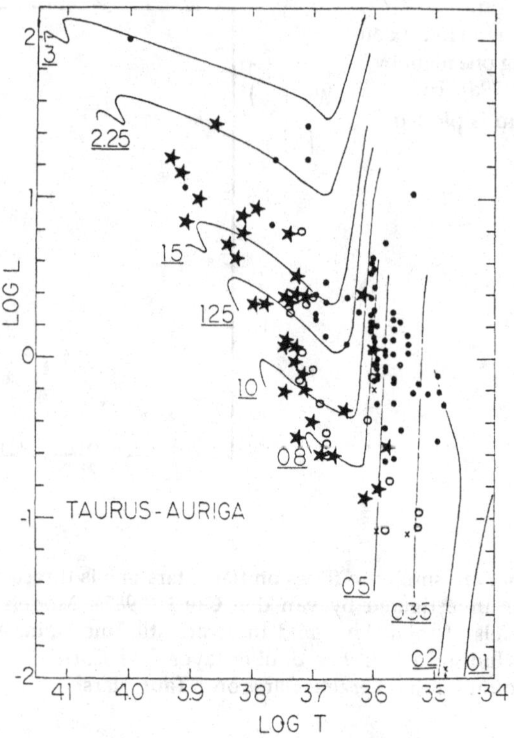

Figure 8. HR-diagram (luminosity versus effective temperature) where the dots represent the T Tauri stars in the Taurus-Auriga region as plotted by Cohen and Kuhi (1979) with evolutionary model tracks for contracting stars of different mass. Superimposed are the post-T Tauri candidates found by Lindroos (1986) - marked as stars - and by Walter et al. (1988) - marked as open circles.

Turning now to the question of how flare activity changes with time as the stars evolve towards the main sequence we will focus at Fig. 8 as a starting point. In this HR-diagram we have plotted the post-T Tauri star candidates found by Lindroos (1986) from studies of young secondary components in double or multiple stars and those found by Walter et al. (1988) from X-ray surveys in the Taurus-Auriga region together with the T Tauri stars in Taurus-Auriga according to Cohen and Kuhi (1979). The post-T Tauri candidates typically fall between the region with T Tauri stars and the main sequence. It is important to notice in relation to the discussion mentioned in Chapter 1 about the relation between T Tauri stars and flare stars in young stellar aggregates, that the post-T Tauri candidates represent a class of stars which is different from the ordinary flare stars. Most likely the late-type flare stars are derivatives of low-mass pre-main sequence objects most of which never had the high luminosities observed for the T Tauri stars in Fig. 8.

What can be said about the flare activity of the post-T Tauri stars, presumably representing ages between $5{\cdot}10^6$ years to zero-age main sequence age? In fact, very little is known about short-term variability on these stars. Tagliaferri et al. (1988) discovered a powerful X-ray flare, with a total energy of 10^{35} erg, on HD 560B - one of the stars in Lindroos'

(1986) sample. This indicates that flares of 10^4 times the energies liberated during prominent solar flares occur also after the T Tauri phase.

On the other hand, Feigelson (1989) reported that on the whole the X-ray post-T Tauri candidates seem to be more quiescent than the T Tauri stars. Furthermore, Hojaev (1989) has reported that during photographic patrole over a total of about 900 hours of observing time in the Taurus-Auriga field described in Chapter 3.1, no flares with $\Delta U > 0.2$ mag. were found on the (few) post-T Tauri candidates of Walter et al. (1988) in the field.

To summerize, we have a clear indication that the general degree of flare activity decays after 10^7 years or so but also that very energetic flares may still occur. Certainly, this is an area to explore further by for instance monitoring post-T Tauri stars photometrically over longer periods of time.

Finally, there is a cosmogonic connection to the question of flare activity evolution with age. If pre-main sequence solar mass stars are in a state of X- and/or U-flaring at levels of 10^4 times the energy release of large solar flares during several percent of the time over 10^8 years, it could mean that an early atmosphere of the planet Earth was exposed to very high doses of X-ray and ultraviolet radiation (~10^9 times that of the present quiescent sun) over a total period of several 10^6 years. Although such fluxes may be of minor importance to the thermal history of our atmosphere it may be of great importance to the chemical evolution of the atmosphere.

References

Ambartsumian, V.A.: 1957, in "Non-stable stars", IAU symp. No 3 (ed. G.H. Herbig) p. 177

André, P., Montmerle, T., Feigelson, E.D., Stine, P.C., Klein, K.L.: 1988, In "Activity in cool star envelopes" (eds. O. Havnes, B.R. Pettersen, H.M.M. Scmitt, J.E. Solheim) Kluwer, p. 293

Askebjer, P.: 1989, Master Thesis (Stockholm Observatory, Sweden)

Bastian, U., Mundt, R.: 1979a, Astron. Astrophys. 78, 181

Bastian, U., Mundt, R.: 1979b, Astron. Astrophys. Suppl. 36, 57

Bertout, C., Basri, G., Bouvier, J.: 1988, Astrophys. J. 330, 350

Bouvier, J.: 1986, in "Protostars and molecular clouds" (eds. T. Montmerle, C. Bertout) Commissariat a l'Energie Atomique, France, p. 189

Bouvier, J., Bertout, C., Benz, W., M. Mayor, M.: 1986, Astron. Astrophys. 165, 110

Bouvier, J., Bertout, C., Boucher, P.: 1989, Astron. Astrophys. Suppl., in press

Butler, C.J., Rodono, M., Foing, B.H.: 1988, Astron. Astrophys. Lett. 206, L1

Calvet, N., Basri, G., Imhoff, C.L., Giampapa, M.S.: 1985, Astrophys. J. 293, 575

Carlqvist, P.: 1986, IEEE Transactions on plasma science, PS-14, p. 794

Cohen, M., Kuhi, L.V.: 1979, Astrophys.J. Suppl. 41, 743

Feigelson, E.D.: 1987, in "Protostars and molecular clouds" (eds. T. Montmerle, C. Bertout) Commissariat a l'Energie Atomique, France, p. 123

Feigelson, E.D.: 1989, private communication

Fischerström, C., Gahm, G.F.: 1989, preprint

Gahm, G.F.: 1986, In "Flares: Solar and Stellar", Rutherford Appleton Laboratory, RAL-86-085, p. 124

Gahm, G.F.: 1988, In "Formation and evolution of low mass stars" (Eds. A.K. Dupree, M.T.V.T. Lago) NATO ASI Series, p. 295

Gahm, G.F.: 1989, in preparation

Gahm, G.F., Fischerström, C., Liseau, R., Lindroos, K.P.: 1989, Astron. Astrophys. 211, 115

Giampapa, M.S.: 1986, In "Flares: Solar and Stellar", Rutherford Appleton Laboratory, RAL-86-085, p. 232

Grinin, V.P., Efimov, Ju. S., Krasnobatsev, V.I., Shakhovskaya, N.I., Shcherbakov, A.G., Zaitseva, G.V.., Kolotilov, E.A., Shanin, G.I., Kiselev, N.N., Gjulaliev, Ch.G., Salmanov, I.R.: 1980, Peremennye Zvezdy (Variable Stars) 21, 247

Haro, G.: 1976, Bol. Inst. Tonantzintla 2, 3

Hartmann, L., Hewett, R., Stahler, S., Mathieau, R.D.: 1986 Astrophys. J. 309, 275

Herbig, G.H.: 1962, Advances Astron. Astrophys. 1, 47

Herbig, G.H., Bell, K.R.: 1988, Lick Obs. Bull., No 1111

Herbst, W., Holtzman, J.A., Klasky, R.S.: 1983, Astron. J. 88, 1648

Herbst, W., Booth, J.F., Koret, D.L., Zaitseva, G.V., Shakhovskaja, N.I., Vrba, F.J., Covino, E., Terranegra, L., Vittone, A., Hoff, D., Kelsey, L., Lines, R., Barksdale, W.: 1987, Astron. J. 94, 137

Hojaev, A.S.: 1987, Astrofizika (russ.) 27, 207

Hojaev, A.S.: 1989, private communication

Kardopolov, V.I., Filip'ev, G.K.: 1985, Peremennye Zvezdy (Variable Stars) 22, 103

Kilyachkov, N.N., Shevchenko, V.S.: 1976, Astron. J. USSR Lett. 2, 494

Kuan, P.: 1976, Astrophys. J. 210, 129

Lacy, C.H., Moffett, T.S., Evans, D.S.: 1976, Astrophys. J. Suppl. 30, 85

Lago, M.T.V.T.: 1988, In "Formation and evolution of low mass stars" (Eds. A.K. Dupree, M.T.V.T. Lago) NATO ASI Series, p. 209

Lindroos, K.P.: 1986, Astron. Astrophys. 156, 223

Lynden-Bell, D., Pringle, J.E.: 1974, Mon. Not. Roy. Astron. Soc. 168, 603

Mosidze, L.N.: 1970, Bull. Obs. Abastumani 39, 21

Montmerle, T.: in "Low mass star formation and pre-main sequence objects" ESO-WS, in press

Montmerle, T., Koch-Miramond, L., Falgarone, E., Grindlay, J.E.: 1983, Astrophys. J. 269,182

Pringle, J.E.: 1981, Ann. Rev. Astron. Astrophys. 19, 137

Pugach, A.F.: 1975, Peremmenye Zvezdy Prilozenie (Variable Stars Suppl.) 2, 195

Rucinsky, S.M.: 1988, Information Bull. Variable Stars (IAU), No. 3146

Rucinski, S.M., Krautter, J.: 1983, Astron. Astrophys. 121, 217

Rydgren, A.E., Vrba, F.J.: 1983, Astrophys. J. 267, 191

Rydgren, A.E., Zak, D.S., Vrba, F.J., Chugainov, P.F., Zaitseva, G.V.: 1984, Astron. J. 89, 1015

Schneeberger, T.J., Worden, S.P., Africano, J.L.: 1979a, Bull. American Astron. Soc. 11, 439

Schneeberger, T.J., Worden, S.P., Africano, J.L.: 1979b, Information Bull. Variable Stars (IAU) No. 1582

Shaimieva, A.F., Shutiomova, N.: 1985, Peremennye Zvezdy (Variable Stars) 22, 176

Shakura, N.I., Sunyaev, R.A.: 1973, Astron. Astrophys. 24, 337

Shevchenko, V.S., Shutiomova, N.A.: 1981, Astrofizika (russ,) 17, 286

Tagliaferri, G., Giommi, P., Angelini, L., Osborne, J.P., Pallavicini, P.: 1988, Astrophys. J. Lett. 331, L 113

van den Oord, G.H.J.: 1987, "Stellar Flares", thesis (Utrecht, Holland)

Vrba, F.J., Rydgren, A.E., Chugainov, P.F., Shakovskaya, N.I., Zak, D.S.: 1986, Astrophys. J. 306, 199

Walter, F.M., Brown, A., Mathieau, R.D., Myers, P.C., Vrba, F.J.: 1988, Astron. J. 96, 297

Whittet, D.C.B., Davies, J.K., Evans, A., Bode, M.F., Robson, E.I., Banfield, R.M.: 1985, South African Obs. Circ. No 9, p. 55

Worden, S.P., Schneeberger, T.J., Kuhn, J.R., Africano, J.L.: 1981, Astrophys. J. 244, 520

Zaitseva, G.V., Kolotilov, E.A., Tarasov, A.E., Shenavrin, V.I., Shcherbakov, A.G.: 1985, Astron. J. USSR Lett. 11, 109

MIRZOYAN: You have found that the ratios of energies in UBV regions of the spectrum are quite different. Is this because of the flares? If it is so then it can be assumed that it is a result of small number of flares observed in the case of T Tauri stars. But if it is real difference, it could be explained by the observational fact that for flare stars we observe comparatively pure flare radiation while for T Tauri stars it is a mixture of flare radiation and star radiation (flares on T Tauri stars do not usually begin from minimum level).

GAHM: My feeling is that the flare activity is high on both type of stars but that on the classical T Tauri stars the rapid variations are dominated by this other process.

BENZ: It is interesting how low the temperature of T Tauri flares is. Given the temperature and luminosity of a black body it is easy to calculate the size of the flare source. Can you tell me how big it is?

GAHM: I have not done this, but of course it is an easy thing to do and it should be made.

MONTMERLE: The distinction you see in U band light curves for NTTS and CTTS is very interesting. Given that NTTS have no (hot) gas around them, and that CTTS on the contrary often have ionized winds with dM/dt up to about 10**(-8) solar masses per year, don't you think the differences might be due essentially to transfer effects of flare radiation through this circumstellar material?

GAHM: Yes definitely, this is certainly one possibility.

APPENZELLER: I would like to comment on your estimate of the minimum time scale of variations caused by changes of the accretion rate. These estimates obviously depend on the assumption about the size of the disk-boundary zone (about one scale height?). As you know the physical nature and hence the extent of these layers are uncertain. In addition, the disks may not be stationary. If the boundary layer is very extended and if the matter is in free fall near the stellar surface, variations as short as a day or perhaps a few hours could be produced by variable accretion.

GAHM: That is interesting. I hope that observations of phenomena that may take place in the interior circumstellar regions of young stars can help to restrict the models. Of course we do not know the actual physical situation. But what is sure is that it is likely to be much more complicated than the simple models predict.

T TAURI STARS AND FLARE STARS: COMMON PROPERTIES AND DIFFERENCES

I. APPENZELLER
Landessternwarte Königstuhl
D-6900 Heidelberg 1

ABSTRACT. T Tauri stars and flare stars are both magnetically active late-type stars of low mass and low to moderate luminosities. The flares observed in these two classes of variables show similar properties and, thus, probably have the same physical origin. On the other hand, at least the majority of the classical T Tauri stars seem to be surrounded by cool, dusty (accretion) disks, which are absent or undetectable in most classical flare stars.

1 Common Properties

Similarities between the flare stars (FSs) and the T Tauri stars (TTSs) have been noted in many earlier papers (see e.g. Haro 1957, 1976, Ambartsumian and Mirzoyan 1975, Mirzoyan 1977). New high-quality observations of TTSs obtained in recent years at many different wavelengths, and summarized, e.g., by Bertout (1989) and Appenzeller and Mundt (1989), confirm the presence of many common properties. However, the new observational data also demonstrate the presence of characteristic differences between these two classes of variable late-type stars.

A characteristic property shared by the TTSs and the FSs is the occurence of flares, i.e. short-term brightness outbursts with a rapid rise and a slower decline. Although in individual TTSs flares seem to occur less frequently than in classical flare stars, the properties of the T Tauri flares observed at UV and visual wavelengths (see e.g. Kilyachkov and Shevchenko 1976), at X-rays (Feigelson and DeCampli 1981, Montmerle et al. 1983, Walter and Kuhi 1984), and at radio wavelengths (Feigelson and Montmerle 1985, André et al. 1987, Stine et al. 1988, Feigelson 1988) agree well with those found for classical flare stars.

Flares are known to occur in all subclasses of the TTSs, including strong-emission classical TTS (such as DG Tau and RW Aur), weak-emission TTSs, and post-TTSs (such as DoAr 21). Additional support for a common origin of the FS and TTS flares is indicated by the occurrence of spectral signatures typical for stellar flares in

L. V. Mirzoyan et al. (eds.), Flare Stars in Star Clusters, Associations and the Solar Vicinity, 209–213.

the short-term emission-line variations of classical TTSs (see e.g. Appenzeller at al. 1983).

Another common property of FSs and TTSs are quasi-periodic light variations, which are generally assumed to be caused by magnetic star spots on the surface of rotating stars, and which provide indirect evidence for extensive photospheric magnetic fields in FSs as well as in TTSs. In the case of the FSs the quasi-periodic variations have been known since many years from the systematic investigation by Krzeminski and Kraft (1967) and many subsequent papers. For the TTSs this phenomenon has been studied, e.g., by Hoffmeister (1965), Rydgren et al. (1984), Bouvier et al. (1986), Herbst et al. (1986), Vrba et al. (1986, 1988, 1989), Herbst and Koret (1988), and Bouvier and Bertout (1989).

2 Differences

Among the empirical properties which are observed for TTSs, but *not* in classical FSs, are strong, cool, and sometimes surprisingly well collimated stellar winds or outflows (cf. e.g. Lada 1985, Mundt 1988, Appenzeller and Mundt 1989) and cool, dusty circumstellar evelopes (cf. e.g. Chini 1989). Line profile studies provide conclusive evidence, that these cool envelopes show a disk structure and extend to at least about 100 AU from the stars (cf. e.g. Appenzeller 1983, Jankovics et al. 1983, Appenzeller 1989).

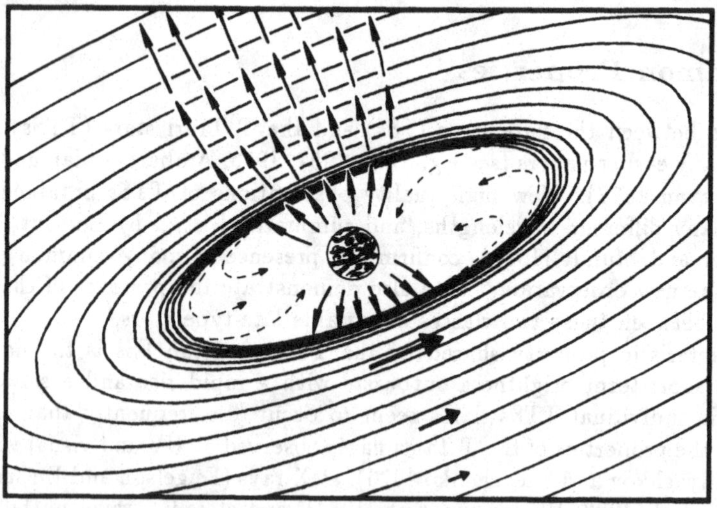

Figure 1: Schematic outline of the accretion disk model of classical T Tauri stars

As shown by Bertout et al. (1988), Basri and Bertout (1989), and others, the observed continuum energy distribution and many other observed properties of the classical TTSs can be explained by assuming that (turbulent or magnetic) angular momentum transfer results in a heating of the disks and in an accretion flow to the central star . Since TTSs are known to rotate much more slowly than the Keplerian orbital velocity near their surfaces, there must be a transition zone or "boundary layer" between the inner edge of the accretion disk and the stellar surface. In this zone at least half of the initial potential energy of the accreted matter is converted into heat. From the presence of strongly redshifted absorption features of the Na I resonance lines (and sometimes also of Fe II, Ca II, Balmer, and other lines) it is clear that at least part of the TTSs have extended regions of (more or less) free falling matter inside the inner disk boundary (cf. Figure 1). The absence of orbiting material in these inner zones may be due to the interaction with the slowly rotating magnetic field of the central star. In some TTSs (such as DR Tau, cf. Appenzeller et al. 1988) observed infall velocities up to to about 400 km s^{-1} (i.e. of the order of the surface escape velocity) indicate that the free fall zone extends over at least several stellar radii. In this case most of the boundary-layer energy dissipation is expected to take place in a shock front close to the stellar surface.

The rotating magnetic field also provides a mechanism for driving and collimating the observed winds along the direction of the rotation axis. Because of its high specific angular momentum this rotationally driven wind efficiently removes the angular momentum of the accreted matter (cf., e.g., Pudritz 1988, Appenzeller 1989).

In the case of the classical FSs, we have no evidence for either cool circumstellar disks or for mass accretion. Very likely the FSs represent a later evolutionary phase of the low-mass stellar evolution, where the disks have been depleted by the accretion flows and other effects (e.g. the formation of planets). An intermediate stage may be the weak-emission line or "naked" TTSs which (at least in part) have cool circumstellar disks (Chini 1989), but show no indications of significant mass accretion.

References

Ambartsuminan, V.A., Mirzoyan, L.V.: 1975, in Proc. IAU Symp. **67**, *Variable Stars and Stellar Evolution*, ed. V.E. Sherwood and L. Plaut, (D. Reidel, Dordrecht), p. 1

André, P., Montmerle, Th., Feigelson, E.D.: 1987 *Astron. J.* **93**, 1182

Appenzeller, I.: 1983, *Rev. Mexicana Astron. Astrof.* **7**, 151

Appenzeller, I.: 1989, in Proc. 4th IAP Astrophysics Meeting on *Modelling the Stellar Environment*, eds. P. Delache et al. (Editions Frontieres), p. 47

Appenzeller, I., Mundt, R.: 1989, *Astron. Astrophys. Rev.* **1**, in press

Appenzeller, I., Östreicher, R., Schiffer, J.G., Egge, K.E., Pettersen, B.R.: 1983, *Astron. Astrophys.* **118**, 75

Appenzeller, I., Reitermann, A., Stahl, O.: 1988, *Publ. Astr. Soc. Pacific* **100**, 815

Basri, G., Bertout, C.: 1989, *Astrophys. J.* **341**, 340

Bertout, C.: 1989, *Ann. Rev. Astron. Astrophys.* **27**, 351

Bertout, C., Basri, G., Bouvier, J.: 1988 *Astrophys. J.* **330**, 350

Bouvier, J., Bertout, C.: 1989, *Astron. Astrophys.* **221**, 99

Bouvier, J., Bertout, C., Benz, W., Mayor, M.: 1986, *Astron. Astrophys.* **165**, 110

Chini, R.: 1989, in Proc. ESO Workshop on *Low Mass Star Formation and Pre-main Sequence Objects*, ed. B. Reipurth (ESO, Garching), in press

Feigelson, E.D.: 1988, in *Cool Stars, Stellar Systems, and the Sun*, eds. J.L. Linsky, R.E. Stencel, *Springer Lecture Notes in Physics* **291**, 455

Feigelson, E.D., DeCampli,W.M.: 1981, *Astrophys. J. (Lett.)* **243**, L89

Feigelson, E.D., Montmerle, Th.: 1985, *Astrophys. J. (Lett.)* **289**, L19
Haro, G.: 1957, in Proc. IAU Symp. **3** on *Non-stable Stars*, ed. G.H. Herbig (Univ. Cambridge Press), p. 26
Haro, G.: 1976, *Bol. Inst. Tonantzintla* **2**, 3
Herbst, W., Booth, J.F., Chugainov, P.F., Zajtseva, G.V., Barksdale, W., Teranegra, W., Vrba, F.J.: 1986, *Astrophys. J. (Lett.)* **310**, L71
Herbst, W., Koret, D.L.: 1988, *Astron. J.* **96**, 1949
Hoffmeister, C.: 1965, *Veröffentl. Sonneberg* **6**, 97
Jankovics, I., Appenzeller, I., Krautter, J.: 1983, *Publ. Astr. Soc. Pacific* **95**, 883
Kilyachkov, N.N., Shevchenko, V.S.: 1976, *Sov. Astron. Lett.* **2**, 193
Krzeminski, W., Kraft, R.P.: 1967, *Astron. J.* **72**, 307
Lada, C.J.: 1985, *Ann. Rev. Astron. Astrophys.* **23**, 267
Mirzoyan, L.V.: 1977, Veröff. Remeis-Sternwarte **XI**, Nr. 121, p. 106
Montmerle, Th., Koch-Miramond, L., Falgarone, E., Grindlay, J.: 1983, *Astrophys. J.* **269**, 182
Mundt, R.: 1988, in NATO-ASI on *Formation and Evolution of Low Mass Stars*, eds. A. Dupree and M.T.V.T. Lago (Kluwer, Dordrecht), p. 257
Pudritz, R.E.: 1988 in *Galactic and Extragalactic Star Formation*, ed. R.E. Pudritz and M. Fich (Kluwer, Dordrecht) p. 135
Rydgren, A.E., Schmelz, J.T., Zak, D.S., Vrba, F.J., Chugainov, P.F., Zajtseva, G.V.: 1984, *Astron. J.* **89**, 1015
Stine, P.C., Feigelson, E.D., André, P., Montmerle, T.: 1988, *Astron. J.* **96**, 1394
Vrba, F.J., Herbst, W., Booth, J.F.: 1988, *Astron. J.* **96**, 1032
Vrba, F.J., Rydgren, A.E., Chugainov, P.F., Shakhovskaja, N.I., Weaver W.B.: 1989, *Astron. J.* **97**, 483
Vrba, F.J., Luginbuhl, C.B.,Strom, S.E., Strom, K.M., Heyer, M.J.: 1986, *Astron. J.* **92**, 633
Walter, F.M., Kuhi, L.V.: 1984, *Astrophys. J.* **284**, 194

BLAAUW: In your presentation I did not hear reference to the interstellar medium within which these stars are located. This question occurred to me particularly when you spoke about the cases with observable inflow of matter. Should this be directly related to the ISM? Do we know a star with inflow situated in a "clean, empty" volume?

GYULBUDAGHIAN: Did you observe the change of duplicity of emission lines during the change of line widths (i.e. sometimes the right component is seen, sometimes the left, sometimes both)?

APPENZELLER: Yes. All T Tauri stars with double-peaked Balmer lines which we looked at showed strong variations of the relative strength of the blue and red peaks.

BASTIEN: I have two comments to make. One must be careful when interpreting line profiles of extreme T Tauri stars, e.g. HL Tauri, which are seen nearly edge on. The light from these stars is reflected by the circumstellar disk and this complicates the interpretation of line profiles. Secondly, in the Monte Carlo results of circumstellar disks which I described earlier,there has to be a "hole" or empty region close to the star as you suggested. The results of these models are sensitive to the size of the hole and the distribution of matter in the inner disk region. Therefore, it may be possible to learn about this interesting region close to the star from these models.

MIRZOYAN: Many years ago in the Bamberg colloquium you tried to show that the majority of the T Tauri stars have anti-P Cygni spectral line profiles. What is your opinion on this problem now?

APPENZELLER: Current disk models of the T Tauri stars (TTS) predict that all classical T Tauri stars (CTTS) have mass accretion. How often this mass accretion manifests itself in inverse P Cygni profiles is unclear. Observationally we know at present between 30 and 40 TTS where inverse P Cygni profiles have been observed at least once. I guess that this is about 50 % of all CTTS which have been observed at sufficient spectral resolution. So the estimate made in Bamberg may still be correct.

CIRCUMSTELLAR PHENOMENA AND THE POSITION OF T TAURI STARS IN THE COLOUR MAGNITUDE DIAGRAM

W. GÖTZ
Academy of Sciences of GDR
Central Institute of Astrophysics
Sonneberg Observatory
PSF 55-27/28
Sonneberg
DDR - 6400

ABSTRACT. On the basis of investigations concerning the character and the behaviour of circumstellar shells at T Tauri stars it is shown that the position of these objects in the colour magnitude diagram, below and above the main sequence, is caused by cosmogonic circumstellar effects originated by the mass loss of the stars.

1. INTRODUCTION

From photometric observations of T Tauri and Flare stars published by several authors it could be shown that these objects are situated below and above the main sequence in the colour magnitude diagram $M_V/(B-V)_0$ though, owing to their evolutionary ages and according to theory, their position should always be above the main sequence.

Aggregates where this phenomenon can be observed are for instance the Orion Nebula, the Taurus dark cloud, NGC 2264 and the Pleiades.

In this report it will be shown that this phenomenon can be explained by the character and the behaviour of circumstellar shells, which in the end are originated by the mass loss of the stars.

2. OBSERVATIONAL CORRELATIONS

From statistical investigations of circumstellar shells at T Tauri stars made at Sonneberg Observatory (1971, 1980, 1984) it could be inferred that the most important observational parameter of these objects is the strength of their Hα emission. It could be shown that the strength of Hα emission is closely connected with the individual mass loss rate of the objects as well as the circumstellar absorption obtained by star counts in their surroundings in the sense that high mass loss rates are correlated with strong Hα emission and strong circumstellar absorption.

L. V. Mirzoyan et al. (eds.), Flare Stars in Star Clusters, Associations and the Solar Vicinity, 215–218.

Small mass loss rates however go with weak $H\alpha$ emission and weak absorption.

From studies in several T associations and Flare star aggregates of different ages it could further be shown that the strength of $H\alpha$ emission is also a cosmogonic parameter. $H\alpha$ emission is strong with stars in extremely young or young aggregates, but it decreases monotonically with age and becomes weak later than the logarithmic age $\log \tau \geq 7.0$. Therefore, in old aggregates $H\alpha$ emission should be very weak or not observable.

In this connection it is worth mentioning that the colour indices (B-V), (U-B) and (U-V), too, go with the strength of $H\alpha$ emission. This relation is characterized by increasing colour indices with decreasing $H\alpha$ emission or, considering the cosmogonic relation, with increasing ages of the objects. A similar behaviour results from index excesses, which are characterized by the differences between the observed colour indices (B-V) and those according to the spectral types. Here, with increasing evolutionary phases of the objects in the pre-main sequence stage the index excesses become smaller.

All those observational facts can be explained by the causal connection between the gas and dust shell phenomenon and the cosmogonic mass loss of the stars, which is the connection link between the stars and their shells and which appears in the early phase of the pre-main sequence stage and decreases, like the accompanying shell phenomena, during the evolution of the stars.

3. CONCLUSION

The correlations mentioned above, especially those that characterize light depressions and colour excesses caused by cosmogonic circumstellar effects, provide a possibility to reduce the observations free of interstellar absorption according to the strength of $H\alpha$ emission to the stage where the circumstellar shell phenomena should be weak or totally absent in the same way as the $H\alpha$ emission. The results obtained in this manner are shown in Fig. 1 and Fig. 2, where the positions of $H\alpha$ objects of NGC 2264 in the colour magnitude diagram before and after the reduction are shown. From Fig. 2 it can be seen that in Fig. 1 most of the stars which are situated below the main sequence now appear above the main sequence, just at that place where we should find them according to their evolutionary phases in the pre-main sequence stage and according to the theory and where we find also stars without $H\alpha$ emission (circles) but with position above the main sequence in the two colour diagram similiar to the $H\alpha$ emission objects.

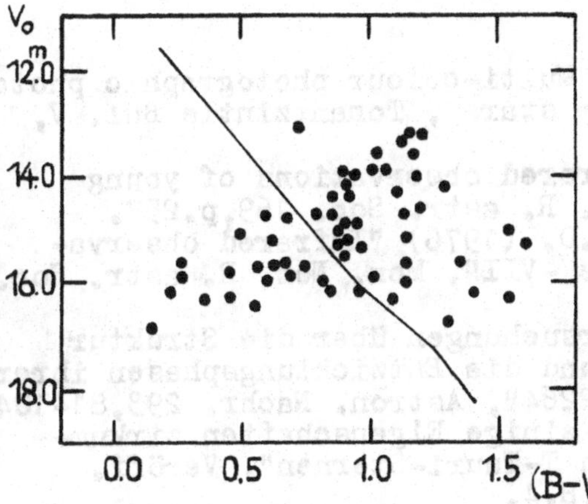

Figure 1. The position of Hα emission stars in the colour magnitude diagram of NGC 2264.

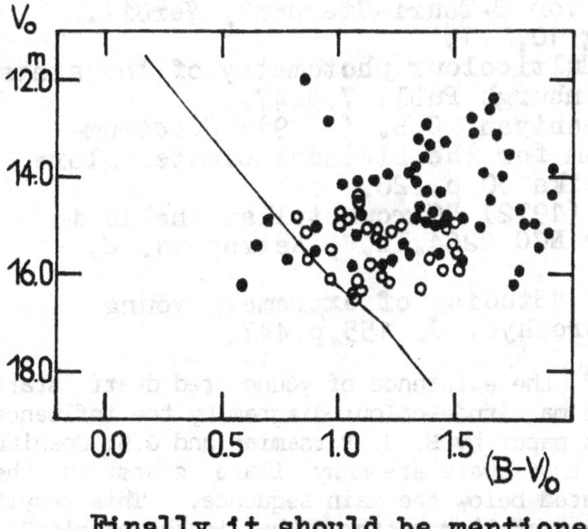

Figure 2. The position of Hα emission stars in the colour magnitude diagram after reducing the cosmogonic circumstellar effects.

Finally it should be mentioned that, as could be shown in a shell model, the circumstellar phenomena responsible for the discussed positions of T Tauri and Flare stars in the colour magnitude diagram are caused by the behaviour of the true shell parameters. These parameters are shell temperature, electron density, flux of Hα emission, specific fluxes in the X-ray and UV range originated by magnetobremsstrahlung, and the flux of dust radiation. They are all decreasing with increasing colour indices (U-V) and with increasing evolutionary phases in the pre-main sequence stage.

REFERENCES

Andrews, A.D. (1970) "Multi-colour photographic photometry of Orion flare stars", Tonantzintla Bul. V, p.195.

Cohen, M. (1974) "Infrared observations of young stars -V", Mon. Not. R. astr. Soc. 169,p.257.

Cohen, M., Schwartz,R.D. (1976) "Infrared observations of young stars -VII", Mon. Not. R. astr. Soc. 174,p.137.

Götz, W. (1971) "Untersuchungen über die Struktur junger Sternhaufen und die Entwicklungsphasen ihrer Mitglieder, I. NGC 2264", Astron. Nachr. 293,81-104.

Götz, W. (1980) "Über einige Eigenschaften zirkumstellarer Hüllen von T-Tauri-Sternen", Veröff. Sternw. Sonneberg 9,H.4.

Götz, W. (1984) "Zum Charakter und Verhalten zirkumstellarer Hüllen von T-Tauri-Sternen", Veröff. Sternw. Sonneberg 10,H.1.

Nandy, K. (1971) "Multicolour photometry of the stars in NGC 2264", Edinburgh Publ. 7,p.47.

Parsamian, E.S., Ohaniyan, G.B. (1989) "Spectrum-luminosity diagram for the Pleiades cluster flare stars", Astrophysika 30,p.220.

Strom, S.E. et al. (1972) "Circumstellar shells in the young cluster NGC 2264.II.", Astrophys. J. 171,p.267.

Walker, M.F. (1969) "Studies of extremely young clusters.V.", Astrophys. J. 155,p.447.

MIRZOYAN: You have explained the existence of young red dwarf stars below the main sequence in the magnitude-colour diagram by the influence of circumstellar matter. In a paper by E. S. Parsamian and G.B. Ohanian in Astrofizika they showed that there are many flare stars in the Pleiades cluster that are located below the main sequence. This result does not seem to depend on circumstellar matter. What is the problem?

GÖTZ: If there is a dust shell around the stars there exists a light depression. In considering this effect the magnitudes become brighter. In the case that circumstellar matter is absent I cannot explain this problem.

LADA: There is another effect which can have the result of moving stars in NGC 2264 from below to above the main sequence, namely that there is variable extinction in the cluster. This cluster actually extends into a giant molecular cloud and many of the faint stars have appreciable extinction. When they are de-reddened properly, they do not lie below the main sequence.

GÖTZ: The circumstellar extinction, which was determined by star counts was considered in agreement with the strength of H-alpha emission.

HIGH RESOLUTION SPECTROSCOPY OF RY TAURI: VARIABILITY OF THE NA D LINE PROFILES

P.P.PETROV
Crimea Astrophysical Observatory
p/o Nauchny,334413,Crimea,USSR

ABSTRACT. Ejection and accretion of gas clouds in the vicinity of RY Tau were discovered. The existense of large scale "stellar prominences" around young stars is suggested.

For a long time the large scale gas flows around TTau-type stars have been a subject for discussion among the astronomers studying the problems of star formation. Both processes - outflow and accretion have been observed through the investigation of line profiles, mainly the hydrogen lines (see, e.g. Grinin et al.,1985).

Spectroscopic CCD observations of the TTau-type star RY Tau in 1986/87 revealed the variability of the NaI D line profiles. Each of the doublet line incorporates three different components:

- broad absorbtion of photospheric origin (its profile corresponds to the rotational velocity of the star $v \sin i = 55$ km/s);
- narrow absorption of interstellar origin;
- additional *variable* absorption, which disturbes the line profile and shows day-to-day variations.

In order to extract the variable component of the NaI lines the average spectrum of RY Tau was subtracted from each of the individual spectrum of the star. The derived differential (residual) spectra contain mainly the NaI D absorptions which are variable in intensity and radial velocity. The radial velocity of the residuals vary from -100 to +100 km/s, sometimes weak absorptions appear at larger radial velocity, up to 200 km/s.

The analysis of the differential spectra of RY Tau leads to a conclusion that the residual NaI lines originate from relatively cool gas clouds, which are moving inside the stellar wind, ascending and descending on a time scale of a few days, at the distances within three stellar radii. This phenomenon resembles the eruptive solar prominences - ejection of cool gas into corona up to the distance of about one solar radius.

Physical conditions in the "prominences" of RY Tau are also close to these of solar prominences. The phenomenon of "stellar prominences"

L. V. Mirzoyan et al. (eds.), Flare Stars in Star Clusters, Associations and the Solar Vicinity, 219–220.

in other young stars can be studied through spectroscopic monitoring of selected spectral features using techniques of high signal-to-noise ratio and high resolution spectroscopy.

More details about this investigation are included in the paper "Prominences of RY Tau" submitted to the "Astrophysics and Space Science".

References

Grinin V.P., Petrov P.P. and Shakhovskaya N.I. (1985) "The result of spectroscopic and photometric patrol observations of RW Aurigae. I. Variability of the Balmer lines", Bull. Crimean Astrophys. Obs. 71, 109-127.

HERBST: Do you find any correlation between the brightness of the star and the spectral features?

PETROV: No such correlation was found.

APPENZELLER: Do you assume your absorbing clouds to move supersonically with respect to their environment or do you expect a very hot surrounding gas so that 100 km/s is subsonic?

PETROV: The clouds are moving inside the hot stellar wind of RY Tau. The presence of the hot wind is evident from the specific Balmer decrement and other lines, like the He I D3.

HERBIG-HARO PHENOMENA ASSOCIATED WITH T TAURI: EVIDENCE FOR A PRECESSING JET?

RICHARD D. SCHWARTZ
Department of Physics
University of Missouri - St. Louis
8001 Natural Bridge Road
St. Louis, Missouri 63121
U.S.A.

ABSTRACT. A [S II] emission-line image of T Tauri and its surrounding nebulosity suggests the presence of a precessing jet in the system.

1. INTRODUCTION

The early observations of Burnham and later work by Herbig established the presence of an extended emission nebula associated with T Tau (see Schwartz 1974 and references therein). Schwartz (1975) demonstrated that Burnham's Nebula had a spectrum characteristic of Herbig-Haro (HH) objects, and reported the discovery of faint HH emission about 30" west and 4" south of T Tau near the inner edge of Hind's reflection nebula (NGC 1555). Burke et al.(1986), with CCD imaging and long-slit spectra, discovered faint high-velocity (-40 to -150 km s^{-1}) gas between T Tau and a relatively bright knot of emission (HH 1555) located 32" west of T Tau. Dyck et al.(1982) discovered an infrared (IR) companion about 0.6" south of T Tau. Subsequent work (see Schwartz et al. 1986) has demonstrated that the IR companion is the source of significant IR and radio continuum, suggesting that it may possess a substantial stellar wind. This paper reports the discovery of additional HH nebulosity around T Tau, and the presence of morphological structure which may result from a precessing jet in the system.

2. OBSERVATIONS AND DATA ANALYSIS

Direct CCD observations of the T Tau system with a [S II] λ6730 filter were obtained with the Kitt Peak National Observatory 2.1-m telescope on 23 Nov 1987. To avoid image saturation, five successive 100 sec exposures of T Tau were coadded. A nearby reference star (SAO 93887) was imaged in similar fashion (five 35 sec exposures) to provide a stellar point spread function normalized to the central intensity of the T Tau image. With T Tau placed 3" outside of the CCD frame, a 2000 sec [S II] exposure of the emission and reflection nebulosity west of T Tau was obtained. A 500 sec I-band exposure was obtained at the same position to provide a continuum image of the region.

Following image registration, the continuum image of the reference

L. V. Mirzoyan et al. (eds.), Flare Stars in Star Clusters, Associations and the Solar Vicinity, 221–224.

star was subtracted from the image of T Tau. Likewise, after normalization of the I-band continuum in Hind's nebula to the continuum observed at the same location with the [S II] filter, the I-band image was subtracted from the [S II] image. The resultant frames, showing basically the [S II] emission around T Tau, were then joined to provide the single image shown in the contour picture of Fig. 1. Burnham's Nebula extending southward from the star is seen to exhibit what appears to be a hint of spiral shape. A new HH knot (A) is located on a continuation of the spiral about 17" SSW of the star.

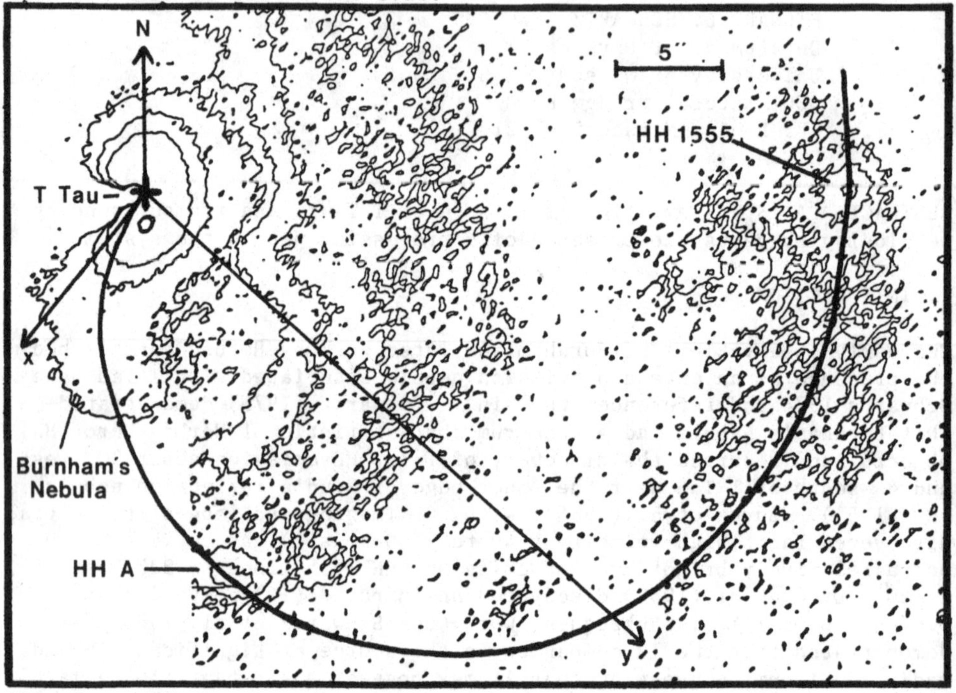

Fig. 1. An intensity contour plot of [S II] emission around T Tau. The x-y system is inclined from an x'-y' system by angles α,β (see text)

3. INTERPRETATION

The radial velocity measurements of Solf et al. (1988) indicate the presence of a bipolar flow from the T Tau system with a jump of more than 100 km s^{-1} from negative values 1.5" south of the star to positive values 1.5" north of the star. If one assumes that a collimated constant velocity flow emanates from a precessing source, a spiral pattern will be traced by the flow according to the relations:

$$x(\phi') = (Pv\phi'/360)[\cos\phi'\sin\psi\cos\alpha + \sin\alpha]$$

$$y(\phi') = (Pv\phi'/360)[\sin\phi'\sin\psi\cos\beta + \sin\beta]$$

where ϕ' is the phase angle of the flow at a given point on the spiral measured from the x'-axis which is in position angle 220°(the present position angle of the flow), P is the period of precession, v the velocity of the flow, ψ is the opening half-angle of the precession cone, α is the inclination of the central axis of the precession cone in the x'-z' plane, and β is the inclination of the central axis of the cone in the y'-z' plane. Without independent estimates of the product Pv or the angle ψ, it is not possible to obtain a unique fit for a curve passing through Burnham's Nebula, HH A, and HH 1555. However, assuming a value $\psi = 20°$ and carrying out a variation of the parameters α and β yields the fit seen in Fig. 1 where $\alpha = \beta = 5°$. For a distance of 140 pc, the fit gives $Pv = 6.2 \times 10^{12}$ km. Therefore jet velocities in the range 50-200 km s^{-1} would yield precession periods in the range 1000-4000 yrs. For smaller values of ψ, the product Pv increases linearly. The small inclination (≅7°) of the present jet axis implied by the fit agrees with the T Tau rotation axis inclination (≅10°) found by Herbst et al. (1986), and if the flow originates from the IR companion, one could conclude that the two objects have comparable inclinations.

Fukue and Yokoo (1986) consider several scenarios in which precessing jets can arise from a young stellar object. Models which invoke torques between T Tau and the IR companion generally yield precession periods $>10^6$ yrs because of their relatively great separation (≅84 A.U.). A more likely source of precession considered by these authors is that of a rapidly rotating, oblate central star which drives the precession of an orbiting torus (disk) misaligned with the equatorial plane of the star. Fukue and Yokoo find $P < 10^4$ yrs for toroids with $r < 1$ A.U. Such a source would probably require a relatively massive accretion disk with matter effectively weighted toward the inner disk, and if identified with the IR companion it would imply that the disk is optically thick even as viewed from the polar direction since no optical component is seen. It is not evident, however, if this is consistent with the presence of a jet which might be expected to have cleared matter from the polar zone. An additional observational feature is the prominent nebular emission immediately west (within a few arcsec) of the star and which may not fit into the bipolar jet model. To assess further the viability of the precessing jet model, it would be useful to obtain high spatial resolution images (<1") of the nebulosity around the star, and to map the velocity field of all material along the putative jet. This work was supported by NSF grant AST 8813917.

4. REFERENCES

Burke, T., Brugel, E.W. and Mundt, R. (1986) Astr. Ap. 163, 83-92.
Dyck, H., Simon, T. and Zuckerman, B. (1982) Ap. J. Lett. 255, L103-106.
Herbst, W., Booth, J., Chugainov, P., Zajtseva, G., Barksdale, W., Covina, E., Terranegra, L., Vittone, A. and Vrba, F. (1986) Ap. J. Lett. 310, L71-75.
Fukue, J. and Yokoo, T. (1986) Nature 321, 841-842.
Schwartz, P.R., Simon, T. and Campbell, R. (1986), Ap. J. 303, 233-238.
Schwartz, R.D. (1974) Ap. J. 191, 419-432.
Schwartz, R.D. (1975) Ap. J. 195, 631-642.
Solf, J., Böhm, K.-H. and Raga, A. (1988) Ap. J. 334, 229-251.

MONTMERLE: Is there any other example of a "precessing jet"? Would the S-shaped feature discovered by S. Strom a few years ago belong to the same category?

SCHWARTZ: I am unaware of any similar spiral jets observed in other young stars. Such jets are seen in some ejections from active galactic nuclei, and of course in the galactic source SS 433. Strom's "twisted filament" near HH 12 appears to be more like a "helix" instead of an opening spiral pattern. Some investigators have suggested that magnetic fields could give rise to helical motion.

SURVEY OBSERVATIONS OF EMISSION LINE STARS AND HERBIG-HARO OBJECTS AT THE KISO OBSERVATORY

K. ISHIDA
Institute of Astronomy, University of Tokyo
2-21-1 Osawa, Mitaka, Tokyo 181 Japan

ABSTRACT. Summary of the works on emission line stars and Herbig-Haro objects done by visiting astronomers at Kiso is presented.

1. Introduction

Kiso Observatory was founded in 1974 to operate Schmidt telescope with a corrector of 105 cm diameter, 150 cm main mirror, and 330 cm focal length. Since then, the Kiso Schmidt took 6000 photographs on plates purchased from Eastman Kodak. All the plates taken by the Kiso Schmidt and the image data processing system including several measuring machines are kept at the Observatory site on the mountain. The number of resident staff member is seven, three astronomers and four technicians. The Kiso Observatory accepts about 45 scientific programs and about 1000 man-day per year visitings of astronomers from domestic and foreign institutes in current years. The objectives distribute in 1) galaxies and clusters of galaxies (15), 2) nebulae and star clusters (10), 3) stars and stellar objects (9), 4) solar system and position astronomy (8), 5) others (3). One third of the visitings is occupied by those of graduate students guided by the resident astronomers at Kiso. The main program of the Kiso astronomers is isophotometry of a large number of galaxies in current years. Nevertheless, survey observations of emission line stars (ELS) and Herbig-Haro (HH) objects have been actively carried out by the two groups of astronomers with the collaboration of the staff members at Kiso. In the following sections, some of the results of survey observations of ELS and HH objects done by the two groups of astronomers are summarized.

2. Works done by Kogure's group

The star forming activity is intimately related to the formation and evolution of galaxies, because the conversion of masses from gas into stellar objects realizes the long lived galaxies existing safely against interaction with other intergalactic objects. In this respect, we can identify three problems that are the mass spectrum at the time of star formation in a particular site, the time dependence of star formation in the Galaxy, and the efficiency of the conversion

L. V. Mirzoyan et al. (eds.), Flare Stars in Star Clusters, Associations and the Solar Vicinity, 225–228.

of gas into stars.

ELS are promising candidates of T Tau type stars, which are pre-main sequence stars of intermediate masses, in contrast to that OB stars are young massive stars. By extending survey observations of ELS to as wide area as those done for OB stars (Warren and Hesser 1977), we may probably look at the difference of mass spectra with which stars form from dense cloud and from less-dense cloud.

The adopted method to list up ELS is to take plates by the Kiso Schmidt attached with 4 degree objective prism on Kodak 103aE (hypersensitized) behind Schott RG610. The spectra on the plates are about 1 mm long between $\lambda\lambda$ 6100-6800 A, which is long enough to detect Hα emission line (Wiramihardja et al. 1986, Wiramihardja et al. 1989, Kogure et al. 1989).

In the Orion region of 100 square degree area, 1070 ELS with magnitude range of V = 13-17 were found. The galactic latitude and distance 500 pc of the Orion OB association (Warren and Hesser 1977) suggest that the ELS are more than 100 pc below the galactic plane, and the ELS can not be farther away than the Orion OB association. The distance modulus m-M = 8.5 and the estimated interstellar extinction Av = 1.0 indicate that most of the ELS are of T Tau type. If so, the number ratio of ELS to OB stars in different places tells us difference of the mass spectrum with which stars form. The surface number densities of Ori OB1 members and of ELS are both higher in regions of high gas density cloud shown in the CO map (Maddalena et al. 1986) and in the Av map (Tomita, private communication). Remarkable thing is that the ratio of ELS to Ori OB members in surface number is about ten times higher in the less-dense part of dark cloud than in the dense part in the Orion region. In another word, the stars of intermediate masses are found in a wide area, whereas the massive stars are apparently found in a more restricted region of high gas density. This difference of number ratio of ELS to OB association member in place to place can simply be interpleted as the difference of mass spectrum correlated to the gas density of its parental cloud, if the ages of Ori OB association members and of ELS are nearly the same, and the life times of the both kinds of stars are again nearly the same.

However, there is a hint to indicate that there is a statistical difference in ages of ELS in its parental cloud of different densities. The ELS are classified into 6 grades (0-5) by relative intensities of Hα emission. Fraction of ELS with 3(medium), 4(strong), and 5(very strong) Hα is much larger in the region of dense cloud. Further study is needed to draw a firm conclusion on the mass spectrum and time dependence of star formation in the Orion region. The catalog of 1070 ELS to list coordinates α, δ, visual magnitude and Hα relative intensity is in preparation and can be obtained from Kogure, Kyoto University.

3. Works done by Ogura's group

ELS were searched on 103aE plates (hypersensitized) taken through RG645 by the Kiso Schmidt attached with the 4 degree objective prism in regions in and around 39 Bok globules of about 11 square degrees (Ogura and Hasegawa 1983, Ogura and Hidayat 1985), in Mon OB1/R1 of about 7 sq deg (Ogura 1984), and in dark cloud Lynds 1228 of about 7 sq deg (Sato et al. 1989). The total number of ELS detected and suspected is 295, of which 220 are ELS newly discovered.

A new technique was employed to search for HH objects in a wide area from homogeneous material. Two exposures are given on a plate with a small displacement behind two different filters one after the other, with a convenient mechanism to exchange a glass filter in front of loaded plate during the telescope is tracking on any sky area. The combinations of emulsion 103aF (hypersensitized) and filters RG645 and RG610 are utilized. The exposure times were 60 min and 35 min in most cases to give equal brightness side by side for ordinary objects, whereas unequal for HH objects with Hα emission. Slit spectroscopic observations or CCD direct image observations behind interference filters is needed to segregate HH objects from reflection nebulae out of the survey listings of the Kiso Schmidt. The number of HH object candidates found in M42, NGC 2264, NGC 7822, and NGC 1499 was 18 (Ogura 1987) and 10 of which were confirmed to be bona fide HH objects (Ogura and Walsh 1989, Walsh and Ogura 1989).

About 900 HH candidates in M42/L1641, about 60 in Ori B/L1630, and about 60 in NGC 2264 were found on deep, direct red plates taken by the SERC Schmidt telescope. The catalog of the HH candidates is being prepared with measuring machines at Kiso and supplementary observations to confirm the HH characteristics are on-going with using Anglo-Australian Telescope attached with a fiber spectrograph (Ogura, private communication).

Acknowledgments. I thank Prof. T. Kogure and Dr. K. Ogura for sending me material in advance of publication.

References

Kogure, T., Yoshida, S., Wiramihardja, S.D., Nakano, M., Iwata, T. and Ogura, K. (1989) Pub. Astr. Soc. Japan **41**, No.6, in press.

Maddalena, R.B., Morris, M., Moscowitz, J. and Thaddeus, P. (1986) Astrophys. J. **303**, 375.

Ogura, K. (1984) Pub. Astr. Soc. Japan **36**, 139.

Ogura, K. (1987) IAU Symposium No. 115, Star Forming Regions, P. 341, ed. Peimbert, M. & Jugaku, J., Reidel, Dordrecht, Holland.

Ogura, K. and Hasegawa, T. (1983) Pub. Astr. Soc. Japan **35**, 299.

Ogura, K. and Hidayat, B. (1985) Pub. Astr. Soc. Japan **37**, 537.

Ogura, K. and Walsh, J.R. (1989) Mon. Not. R. astr. Soc., to be submitted.

Sato, F., Ogura, K. and Matsumoto, A. (1989) Pub. Astr. Soc. Japan, to be submitted.

Walsh, J.R. and Ogura, K. (1989) Mon. Not. R. astr. Soc., to be submitted.

Warren, Jr., W.H. and Hesser, J.E. (1977) Astrophys. J. Suppl. **34**, 115.

Wiramihardja, S.D., Kogure, T., Nakano. M. and Yoshida, S. (1986) Pub. Astr. Soc. Japan **38**, 395.

Wiramihardja, S.D., Kogure, T., Yoshida, S., Ogura, K. and Nakano, M., (1989) Pub. Astr. Soc. Japan **41**, 155.

BLAAUW: Among the transparancies you showed was one of the Orion region exhibiting the molecular clouds and the various OB subgroups, and there was also one showing equi-surface density curves of the emission line stars. I noted that there appears to be a correlation between the location of subgroups Ia and a feature in the emission line star density. This is interesting because this subgroup is almost void of interstellar matter and this implies that the subgroup Ia also contains low mass stars.

FU Orionis eruptions and early stellar evolution

Bo Reipurth
European Southern Observatory
Casilla 19001
Santiago 19
Chile

Abstract The FU Orionis phenomenon has attracted increasing attention in recent years, and is now accepted as a crucial element in the early evolution of low mass stars. The general characteristics of FUors are outlined and individual members of the class are discussed. The discovery of a new FUor, BBW 76, is presented, together with a discussion of photometric and spectroscopic observations of the star. The evidence for circumstellar disks around T Tauri stars is briefly outlined, and the FUor phenomenon is discussed in the context of a disk accretion model. A large increase in the accretion rate through a circumstellar disk makes the disk self-luminous with a luminosity two or more orders larger than that of the star. Massive cool winds rise from FUors, and it is conceivable that they are related to the initiation of Herbig-Haro flows. The FUor phenomenon appears to be repetitive, and newborn low-mass stars may be cycling between the FUor state and the T Tauri state.

1 The FUor phenomenon

The story of FU Orionis eruptions began in 1936, when a faint red variable star, FU Orionis, located in the dark cloud B35 in the λ Ori region, brightened by 6 magnitudes in a period possibly as short as four months. Early photometry has been discussed by Wachmann (1939, 1954) and a detailed photometric and spectroscopic analysis is presented by Herbig (1966, 1977).

There are now five such objects known, dubbed FUors by Ambartsumian (1971), which all have in common that they were observed to brighten considerably, although on rather different time scales. In their bright states, these stars share a number of spectroscopic properties, notably F-G type supergiant spectra with abnormally strong Balmer absorption lines, strong Li I 6707 lines, pronounced P Cygni profiles at Hα and several other lines, and broad lines suggesting rapid rotation. They are all located in regions of recent star formation, and are associated with reflection nebulae. After

L. V. Mirzoyan et al. (eds.), Flare Stars in Star Clusters, Associations and the Solar Vicinity, 229–251.

reaching maximum brightness they presumably gradually decay, although only one object (V1057 Cyg) has faded substantially since its eruption (Fig. 1); since we are not seeing numerous objects in high states it follows that all must eventually decay.

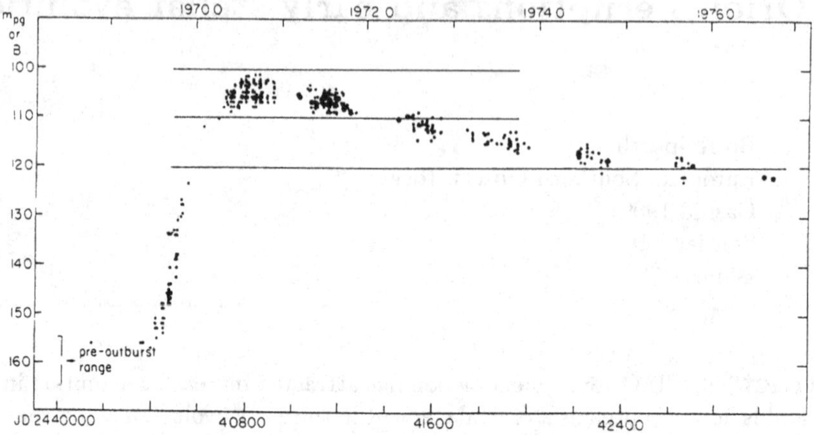

Fig. 1. The B-light curve of V1057 Cyg showing the eruption in 1969/70 and the subsequent decay. From Herbig (1977).

The progenitors of FUors are faint, red, slightly variable stars, all presumably T Tauri stars, although for only V1057 Cyg is a pre-outburst spectrum available (Herbig 1958); this spectrum shows the emission lines characteristic of T Tauri stars.

Based on event statistics and the estimated number of observable T Tau stars, Herbig (1977) demonstrated that FUor eruptions almost certainly occur repetitively in all T Tau stars in the course of their evolution. (see Section 8).

Perhaps the most outstanding spectroscopic characteristic of FUors is their distinct P Cygni profiles, notably at Hα and at the Na I 5890/5896 lines (Bastian and Mundt 1985). Fig. 2 shows high-resolution spectra of these lines in FU Ori. Deep blueshifted absorption troughs testify to the presence of massive stellar winds in these objects (for further details see Sec. 6). Another notable feature of FUors is their gradual change of spectral type with wavelength: in the red spectral region they have a later spectral type than in the blue, and in the near-infrared they show the CO bands typical of K-M stars (Cohen 1975, Mould et al. 1978).

On the basis of such spectroscopic characteristics, several stars have been suggested as FUors, without evidence for an eruption, in particular Z CMa and L1551 IRS 5. These two objects are discussed, together with the five known classical FUors, in the following.

• FU Orionis

The photometric behaviour of FU Ori from 1936 to 1977 has been compiled by Herbig (1977), and from 1978 to 1985 by Kolotilov and Petrov (1985). During that time the star has remained in its elevated state, with only a slight overall fading. Two years after the eruption the star suddenly dimmed by a magnitude, and subsequently started a slow oscillation which ceased after 20 years. Overall the star has faded by about a magnitude in the blue since the eruption.

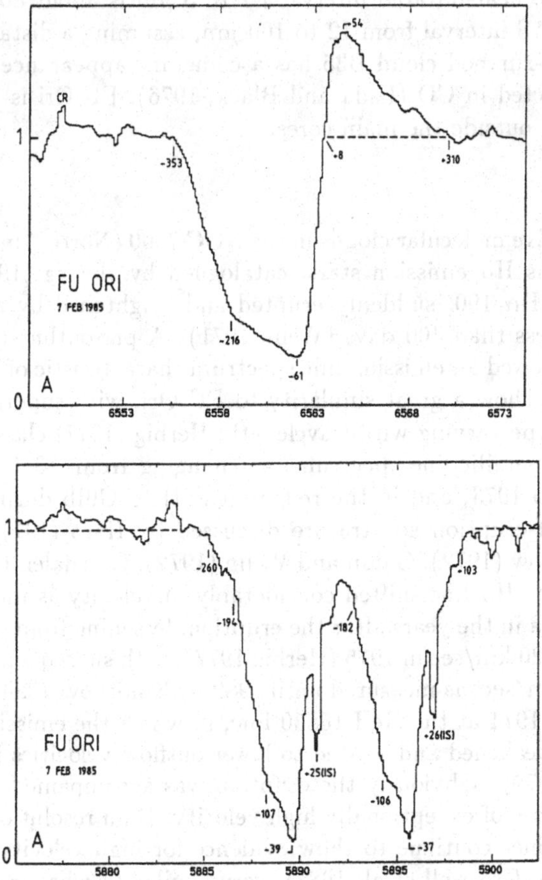

Fig. 2. High-resolution line profiles of the Hα and the Na I lines in FU Orionis, showing evidence for massive winds. From Reipurth (1990).

The peculiar composite spectrum of FU Ori has been described in detail by Herbig (1966), who also gives references to the early work on this object. At low resolution the star appears as an F-G-type supergiant, but with unusually prominent Balmer lines. At higher resolution one sees two components, the above mentioned supergiant spectrum as well as a set of blueshifted components particularly strong at the Balmer lines and at low-level lines in both neutral and ionized metals. The best compromise classification of the supergiant spectrum is F2:p I-II (Herbig 1966).

The far-red spectrum has been described by Shanin (1979), and ultraviolet spectra by Ewald et al. (1986) and Kenyon et al. (1989).

FU Ori is a bright near-infrared source (e.g. Cohen 1973) and is also detectable at far-infrared wavelengths with a dust-temperature of 15 K (Smith et al. 1982). The luminosity in the near-infrared interval 1.2 to 5 μm is about 80 $L_\odot$ and it is about 20 $L_\odot$ in the IRAS interval from 12 to 100 μm, assuming a distance of 460 pc.

The bright-rimmed cloud B35 has a cometary appearance. It harbors several cloud cores detected in CO (Lada and Black, 1976); FU Ori is located towards the tail of the cloud, outside the main cores.

• V1057 Cyg

The extensive molecular clouds in the NGC 7000 (North America Nebula) region contain numerous Hα emission stars, catalogued by Herbig (1958). In 1969–1970, one of these, LkHα 190, suddenly erupted and brightened by about 5 magnitudes in a time-span less than 400 days (Welin 1971). A pre-outburst spectrum taken by Herbig (1958) showed an emission line spectrum characteristic of T Tauri stars. Post-eruption spectra show a great similarity to FU Ori, with supergiant characteristics and a spectral type varying with wavelength; Herbig (1977) classified the star in the blue (from the metallic-line spectrum) as changing from A3–5 II to early F, II or III, from 1971 to 1973, and in the red from F5II to G0Ib during the same period. Other early post-eruption spectra are discussed by Haro (1971), Mendoza (1971), Schwartz and Snow (1972), Gahm and Welin (1972), Grasdalen (1973) and Chalonge et al. (1983). The Hα line shifted considerably in velocity as measured on medium-resolution spectra in the years after the eruption, lessening from –560 km/sec in early 1971 to about –220 km/sec in 1975 (Herbig 1977), with subsequent variations between –200 and –300 km/sec, as measured until 1982 by Kolotilov (1984). A P Cygni profile was observed in 1974 at the He I 10830 line, in which the emission disappeared and the absorption weakened and shifted to lower outflow velocities in the following two years (Shanin 1979). Obviously the eruption was accompanied by a brief period of enhanced mass loss of exceptionally high velocity. High resolution spectra of the Hα and Sodium D lines continue to show evidence for high velocity mass loss (Bastian and Mundt 1985, Croswell et al. 1987), very similar to what is observed in FU Ori (see Fig. 2 and Section 6).

The optical light curve of V1057 Cyg is well observed. Fig. 1 shows the monotonic decline after the eruption until 1977 (Herbig 1977); further photometry up to 1985 has been obtained by Kopatskaya (1984) and Ibragimov and Shevchenko (1987). The star is now almost halfway back to its average pre-outburst level, but the rate of decrease in brightness appears to be slowing down. Ibragimov and Shevchenko (1987) suggest that there could be a cyclic variation in their photometry with a period around 12 days. The decline in brightness of V1057 Cyg has been accompanied by a corresponding fading of the extensive reflection nebulosity around the star (Duncan et al. 1981)

Infrared observations of V1057 Cyg obtained in the seventeen years following outburst have been analyzed by Simon and Joyce (1988), who also give reference to earlier infrared work. The infrared fluxes have decreased at all wavelengths observed between 1.2 and 20 μm, but much more rapidly at the shorter wavelengths. At 10 and 20 μm an initial very rapid fading subsequently leveled off. To explain the rate of decrease as a function of wavelength is an essential test for any model of the FUor phenomenon.

In 1973 a 1720 MHz OH maser was observed at V1057 Cyg, which gradually faded beyond detection towards the end of 1974 (Lo and Bechis 1973, 1974; Andersson et al. 1979). A second OH maser outburst was reported by Winnberg et al. (1981) to occur in 1979, also with a subsequent fading.

V1057 Cyg was detected at 450, 800 and 1100 μm by Weintraub et al. (1989) providing direct evidence for the presence of circumstellar material; they suggest a lower limit to the mass of this material of 10^{-1} $M_{\odot}$.

• V1515 Cyg

The third FUor, V1515 Cyg, was discovered by Herbig (1977) in the Lynds 897 cloud, a complex region with various signs of recent star formation (Herbig 1960). In contrast to the rapid rise displayed by FU Ori and V1057 Cyg, V1515 Cyg took several decades to rise: from 1944 to 1952 it brightened by about 3 mag from m_B around $17^m.5$, followed by a slower rise until the mid-seventies when it reached $V \simeq 12^m$ (Herbig 1977, Gottlieb and Liller 1978). Since then it appears to have started to fade (Kolotilov and Petrov 1983). It displays the various characteristics of FUors, such as an early G-type high-luminosity spectrum, a bow-shaped reflection nebula (see Fig. 4), deep blueshifted absorptions at Hα and the Sodium lines (Bastian and Mundt 1985), and a strong infrared excess (Cohen 1980).

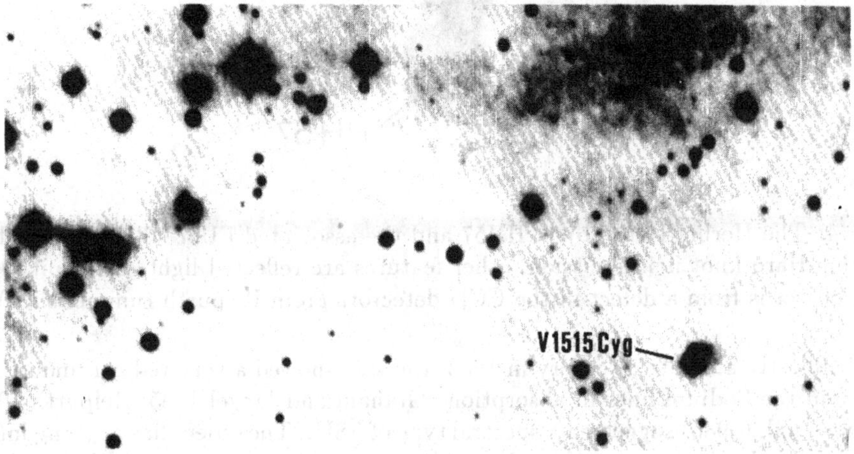

Fig. 3. The FUor V1515 Cyg and its environment, as seen on the red Palomar atlas. The characteristic FUor reflection nebula is evident. Another young star, Parsamian 22, is seen further east in an extended reflection nebula. North is up and east is left.

• V1735 Cyg (=Elias 1-12)

This object was found as a strong infrared source in an infrared survey of the IC 5146 dark cloud complex (Elias 1978). It is a bright near-infrared source rising steeply towards the mid-infrared. It is also a bright IRAS source, with an IRAS luminosity of 75 $L_\odot$ at the assumed distance of 900 pc. On the red Palomar atlas plate from 1952, no object brighter than 20^m was visible, but a similar plate from 1965 showed a star of about 15^m with a nearby faint reflection nebula. An optical low-resolution spectrum shows a red continuum with Hα in absorption, displaced 4 Å bluewards, suggesting a P Cygni profile. The near-infrared spectrum shows strong H_2O and CO absorption features, as in other FUors. CO line profiles of the surrounding cloud show broad wings, suggesting outflow from the star (Levreault 1983).

• V346 Nor (= HH 57 IRS)

The most recent FUor to be discovered is V346 Nor, which appeared in a molecular cloud in the southern constellation Norma, next to the Herbig-Haro object HH57 (Graham 1983, Graham and Frogel 1985). Fig. 4 shows a red broadband CCD image of the region, which demonstrates the close association between HH object and FUor: a reflection nebula stretches out from the star to reach the HH object. The proper motion vector of the HH object points away from the FUor (Schwartz et al. 1984).

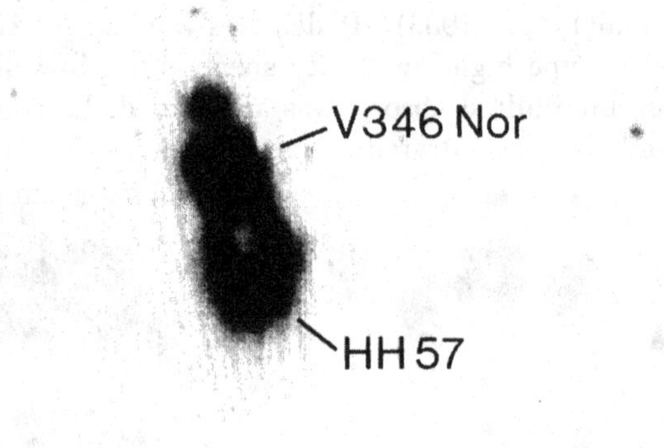

Fig. 4. The Herbig-Haro object HH57 and the associated FUor V346 Nor. Only the Herbig-Haro knot is in emission, other features are reflected light. The spike in the HH object is from a defect in the CCD detector. From Reipurth (unpublished).

Shortly after its discovery in 1983, the star showed a very red continuum, with Hα and the Sodium lines in absorption (Graham and Frogel 1985, Reipurth 1985a). Cohen et al. (1986) suggested a spectral type of F8III. The object has a strong infrared excess. In the far-infrared there is an elongated structure perpendicular to the line between HH object and star (Cohen et al. 1985). A molecular outflow from the FUor has been detected along the axis to the HH object (Reipurth et al. 1990).

Little is known about the rise time. Examination of Schmidt plates suggest a slower rise than FU Ori and V1057 Cyg, more reminiscent of V1515 Cyg (Reipurth 1985a).

The luminosity is poorly determined, mainly because the distance is very uncertain. Various distance estimates are 300 pc (Reipurth 1985a), 700 pc (Graham and Frogel 1985) and 940 pc (Cohen et al. 1986). The last estimate is derived assuming the FUor has the same luminosity as FU Ori at peak brightness.

• Z Canis Majoris

This object has long been considered a Herbig Ae/Be star, but it possesses a number of characteristics which suggest it may be a FUor (Hartmann et al. 1989): it exhibits strong blueshifted Balmer and Sodium absorption troughs (Finkenzeller and Jankovics 1984, Finkenzeller and Mundt 1984), a unique spectral type cannot be assigned (Herbig 1960), CO absorption at 2.3 μm (Hartmann et al. 1989), and a characteristic reflection nebula (Herbig 1960, Goodrich 1987). It also shows double-peaked absorption line profiles (Hartmann et al. 1989), as seen in other FUors and predicted by the accretion disk model (see Sect. 5).

No outburst has been observed, neither is Z CMa known to fade gradually, instead it shows irregular variability with amplitudes up to $\Delta V \sim 2$ mag (e.g. Covino et al. 1984).

Near-infrared speckle observations have shown Z CMa to be highly elongated with a size of about $0.1''$ at 2.2 μm, either due to scattering in a circumstellar disk (Leinert and Haas 1987) or because of a close companion (Koresko et al. 1989). It has also been resolved at radio continuum wavelengths (Cohen and Bieging 1986). The close visual companion reported by Finkenzeller and Mundt (1984) appears to be spurious.

It is likely that Z CMa is a FUor. With a luminosity of 3500 $L_\odot$ it is more luminous than any of the other known FUors.

• L1551 IRS 5

This object is an infrared source embedded in the L1551 cloud (Strom et al. 1976) with a luminosity of 36 $L_\odot$ (Cohen and Schwartz 1987). It is presumably very young, drives a major bipolar molecular outflow (Snell et al. 1980), and a series of Herbig-Haro objects are moving supersonically away from it (Cudworth and Herbig 1979, Neckel and Staude 1987). High resolution $C^{18}O$ observations have revealed an elongated concentration of gas, with a long axis of 1400 AU and a mass of $\sim$ 0.1 $M_\odot$, oriented perpendicular to the outflow-axis (Sargent et al. 1988). A small optical and infrared nebula outlines a cavity reflecting the light from the source (e.g. Moneti et al. 1988, Campbell et al. 1988). The optical spectrum of the source has been observed by Mundt et al. (1985) and Stocke et al. (1988) via this reflection nebula. They find a P Cygni profile at Hα, with blueshifted absorption extending to more than 500 km/sec, and only weak Hα emission. This is unlike the Hα profiles of T Tauri stars, but similar to what is observed in FUors. There is some tentative evidence for a modest change in spectral type as one goes from blue to red, from G to K. Carr et al. (1987) observed the 2μm spectrum of L1551 IRS 5 and detected the CO bands in absorption at a strength suggesting a spectral type of K2III. They also find that H_2O absorption is possibly present. The sum of the observational evidence thus points to L1551 IRS 5 being in a post-eruption FUor state.

2 BBW76 – a new southern FUor

The photometric behaviour of the FUors discussed in the previous section show large differences. In order to better discern common properties from individual characteristics, it would be very useful to enlarge the small sample of known FUors. This can be done in two ways: either by searching for major stellar eruptions in molecular cloud regions in new or older plate material, or by closer examination of individual nebulous stars in star forming clouds. This latter approach presupposes that FUors have some sufficiently well defined characteristics, which allow them to be identified without evidence for an outburst. As discussed in the previous section, this appears in certain cases to be possible.

Both types of studies have been carried out at the European Southern Observatory at La Silla over the last 10 years, and one object has been identified as an apparently bona fide FUor (Reipurth 1985b, 1990). The star, BBW 76, is a 12th magnitude object associated with a small, anonymous cloud in the direction of the constellation Puppis at a position α_{1950}: $7^h48^m40^s.4$, δ_{1950}: $-32°58'43''$. The star and cloud are apparently associated, because the star displays a distinct, curved reflection nebula. Also, the velocity of the cloud as measured in the ^{12}CO transition equals the velocity of the star, as well as of the interstellar absorption lines seen in high resolution Na I spectra.

Low resolution spectra of the star reveal no emission lines, but a composite of several types of features, mixing a host of metallic lines, strong Balmer lines and molecular bands. Based on a blue spectrum, and excluding the Balmer lines, the best spectral type is G0-2 I. Fig. 5 shows a low-resolution spectrum of BBW 76 compared to one of FU Ori itself. The similarity of the two stars is striking. Further to the red the two stars are equally similar, and both show strong Li I 6707 absorption. In the near-infrared region, BBW 76 shows weak 2.3 μm CO absorption, suggestive of a much later spectral type.

At much higher resolution, the similarities between BBW 76 and FU Ori continue. Fig. 6 shows high-resolution echelle spectra of Hα and the Na I 5890/5896 lines in BBW 76. Comparison with similar spectra for FU Ori shown in Fig. 2 immediately suggests that similar processes operate in both stars. Hα displays a P Cygni profile with weak emission and a deep absorption trough several hundred km/sec wide. The sodium lines also show massive blueshifted absorption features, which are structured, suggesting spatial/temporal variability. The winds responsible for these features are further discussed in section 6.

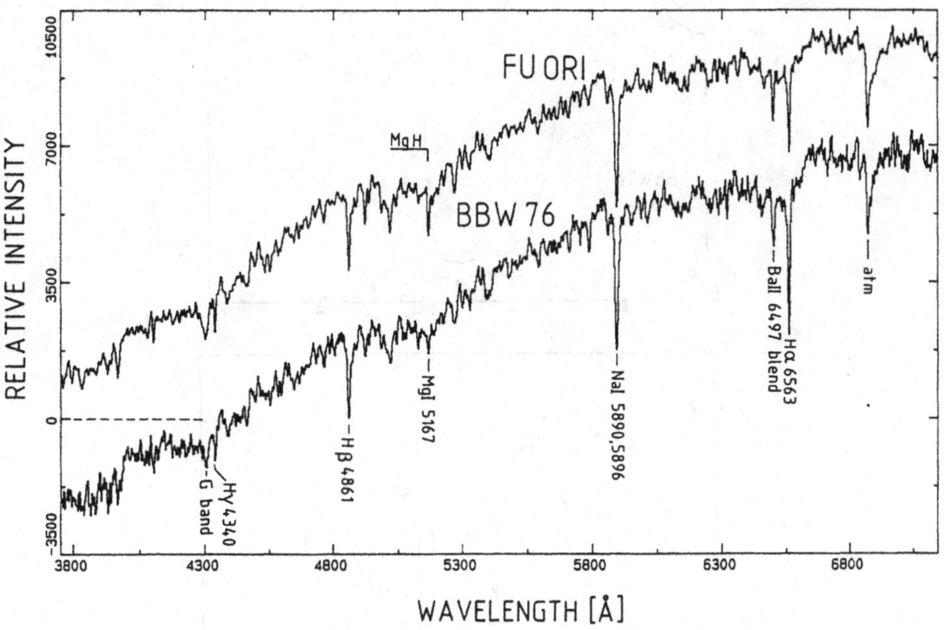

Fig. 5. Low resolution spectra of BBW 76 and FU Orionis, showing the remarkable similarity between the two objects. From Reipurth (1990).

If BBW 76 is a FUor decaying from an earlier eruption, it should exhibit a gradual fading at all wavelengths. Photometric observations, at optical as well as infrared wavelengths, demonstrate that this is indeed the case. Fig. 7 shows the photometric behaviour in the V-band of BBW 76 between 1983 and 1988. To within the photometric accuracy the star is fading steadily at a rate of about $0\overset{m}{.}025\ yr^{-1}$. This is a rate almost twice as fast as FU Ori, but five times slower than V1057 Cyg.

The evidence presented so far strongly suggests that BBW 76 is a FUor. The piece of information missing is the date of its initial eruption. A still ongoing survey of old plate archives has revealed no major variability going back as early as 1927.

In summary, BBW 76 is a young star, with strong lithium and associated with a molecular cloud, which spectroscopically shares all the features of FU Ori itself, and which is gradually fading at optical and infrared wavelengths. If a major brighteningof the star has occurred, it must have taken place more than 60 years ago.

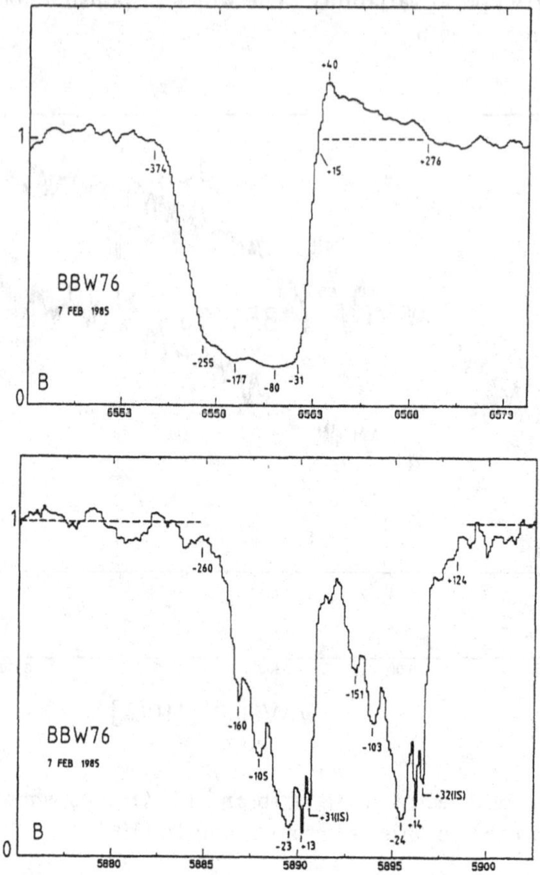

Fig. 6. High resolution line profiles of the Hα and Na I lines in the new FUor BBW 76, showing the same signatures of mass loss as in FU Orionis. From Reipurth (1990).

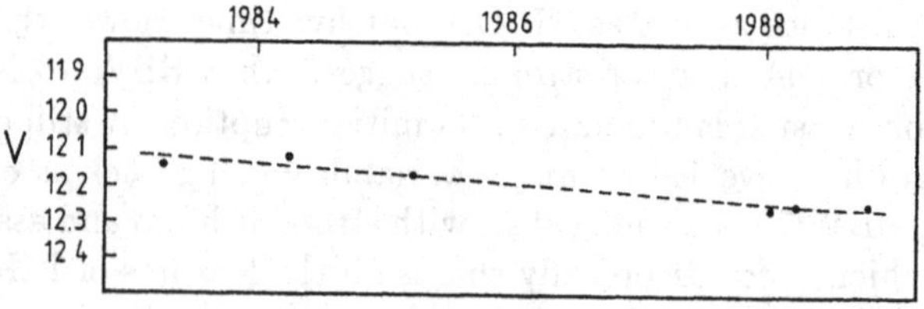

Fig. 7. The V-light curve of the new FUor BBW 76 from 1983 to 1989. The object is clearly fading; a similar behaviour is found at near-infrared wavelengths. From Reipurth (1990).

3 EXors

Herbig (1977) pointed out that if FU Ori eruptions "occur in T Tauri stars but are absent on the main sequence, then in a given object they probably weaken in intensity, duration, or frequency as the star evolves toward the main sequence, instead of ceasing abruptly. If so, minor activity reminiscent of the FU Ori phenomenon might be present in older T Tau stars or post-T Tau objects." It is also possible that the mechanism which drives FUors may operate at lower levels, also in very young stars.

A number of objects are known that fit this description; they are known as EXors, and a list of nine members of the class is given by Herbig (1989).

The EXors, named after the prototype EX Lupi, show outbursts with an amplitude of up to 5 magnitudes, which last for months, in some cases more than a year. They do, however, appear to retain their emission line characteristics also during outbursts, although spectroscopic data are limited. A light curve of the 1955-56 outburst in EX Lupi is shown in Fig. 8.

The best-studied EXor to date is probably DR Tau, a T Tauri star with a rich emission line spectrum and rapid high-amplitude photometric variability superposed on a long term rise in mean brightness (Bertout et al. 1977, Chavarria 1979, Götz 1980, Kolotilov 1987).

The two most recently found EXors are V1118 Ori (= Chanal's object) and V1143 Ori (= Sugano's object). They have had several outbursts in the 1980's, which are discussed by Parsamian and Gasparian (1987), Gasparian and Ohanian (1989), Mirzoyan et al. (1988); see also Parsamian (this volume). Photometric and spectroscopic monitoring of these stars is important for attempts to understand their relationship with the FUor phenomenon.

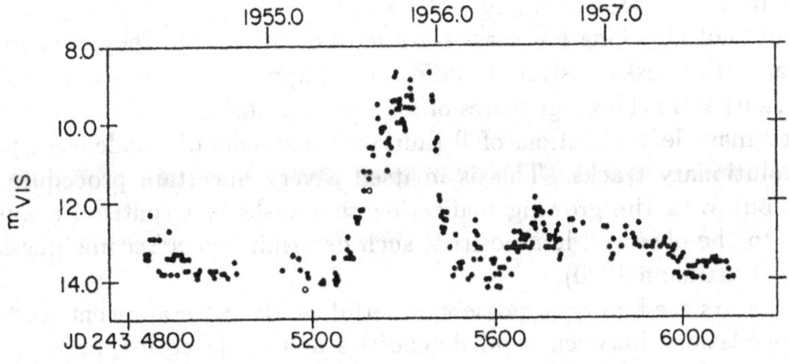

Fig. 8. The 1955/56 eruption of the EXor EX Lup. The data points are visual estimates compiled by Bateson and Jones. From Herbig (1977).

4 Disks around T Tauri stars

It is now generally accepted that circumstellar disks not only are present around the younger of low-mass pre-main sequence stars, but play an essential part in defining a variety of their observable properties. This section briefly summarizes properties of disks around the T Tauri stars, in their role as precursors to FUors (see also the reviews of Appenzeller and of Lada in this volume).

T Tauri disks are divided into *passive* and *active* disks. The passive disks merely reprocess light emitted from the central star, re-emitting it in the infrared. For certain geometries and inclinations, up to 50% more light than emitted by the star can actually reach the observer via scattering. On the other hand, active disks have an additional component to their luminosity from viscous dissipation of energy in the disk, and their infrared excesses can therefore occasionally be very large. As discussed by Lynden-Bell and Pringle (1974), energy dissipation must lead to inflow of material through the disc; active disks are therefore accretion disks. Stars with active disks are likely to have a hot boundary layer between star and disk, where energy is released as disk material comes to rest on the (slower rotating) star.

The spectral shape of infrared energy distributions is almost identical for T Tauri stars with passive and active disks. For a detailed modelling, optical and ultraviolet data must therefore be incorporated, even though the effect of extinction then becomes much more pronounced. Bertout et al. (1988) and Basri and Bertout (1989) have rather successfully modelled the energy distributions of a number of classical T Tauri stars from the ultraviolet to the infrared in terms of a central star surrounded by an active disk. Such accretion disk models are successful in matching observations, because the infrared radiation can be produced by dust near the star without causing an excessive extinction, and because accretion is a convenient source of energy to explain a possible large infrared luminosity. The accretion rates they derive range from $10^{-9} M_{\odot} yr^{-1}$ to $10^{-7} M_{\odot} yr^{-1}$.

Certainly not all young low mass stars show evidence that they are surrounded by actively accreting disks. Estimates differ, but it appears that less than 50% of the classical T Tauri stars show signatures of an accretion disk.

Age and mass determinations of T Tauri stars are generally made using pre-main sequence evolutionary tracks. This is in itself a very uncertain procedure (Mazzitelli 1989), but with the growing realization that disks can contribute sometimes significantly to the observed luminosities, such determinations become questionable (Kenyon and Hartmann 1990).

T Tauri stars are known to possess powerful winds. Several recent studies have suggested correlations between wind diagnostics and evidence for disks. Strom et al. (1988) found that most of young stellar objects with signatures of mass loss are characterized by large infrared luminosity excesses. Cohen et al. (1990) found that stars with strong [OI] emission have bolometric luminosities much larger than their expected stellar luminosities. Cabrit et al. (1990) found [OI] and Hα emission correlated with infrared excess, but no correlation to photospheric luminosity, suggesting that it is the disk, and not the star, which primarily determines the stellar wind strength. It thus appears that mass loss in young low-mass stars is powered by the energy released as material in circumstellar disks accretes onto the stars.

5 Disk accretion in FUors

A number of authors have suggested over the years that the large flare-ups of FUors could represent events in which material accretes onto the central stars (Paczynski 1976, Herbig 1977, Larson 1983, Lin and Papaloizou 1985). Recently, Hartmann and Kenyon (hereafter HK) have developed the details of an accretion disk model for FUors, in which a hot, optically-thick accretion disk dominates the system light at maximum (Hartmann and Kenyon 1985, 1987a, b, Kenyon et al. 1988, Kenyon and Hartmann 1989).

The relations between mass accretion, angular momentum transport and energy dissipation in a viscous disk have been discussed by Shakura and Sunyaev (1973) and Lynden-Bell and Pringle (1974). The effect of viscosity is to transport angular momentum from the inner to the outer part of the disk and thus to control the mass-flow through the disk. While material slowly drifts inwards in the disk, half of the potential energy decrease is used to heat the disk locally. Ultimately the speeding material accretes onto the star, releasing the equivalent of the other half of the potential energy difference in a hot boundary layer. The local temperature and luminosity in the disk is a sensitive function of the mass accretion rate and distance to the star.

Based on these concepts, HK developed a time-independent model of an optically thick disk, consisting of a set of concentric annuli, each of them radiating as a star with a certain effective temperature, or for the outer cool regions, as a black body.

The interpretation of FUors in terms of such a hot, optically thick accretion disk has had some notable successes. It naturally accounts for the otherwise curious change of spectral type with wavelength, wherein a FUor has an F-G type spectrum in the blue range, but a K-M type spectrum in the near-infrared. It further explains the supergiant spectra of FUors, because disks are low-gravity environments. It also makes two predictions, which can be tested by observations.

The first prediction is that absorption-line profiles should be double-peaked if the object observed is a rotating flattened disk. High resolution optical spectra do not show this effect unambiguously in individual lines. But cross-correlation between FUor spectra and spectra of template stars, of similar spectral type in the relevant wavelength range, indeed shows such double peaked profiles in FU Ori, V1057 Cyg and Z CMa (HK, see previously listed references). The linewidths are large (~ 100 km s^{-1}) and the separation of the double peaks can be well matched by disk models.

The second prediction is that lines from optical spectra, emitted in hot, inner,

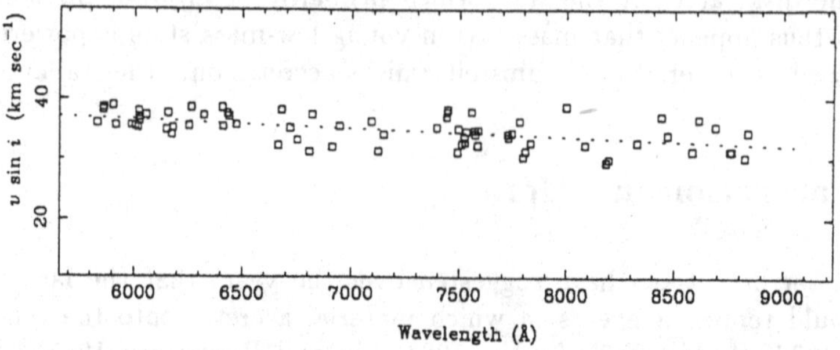

Fig. 9. The rotational velocity *v sin i* determined for a large number of lines in the red spectral region (6000–9000 Å) of V1057 Cyg. There is a clear trend towards lower rotational velocities at longer wavelengths. From Welty et al. (1990).

faster rotating regions, should show larger rotational broadening than lines from infrared spectra, emitted in cooler, more distant and slower rotating regions. Using infrared Fourier transform spectra, HK (Kenyon and Hartmann 1989 and previously listed references) have shown that there is indeed such an effect, and that the observed ratio of optical to infrared rotational broadening can be well reproduced by the disk model. In a more recent study, Welty et al. (1990) obtained high resolution, high signal-to-noise spectra of V1057 Cyg covering the spectral range from 5820 Å to 9370 Å. Fig. 9 shows *v sin i* measured for a number of lines in this wavelength region. The trend towards lower rotational velocity as one goes to the red is evident, and further supports the disk accretion hypothesis. In the same study, Welty et al. (1990) also found a correlation between line width and line transition lower excitation potential, such that higher excitation potential lines are broader, as one would expect in a Keplerian disk with a temperature gradient.

In order to achieve a match between models and observations, very large mass accretion rates, of the order of $10^{-4}\ M_\odot\ \mathrm{yr}^{-1}$, are needed. Assuming an average duration of a FUor eruption of 100 years, thus implies that of the order of $10^{-2} M_\odot$ is accreted in an eruption. Although these numbers obviously are nothing more than simple estimates, they do suggest that if FUor eruptions occur regularly through the pre-main sequence phase, the accreted material may add significantly to the initial stellar mass, and thus influence the evolution of the star.

Despite the ease with which a number of observations can be interpreted in this framework, the accretion disk model is still in its infancy, and it is likely to need considerable elaboration before it can explain a number of more subtle observational facts. Herbig (1989) has drawn attention to some of the shortcomings of the model as it stands at the moment, and suggests that it is still too early to discard other alternatives.

A point of considerable interest is the question of stability of a hot actively accreting circumstellar disk. Clarke et al. (1989) note that the mid-plane temperature in the inner disk is at least 40 000 K, and therefore must contain partially ionized regions. But such regions are prone to local thermal instability, so that a front separating ionized and neutral regions should sweep radially back and forth in the disk, modulating the otherwise steady accretion on timescales of months to years. This is contrary to the observed steady decline of FUors. Clarke et al. (1989) proceed to show that the disk may actually be stabilized against this thermal ionization instability by the effect of advective heat transport.

A problem not yet understood is what initially triggers a FUor outburst. It is difficult to envisage how a disk around a T Tauri star, with an accretion rate of less than $10^{-7}\ M_{\odot}\ yr^{-1}$, can switch into a structure with an accretion rate 3 orders of magnitude larger within a few years. An external triggering mechanism may be needed, such as the close passage of a companion in a highly eccentric orbit (Clarke et al. 1989).

6 The winds of FUors

High-resolution spectra of Hα and the Sodium D lines demonstrate that FUors possess massive, high velocity winds (Bastian and Mundt 1985, Croswell et al. 1987, Reipurth 1990). Profiles of these lines are shown for FU Ori and the new FUor BBW76 in figs. 2 and 6, respectively. In Hα the emission component extends to only low positive velocities, in contrast to the absorption trough which extends to blueshifted velocities of several hundred km/sec. In the Sodium lines, the absorption also extends to very high negative velocities, whereas only weak redshifted absorption of low velocity can sometimes be discerned. Apparently the redshifted material is somehow occulted. This is readily understood if an optically thick disk surrounds the star.

Pudritz and Norman (1983, 1986) suggested that winds from young stellar objects originate in their circumstellar disks. A disk is an attractive wind-source, because its surface-gravity is much lower than that of the central star, and during a FUor eruption it is also many times more luminous than the star. The Pudritz-Norman model assumes that poloidal magnetic fields pervade the disk; gas is accelerated because the magnetic field enforces its corotation with the disk out to relatively large distances. Such centrifugally driven, hydromagnetic winds are very efficient in extracting and removing angular momentum and rotational energy from the disk, although it is not clear that winds can be formed at the large distances from the central source assumed in this scenario. Another disk-wind model assumes that Alfvén waves are generated by turbulence in the disk (Hartmann and Kenyon 1985), and that the FUor winds are driven essentially by the same mechanism suggested for T Tauri winds (Lago 1979, 1984, DeCampli 1981, Hartmann et al. 1982). It is indeed possible that both the magnetic wave model and the rotating magnetic field acceleration mechanism is at play simultaneously.

Detailed analysis of high-resolution line profiles can provide information on the geometry of the wind-forming region, the mass loss-rate and the temperature of the wind. Such a study was performed by Croswell et al. (1987), who did radiative transfer calculations to match observed Hα and Na I profiles; each of their isothermal steady-flow wind models were specified by a mass-loss rate, a wind temperature and a wind-velocity as function of radius, and were joined to a photosphere of T_e = 5000 K and log g = 1. A lower limit to the mass-loss rate of $10^{-5} M_\odot\ yr^{-1}$ was found to be required in order to match the extensive blueshifted Sodium absorption troughs. This was derived assuming a slowly accelerating velocity field; a rapidly accelerating velocity law would demand even higher mass loss rates, approaching the about $10^{-4} M_\odot\ yr^{-1}$ in mass *accretion* required by Hartmann and Kenyon (1985) to account for the observed outburst-luminosity. But rapid wind acceleration is only needed in a spherically-symmetric geometry, where the star is required to occult the redshifted material. If instead one assumes a disk-like geometry, a slow acceleration geometry and thus a lower mass-loss rate can still be adopted. The models which simultaneously can match the observed Hα and Na I absorptions all indicate cool wind temperatures of about 5000 K; at higher temperatures Hα goes into strong emission for mass-loss rates sufficient to produce the observed Na I absorption. The work of Croswell et al. (1987) thus suggests that FUor winds are cool, very massive and arise from a disk-geometry. One consequence of the low wind temperature is that thermal pressure almost certainly is not important in driving the outflow, further supporting the magnetic field models previously discussed.

There are two observations which suggest that the wind rising from the disk has radial structure. Firstly, monitoring of the Hα and Na I line profiles have revealed variations of the high-velocity wings on timescales of a day (Croswell et al. 1987, Pasquini and Reipurth 1990). For wind velocities up to several hundred km/sec, this is likely to imply that this fast component of the wind originates in a region only a few stellar radii in size, presumably located around the central star . Secondly, Kenyon and Hartmann (1989) observed slightly blueshifted CO $v'-v'' = 2-0$ absorption in FU Ori and V1057 Cyg in their Fourier Transform spectra, and concluded that mass ejection from the near-infrared emitting disk regions does not exceed $10^{-7} M_\odot\ yr^{-1}$. Compared with the mass loss rates derived for the hotter, inner, optically emitting regions (Croswell et al. 1987), it appears that the mass loss rates decrease by a factor of roughly 100 as one goes from about 1–2 R_* to about 10 R_*.

There are theoretical arguments why the main mass loss should be concentrated towards the inner disk regions. Pringle (1989) shows that if the wind is driven ultimately by accretion energy, then the wind must come from regions of sufficient depth in the gravitational well, so that enough energy per gram becomes available. He suggests that the wind actually originates in the boundary layer, where the accretion disk feeds into the star, and is driven by powerful toroidal magnetic fields produced by the strong shear in the boundary layer.

Little is known with certainty about the properties of the boundary layer, but it is likely to be the hottest region of the disk. Ultraviolet spectra of several FUors obtained with the IUE satellite have been reported by Ewald et al. (1986) and Kenyon et al. (1989). Ultraviolet absorption features suggest spectral types for FU Ori of A5 I and for Z CMa of F5 I. An unexpected result is that the ultraviolet absorption features are not significantly veiled, implying that a hot boundary layer contributes less than 20% of the radiation between 2600 Å and 3200 Å. An understanding of the properties of the boundary layer is likely to be one of the prime goals of FUor research in the coming years.

7 Do FUors power the Herbig-Haro flows?

The massive, supersonic flows of material lost from a FUor during an eruption are likely to have a profound effect on the ambient medium. It has been proposed that FUor eruptions are responsible for the Herbig-Haro flows (e.g. Dopita 1978, Reipurth 1985a, 1989a).

Herbig-Haro energy sources are, when known, mostly embedded objects, detectable only at infrared wavelengths. In those cases where the source is a visible star, it is generally highly reddened. This suggests that Herbig-Haro energy sources, as a class, belong among the youngest stars known.

The Herbig-Haro jets, the best collimated subset of the Herbig-Haro flows, provide an intriguing glimpse into the recent past of their driving sources. A number of jets show multiple bow shocks (e.g. Reipurth 1989b) testifying to intermittent mass-loss episodes in their driving sources. Viewed as a group, the energy sources also show considerable variety in terms of luminosity, type and visibility. It is here of particular interest that 3 out of the 8 FUors discussed in Section 1 are associated with Herbig-Haro objects. The close association between HH 57 and an emerging FUor has already been discussed in Section 1 (see Fig. 4). The embedded source L1551 IRS 5 produces a jet, seen in Fig. 10 in a red CCD image from Campbell et al. (1988), which has a large proper motion away from the embedded source (Neckel and Staude 1987, Campbell et al. 1988). Z CMa possesses a bipolar Herbig-Haro jet with a total extent of 3.6 pc and large radial velocities of up to –600 km/sec (Poetzel et al. 1989). It is important to note that no other class of young stars can boast a similar high percentage ($\simeq$ 30%) of association with Herbig-Haro objects.

If FUors initiate Herbig-Haro flows, then three questions arise. Firstly, why are the remaining FUors not associated with HH objects? A possible explanation is that the close environment of the star plays a role in focusing the outflow sufficiently to form shocks; note that the two deepest embedded FUors, V346 Nor and L1551 IRS 5, are both associated with Herbig-Haro objects. Secondly, why are most Herbig-Haro energy sources T Tauri stars? That is likely to be due to differences in the timescales of the phenomena involved. Dynamical timescales for Herbig-Haro flows are typically a few thousand years, whereas timescales for the decay of FUors are likely to be of the order of a few hundred years. At any given time, the majority of

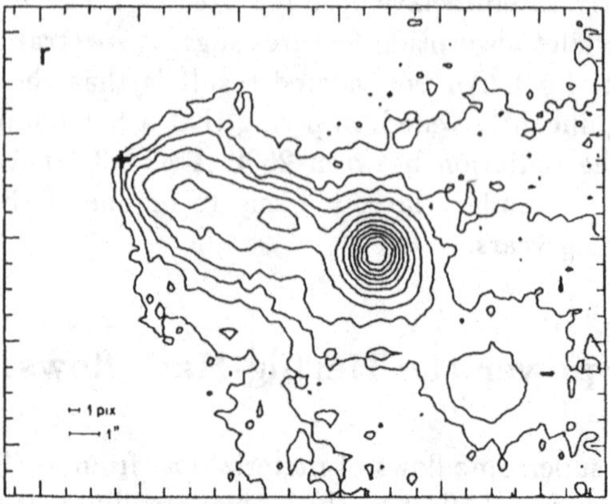

Fig. 10. The Herbig-Haro jet emanating from the embedded FUor L1551 IRS 5, whose position is indicated by a cross, as seen in a CCD image through a red broadband filter. From Campbell et al. (1988).

energy sources have therefore reverted to a stage in which the star again dominates the luminosity. Thirdly, since some Herbig-Haro jets are seen to emanate from T Tauri stars, which therefore still have winds sufficiently powerful and collimated to form shocks, what distinguishes these stars from the average T Tauri star? Essentially all known T Tauri-type Herbig-Haro energy sources belong to the rich emission line variety, Herbig's class 4 and 5 (Herbig 1962), suggesting that they are in a higher state of activity. Perhaps these stars are in EXor-like phases towards the end of their decay from an elevated FUor phase.

8 FUors and pre-main sequence evolution

Based on the number of known FUors, together with an estimate of the population of T Tauri stars that could give rise to FUors, and the time interval in which photographic surveys have been made, Herbig (1977) concluded that the FUor phenomenon is repetitive and occurs many times in each T Tauri star. Deriving a rough mean interval of 10 000 years between eruptions, and assuming a T Tauri phase of 10^6 years duration, Herbig suggested that a T Tauri star might experience as many as 100 eruptions. Estimates made in a different way by Hartmann and Kenyon (1985, 1987a) support Herbig's finding that FUors are repetitive.

It is probable that correct estimates of the FUor frequency cannot be derived by searching for brightenings only in visible stars. If FUors are caused by disk accretion, it seems likely that eruptions would be more frequent in the earliest phases of evolution, during which the stars are still embedded in molecular clouds and their disks are less depleted. Infrared sources could therefore be the most likely reservoir of future FUor eruptions. It is indeed intriguing that embedded Herbig-Haro energy sources have a very large spread in luminosity.

The early evolution of a young low-mass star may be cyclic, with occasional accretion events elevating the star into a FUor stage from which it gradually decays into a T Tauri phase. It has been proposed by a number of authors that even the T Tauri characteristics are powered by accretion. The true unperturbed underlying star may first be seen when the star reaches the weak-line T Tauri phase. The full evolutionary cycle would then be from FUor to T Tauri star to weak-line T Tauri star, except that in the earliest phases the star never reaches the weak-line state because of accretion, and in the late phases the massive accretion events seen as FUors become very rare or non-existent, and the star then basically cycles between the T Tauri and the weak-line T Tauri phase. This cyclic way of looking also at the later phases of pre-main sequence evolution can explain why some weak-line T Tauri stars still can have considerable circumstellar material.

If mass accretion rates reach $10^{-4} M_{\odot}\ yr^{-1}$ during a FUor event, which may last on average 10^2 years, then as much as 0.01 $M_{\odot}$ could be accreted in a single event. If the FUor phenomenon is repetitive, a substantial part of the stellar mass might be added from the disk during the early evolution of the star, as proposed by Mercer-Smith et al. (1984). Even the smaller accretion-rates occurring during the T Tauri phases may, because of the much longer duration of these phases, have an effect on the stellar mass. Accretion may thus alter the evolution in the HR diagram significantly (Hartmann and Kenyon 1990).

FU Orionis eruptions may also have an impact on our understanding of the early solar nebula. Herbig (1977) pointed out that dust grains in the inner regions of the disk could be significantly heated, and speculated that the chondrules found in chondritic meteorites might have formed from dust melted during FUor outbursts. Models of the early solar nebula (e.g. Wood and Morfill 1988) may need to take into account the effects of cyclic heating of disk material by FUor events.

Acknowledgements

I am grateful to Luiz Paulo Vaz for hospitality at the Institute of Physics and Astronomy/UFMG in Belo Horizonte, where this review was written, and to W. Seitter and H. Duerbeck, whose efforts to secure a visum enabled me to participate in this symposium.

References

Ambartsumian, V.A.: 1954, *Comm. Byurakan Obs.* **No. 13**

Ambartsumian, V.A.: 1971, *Astrofizika* **7**, 557 (Astrophysics, **7**, 331)

Andersson, C., Johansson, L.E.B., Winnberg, A., Goss, W.M.: 1979, *Astron. Astrophys.* **80**, 260

Basri, G., Bertout, C.: 1989, *Astrophys. J.* **341**, 340

Bastian, U., Mundt, R.: 1985, *Astron. Astrophys.* **144**, 57

Bertout, C., Krautter, J., Möllenhoff, C., Wolf, B.: 1977, *Astron. Astrophys.* **61**, 737

Bertout, C., Basri, G., Bouvier, J.: 1988, *Astrophys. J.* **330**, 350

Cabrit, S., Edwards, S., Strom, S.E., Strom, K.M.: 1990, in press

Campbell, B., Persson, S.E., Strom, S.E., Grasdalen, G.L.: 1988, *Astron. J.* **95**, 1173

Carr, J.S., Harvey, P.M., Lester, D.F.: 1987, *Astrophys. J.* **321**, L71

Chavarria-K., C.: 1977, *Astron. Astrophys.* **79**, L18

Chalonge, D., Divan, L., Mirzoyan, L.V.: 1982, *Astrofizika*, **18**, 263 (*Astrophysics* **18**, 263)

Clarke, C.J., Lin, D.N.C., Papaloizou, J.C.B.: 1989, *Monthly Notices Roy. Astron. Soc.* **236**, 495

Cohen, M.: 1973, *Monthly Notices Roy. Astron. Soc.* **161**, 105

Cohen, M.: 1975, *Monthly Notices Roy. Astron. Soc.* **173**, 279

Cohen, M.: 1980, *Astron. J.* **85**, 29

Cohen, M., Harvey, P.M., Schwartz, R.D.: 1985, *Astrophys. J.* **296**, 633

Cohen, M., Bieging, J.H.: 1986, *Astron. J.* **92**, 1396

Cohen, M., Dopita, M.A., Schwartz, R.D.: 1986, *Astrophys. J.* **302**, L55

Cohen, M., Schwartz, R.D.: 1987, *Astrophys. J.* **316**, 311

Cohen, M., Emerson, J.P., Beichman, C.A.: 1989, *Astrophys. J.* **339**, 455

Covino, E., Terranegra, L., Vittone, A.A., Russo, G.: 1984, *Astron. J.* **89**, 1868

Croswell, K., Hartmann, L., Avrett, E.H.: 1987, *Astrophys. J.* **312**, 227

Cudworth, K.M., Herbig, G.H.: 1979, *Astron. J.* **84**, 548

DeCampli, W.M.: 1981, *Astrophys. J.*, **244**, 124

Dopita, M.A.: 1978, *Astron. Astrophys.* **63**, 237

Duncan, D.K., Harlan, E.A., Herbig, G.H.: 1981, *Astron. J.* **86**, 1520

Elias, J.H.: 1978, *Astrophys. J.* **223**, 859

Ewald, R., Imhoff, C.L., Giampapa, M.S.: 1986, *In New Insights in Astrophysics*, ESA SP-263, p. 205

Finkenzeller, U., Jankovics, I.: 1984, *Astron. Astrophys. Suppl.* **57**, 285

Finkenzeller, U., Mundt, R.: 1984, *Astron. Astrophys. Suppl.* **55**, 109

Gahm, G.F., Welin, G.: 1972, IBVS No. 741

Gasparian, K.G., Ohanian, G.B.: 1989, *Inf. Bull. Var. Stars* **3327**

Goodrich, R.W.: 1987, *Publ. Astron. Soc. Pacific* **99**, 116

Götz, W.: 1980, *Mitt. Ver. St.* **8**, 143

Gottlieb, E.W., Liller, Wm.: 1978, *Astrophys. J.* **225**, 488

Graham, J.A.: 1983, *IAU Circ.* **3785**

Graham, J.A., Frogel, J.A.: 1985, *Astrophys. J.* **289**, 331

Grasdalen, G.: 1973, *Astrophys. J.* **182**, 781

Haro, G.: 1971, IBVS No. 565

Hartmann, L., Edwards, S., Avrett, E.H.: 1982, *Astrophys. J.* **261**, 279

Hartmann, L., Kenyon, S.J.: 1985, *Astrophys. J.* **299**, 462

Hartmann, L., Kenyon, S.J.: 1987a, *Astrophys. J.* **312**, 243

Hartmann, L., Kenyon, S.J.: 1987b, *Astrophys. J.* **322**, 393

Hartmann, L., Kenyon, S.J.: 1990, *Astrophys. J.* **349**, 190

Hartmann, L., Kenyon, S.J., Hewett, R., Edwards, S., Strom, K.M., Strom, S.E., Stauffer, J.R.: 1989, *Astrophys. J.* **338**, 1001

Herbig, G.H.: 1958, *Astrophys. J.* **128**, 259

Herbig, G.H.: 1960, *Astrophys. J. Suppl.* **4**, 337

Herbig, G.H.: 1962, *Advances in Astron. Astrophys.* **1**, 47

Herbig, G.H.: 1966, *Vistas in Astronomy* **8**, 109

Herbig, G.H.: 1977, *Astrophys. J.* **217**, 693

Herbig, G.H.: 1989, in ESO Workshop on *Low Mass Star Formation and Pre-Main Sequence Objects*, ed. B. Reipurth, p. 233

Ibragimov, M.A., Shevchenko, V.S.: 1987, *Astrophysics* **27**, 337

Kenyon, S.J., Hartmann, L., Hewett, R.: 1988, *Astrophys. J.* **325**, 231

Kenyon, S.J., Hartmann, L.: 1989, preprint

Kenyon, S.J., Hartmann, L.: 1990, *Astrophys. J.* **349**, 197

Kenyon, S.J., Hartmann, L., Imhoff, C.L., Cassatella, A.: 1989, *Astrophys. J.* **344**, 925

Kolotilov, E.A.: 1984, *Sov. Astron. Lett.* **9**, 324

Kolotilov, E.A.: 1987, *Sov. Astron. Lett.* **13**, 16

Kolotilov, E.A., Petrov, P.P.: 1983, *Sov. Astron. Lett.* **9**, 92

Kolotilov, E.A., Petrov, P.P.: 1985, *Sov. Astron. Lett.* **11**, 358

Kopatskaya, E.N.: 1984, *Astrophysics* **20**, 138

Koresko, C.D., Beckwith, S.V.W., Sargent, A.I.: 1989, *Astron. J.* **98**, 1394

Lada, C.J., Black, J.H.: 1976, *Astrophys. J.* **203**, L75

Larson, R.B.: 1983, *Rev. Mexicana Astron. Astrof.* **7**, 219

Lago, M.T.V.T.: 1979, Ph.D. thesis, Univ. of Sussex, Brighton

Lago, M.T.V.T.: 1984, *Monthly Notices Roy. Astron. Soc.* **210**, 323

Leinert, Ch., Haas, M.: 1987, *Astron. Astrophys.* **182**, L47

Levreault, R.M.: 1983, *Astrophys. J.*, 265, 855

Lin, D.N.C., Papaloizou, J.: 1985, in *Protostars and Planets II*, eds. D.C. Black and M.S. Matthews, Univ. of Arizona Press, Tucson, p. 981

Lo, K.Y., Bechis, K.D.: 1973, *Astrophys. J.* **185**, L71

Lo, K,Y., Bechis, K.D.: 1974, *Astrophys. J.* **190**, L125

Lynden-Bell, D., Pringle, J.E.: 1974, *Monthly Notices Roy. Astron. Soc.* **168**, 603

Mazzitelli, I.: 1989, in ESO Workshop on *Low Mass Star Formation and Pre-Main Sequence Objects*, ed. B. Reipurth, p. 433

Mendoza, E.E.: 1971, *Astrophys. J.*, **169**, L117

Mercer-Smith, J.A., Cameron, A.G.W., Epstein, R.I.: 1984, *Astrophys. J.* **279**, 363

Mirzoyan, L.V., Melikian, N.D., Natsvlishvili, R.Sh.: 1988, *Astrofizika* **28**, 540 (Astrophysics **28**, 320)

Moneti, A., Forrest, W.J., Pipher, J.L., Woodward, C.E.: 1988, *Astrophys. J.* **327**, 870

Mould, J.R., Hall, D.N.B., Ridgway, S.T., Hintzen, P., Aaronson, M.: 1978, *Astrophys. J.* **222**, L123

Mundt, R., Stocke, J., Strom, S.E., Strom, K.M., Andersson, E.R.: 1985, *Astrophys. J.* **297**, L41

Neckel, J., Staude, H.J.: 1987, *Astrophys. J.* **322**, L27

Pasquini, L., Reipurth, B.: 1990, in preparation

Paczynski, B.: 1976, *Qly. J. Roy. Astr. Soc.* **17**, 25

Parsamian, E.S., Gasparian, K.G.: 1987, *Astrofizika* **27**, 447 (Astrophysics **27**, 598)

Pringle, J.E.: 1989, *Monthly Notices Roy. Astron. Soc.* **236**, 107

Poetzel, R., Mundt, R., Ray, T.P.: 1989, *Astron. Astrophys.* **224**, L13

Pudritz, R.E., Norman, C.A.: 1983, *Astrophys. J.* **274**, 677

Pudritz, R.E., Norman, C.A.: 1986, *Astrophys. J.* **301**, 571

Reipurth, B.: 1985a, *Astron. Astrophys.* **143**, 435

Reipurth, B.: 1985b, in ESO-IRAM-Onsala Workshop on *(Sub)-millimeter Astronomy*, ed. P.A. Shaver and K. Kjär, p. 459

Reipurth, B.: 1989a, in ESO workshop on *Low Mass Star Formation and Pre-Main Sequence Objects*, ed. B. Reipurth, p. 249

Reipurth, B.: 1989b, Nature **340**, 42

Reipurth, B.: 1990, in preparation

Reipurth, B., Olberg, M., Booth, R.: 1990, in preparation

Sargent, A.I., Beckwith, S., Keene, J., Masson, C.: 1988, *Astrophys. J.*, **333**, 936

Schwartz, R.D., Jones, B.F., Sirk, M.: 1984, *Astron. J.* **89**, 1735

Schwartz, R.D., Snow, T.P.: 1972, *Astrophys. J.*, **177**, L85

Shakura, N.I., Sunyaev, R.A.: 1973, *Astron. Astrophys.* **24**, 337

Shanin, G.I.: 1979, *Sov. Astron.* **23**, 158

Simon, T., Joyce, R.R.: 1988, *Publ. Astron. Soc. Pacific* **100**, 1549

Smith, H.A., Thronson, H.A., Lada, C.J., Harper, D.A., Loewenstein, R.F., Smith, J.: 1982, *Astrophys. J.* **258**, 170

Snell, R.L., Loren, R.B., Plambeck, R.L.: 1980, *Astrophys. J.* **239**, L17

Stocke, J.T., Hartigan, P.M., Strom, S.E., Strom, K.M., Andersson, E.R., Hartmann, L.W., Kenyon, S.J.: 1988, *Astrophys. J. Suppl.* **68**, 229

Strom, K.M., Strom, S.E., Kenyon, S.J., Hartmann, L.: 1988, *Astron. J.* **95**, 534

Strom, S.E., Strom, K.M., Vrba, F.J.: 1976, *Astron. J.* **81**, 320

Wachmann, A.A.: 1939, *Beob. Zirk.* **21**, 12

Wachmann, A.A.: 1954, *Zeitschrift f. Astrophysik* **35**, 74

Welin, G.: 1971, *Astron. Astrophys.* **12**, 312

Welty, A.D., Strom, S.E., Strom, K.M., Hartmann, L., Kenyon, S.J., Grasdalen, G.L., Stauffer, J.R.: 1990, *Astrophys. J.* **349**, 328

Winnberg, A., Graham, D., Walmsley, C.M., Booth, R.S.: 1981, *Astron. Astrophys.* **93**, 79

Wood, J.A., Morfill, G.E.: 1988, in *Meteorites and the Early Solar System*, eds. J.F. Kerridge, M.S. Matthews, University of Arizona Press, p. 329

GAHM: Just a comment on your note on the unusual strength of Ba II at 6496 A. This feature is very strong in red supergiants, so may be this is in line with your picture of line formation in a low pressure disk.

SUBFUORS IN ORION ASSOCIATION

L.G.GASPARIAN, A.S.MELKONIAN, G.B.OHANIAN,
E.S.PARSAMIAN
Byurakan Astrophysical Observatory
Armenian Academy of Sciences, USSR

ABSTRACT. Results of spectral and photometric observations of Sugano star = V1143 Ori in brightness minimum and near it are given. Emission lines of HI, CaII, FeI, TiI, TiII and TiO absorption bands are detected. The appearing envelope is observed also in minimum. A brightness increase of Shanalstar V1118 Ori is observed. In its spectrum lines of HI, CaII, FeI, FeII are found, testifying to formation of an envelope.

During the last ten years data on T Tau type stars which have fuor-like outburst-subfuors were obtained. Two such stars were found in Orion association. These are V1118 Ori [1] and V 1143 Ori [2,3]. In recent years, in Byurakan observatory observations of these stars have been made.

1. OBSERVATIONS

Observational material were obtained by the SAO 6m telescope, 2.6m Cassegrain and 40" Schmidt system telescopes of Byurakan observatory. Observations by 6m telescope were made with scanner with dispersion D=1.8 A/canal, resolution is ≃ 4 A. Observations by 2.6m telescope were made on UAGS spectrograph with inverse dispersion of 101 A/mm, resolution is ≃ 4 A.

2. V1118 ORI

The first outburst of this star was observed in 1982 [1]. We have no information about the rise time. Star was in maximum for four months,the decrease lasted about a year [4]. We have no data about the existence of any spectral observation. The next brightness increase was observed by us in December 1988 [5]. The star spectrum during second outburst on 11.1.89 observed by 6m telescope, is given on Figure 1. Spectrum is moderate intensity with emission lines

L. V. Mirzoyan et al. (eds.), Flare Stars in Star Clusters, Associations and the Solar Vicinity, 253–256.

of HI, CaII, FeI, FeII. Presence of P Cyg type profiles in Balmer series high members are suspected.

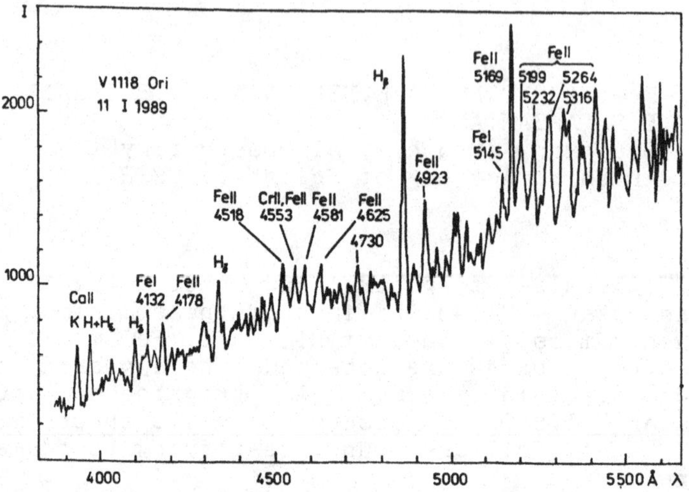

Figure 1. The spectrum of V1118 Ori on 11.1.89.

3. V1143 ORI

The first outburst occured at the end of 1982. Sugano's observations made it possible to define the rise time, which lasted about three months. For more than four months the star was in the maximum with brightness variation about $0^m.5$. Duration of the flare was about 18 months [4,6]. The next outburst occured in 1984 [6] and 1986 [7]. One more brightening perhaps was observed in 1988. According to Natsvlishvili, on April 7 1988, the magnitude of the star reached m_{pg}=15.5 [8].At the same time, by the observations in Crimea observatory on April 13 and 14, the star had m_{pg}=17.3-17.4 [9]. In December 1988 the star was near minimum with m_{pg}=17.6, and in october 1989 m_{pg}~15.5.

Spectral observations of V1143 Ori, carried out in February 1983 during the first outburst, exhibit strong emission line spectrum with the lines of HI, CaII, FeI, FeII, TiII, CrII [10,11]. The spectral type on the decrease stage was about K7-M0 [12]. The spectrograms during the second outburst on decrease stage in 1987-1989 are dominated by emission lines of HI, CaII, FeI, FeII, TiII and by strong absorbtion bands of TiO. Spectral type changed from K7-M0 to M2. The variation of hidrogen lines was observed. Thus, spectral features, both in the beginning of the outburst and after, show that this is a T Tau type star with weak emission in minima.

According to Herbig [10], the spectrum of V1143 Ori near the maximal brightness was similar to subfuors DR Tau and VY Tau. In addition this star shows fast flare with Δm=0.6. So the stage of subfuors must take place in some T Tau stars with weak emission, which shows flare activity also.

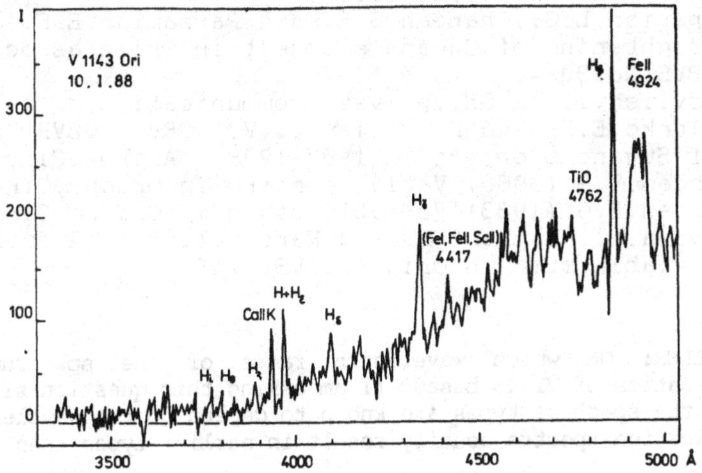

Figure 2. The spectrum of V1143 Ori on 10.01.1988.

4.CONCLUSION

The photometric data of V1143 Oriand V1118 Ori during the outburst show ultraviolet excess which reduce with the decrease of flare [4]. Spectra of these stars during the outburst and after it are similar to T Tau stars with moderate intensity. The solution of the problem, why some T Tau stars become fuors, while the others under the same conditions become only subfuors (perhaps not realised fuors), will help us to understand one of the evolution stage of T Tau stars.

REFERENCES

1. Marsden,B.G.(1984) 'Two variables in the Orion nebula', Circ.IAU,No.3924.
2. Marsden,B.G.(1983)'Variable star in Orion',Circ.IAU,No.3763.

3. Natsvlishvili,R.Sh.(1984) 'New non-stable stars in Orion' IBVS, 2565.
4. Parsamian,E.S. and Gasparian,L.G.(1987) 'On fuorlike variations of the Orion association stars',Astrofizika, 27,447-458.
5. Gasparian,L.G. and Ohanian,G.B.(1989) 'New brightening of Shanal's object in the Orion association',IBVS,3327.
6. Mirzoyan,L.V.,Melikian,N.D. and Natsvlishvili,R.Sh.(1988) 'Unusual light curves of stellar flares in Orion' Astrofizika, 28,540-550.
7. Gasparian,L.G.,Ohanian,G.B. and Parsamian,E.S.(1987) 'New brightening of Sugano's object in Orion association', IBVS,No.3024.
8. Natsvlishvili,R.Sh.,private communication.
9. Pavlenko,E.P. and Prokofieva,V.V.(1988) 'UBVR photometry of Sugano's object in 1987-1988', Astron.Circ.No.1530,7
10. Marsden,B.G.(1983)'Variable stars in Orion',Circ.IAU.No.37
11. Marsden,B.G.(1983)'Variable stars in Orion',Circ.IAU,No.37
12. Chavira,E.,Peimbert,M. and Haro,G.(1985) 'A fuor-like new variable star in Orion',IBVS,2746.

APPENZELLER: On which wavelength range of the spectrum was the K7 classification of V1143 based? (I am asking this question since in T Tau spectra the spectral types are known to depend on the wavelength. Blue classification spectra usually result in earlier types than red spectrograms.)

PARSAMIAN: Classification was made by TiO absorption bands in the long wavelength part of the spectrum.

SLOW FLARES AND ERUPTIVE PHENOMENA IN EARLY STAGES OF STELLAR EVOLUTION

M.A.IBRAHIMOV, V.S.SHEVCHENKO
Astronomical Institute
Uzbek Academy of Sciences
USSR

ABSTRACT. Correlations between different eruptive phenomena (fast and slow flares, large-scale eruptive variations up to fuor phenomenon) in stellar instability during stage of evolution are analysed A conclusion is drawn on the difference in natures of eruptive phenomena lasting some ten minutes and slower variations lasting more than one day.

Analysis of the sample of about 40000 photoelectric observations of the T Tauri stars and related objects, obtained in 1978-88 at the Mount Majdanak, revealed the absence of eruptive phenomena (including the brightness drops or "anti- flares") on a time scale of 10 to 100 days. This fact initiated us to analyse the time distributions of the faster flares and a slower brightness variations.

Five common type of the eruptive phenomena at the early stages of stellar evolution can be characterized by amplitude, duration and morphology.

1) The rapid flares of the UV Ceti-type stars in the solar vicinity and of the flare stars in stellar clusters and associations. The total number of these stars is ower 1400 [1].

2) The slow flares of the flare stars mentioned above. 50 slow flares of 38 flare stars were mentioned in the publications before 1986 [1-6]. 11 "relatively slow" flares with the rising time of about 15-20 minutes were reported by Hojaev [7] for Tau T3-association and by Tsvetkova [8] for γ Cyg region.

3) The flares of T Tau-type stars, which originally were reported by Haro and Chavira [9] for VY Ori in 1953. 17 flares of 13 T Tau stars were detected by Hojaev [10] in Tau T3- assotiation during the 750 hours of observations. The evidences of the flares were reported by Zajtseva [11] for DF Tau, Shevchenko [12] for T Tau, Furtig and Wenzel [13] for RW Aur. Inspite of the finding by Hojaev [10] and

L. V. Mirzoyan et al. (eds.), Flare Stars in Star Clusters, Associations and the Solar Vicinity, 257–260.

Parsamian [14], one may conclude that the flares of T Tau stars are quite rare events: about one flare over 100 hours of observations [11,12]. Our analysis of the Majdanak data sample suggest that this estimate may be too optimistic. The "flares" of DF Tau with the duration of 3 to 5 days may be caused by modulation of light of the rotating star by a hot spot [15,16]. Apparently, the periodical "flares" of other objects, e.g.TZ Ori [5], may be of the same origin. It seems that all the available data do not provided the information about the flares of T Tau stars longer than 1 day.

Table 1. Classification of the eruptive phenomena according to the specific properties of the process

Type of events (examples)	Number of stars	Rising time (day)	Overall duration (day)	Amplit. Δm_{pg}, ΔB
1. Rapid flares (UV Cet type and flare stars in aggregates).	1400	$10^{-5}<t<10^{-2}$	$10^{-4}-10^{-2}$	≤ 4.5
2. Slow flares (the same objects).	40	$10^{-2}<t<10^{-1}$	0.1 - 0.5	≤ 2
3. T Tau star flares (T,DF Tau RW Aur,VY Ori)	~ 40	$10^{-2}<t<$ 1?	<0.5 - 5	$\leq 0.5-1$
?	-	-	10 - 100	-
4. Subfuor erruptions (Shanal and Sugano objects, EX Lup, Z CMa. DR,RY,UZ,VY Tau)	~ 10	~ 100 - 150	$\sim 10^3$	$\leq 3-3.5$
5. Fuor eruptions (FU Ori,V1057,V1515, V1735 Cyg, V350 Cep	5	~ 250 - 7000	$\sim 7\times 10^4$	$\leq$ 4-6

4) The fuor-like flares of the "subfuors" [17]. The light curves of these objects resemble the small scale fuor eruption with the decay time of about 2.5 yrs. The premaximum luminosity of the objects differs of that of fuors and their spectrum at maximum light is also differs of that of fuors.

5) The large scale eruptive phenomena of the FU Ori type stars (fuors). Three objects of this type is known up today

and 2 or 3 objects are suspected to be the fuors.

In the Table 1 we have collected and classified all the eruptive phenomena according to the numbers of events and the specific properties of the process (rising time, duration and amplitude).

The absence of the eruptive phenomena with the duration of 10 to 100 days cannot be explained by any observational selection. The mean property of Majdanak data sample show that the flarelike eruptive phenomena with the duration of 10 to 100 days can more easily be observed. Therefore it is difficult to explain the existing of such rift.

REFERENCES

1. Mirzoyan,L.V. and Ohanian,G.B.(1986) 'Flare stars in stellar clusters and associations ',in L.V.Mirzoyan (ed.), Flare stars and Related Objects, Armenian Academy of Sciences, Yerevan, pp.68-78.
2. Parsamian, E.S. (1980) 'Slow flares in aggregates. I', Astrofizika 16, 87-96.
3. Parsamian, E.S. (1980) 'Slow flares in aggregates. II', Astrofizika 16, 231-241.
4. Mirzoyan,L.V. (1981) Instability and stellar evolution, V.A.Ambartsumian (ed.), Armenian Academy of Sciences, Yerevan.
5. Parsamian, E.S. and H.A.Pogosian(1986) 'Peculiar stars in the Orion association region ', in L.V.Mirzoyan (ed.), Flare stars and Related Objects, Armenian Academy of Sciences, Yerevan, pp. 130-134.
6. Shevchenko V.S.(1986) 'Slow flares on flare dwarfs in the solar neighbourhud ', in L.V.Mirzoyan (ed.), Flare stars and Related Objects, Armenian Academy of Sciences, Yerevan, pp.135-137.
7. Hojaev,A.S. (1986) 'Flare stars in the TDC', in L.V.Mirzoyan (ed.), Flare stars and Related Objects, Armenian Academy of Sciences, Yerevan, pp. 91-100.
8. Tsvetkova, K. (1986) 'Flare stars in the region near γ Cygni', in L.V.Mirzoyan (ed.), Flare stars and Related Objects, Armenian Academy of Sciences, Yerevan, pp.84-90.
9. Haro, G. and Chavira, E. (1965) Vistas in Astronomy 7, 89.
10. Hojaev, A.S. (1987) ' Flares of Orion population variables in the association Taurus T3', Astrofizika, 27, 207-217.
11. Zajtseva,G.V. (1980) 'Flares of the T Tauri-type stars', in L.V.Mirzoyan (ed.),Flare Stars,Fuors and Herbig-Haro Objects, Armenian Academy of Sciences ,Yerevan, pp.61-68.

12. Shevchenko V.S. (1980) 'A comparative analysis of T Tau and red dwarf flares'. in L.V.Mirzoyan (ed.),Flare Stars,Fuors and Herbig-Haro Objects, Armenian Academy of Sciences ,Yerevan, pp.69-75.
13. Furtig, W. and Wenzel. W. (1963) Mitt. Veranderl.Sterne 2, 11.
14. Parsamian. E.S. (1982) Philosophy Doct. Thesis, Yerevan.
15. Bertout, C.,Basri,G. and Bouvier, J.(1988) Astrophys.J. 330, 350.
16. Shevchenko V.S.(1986) 'Quasi-periodic activity T Tau, the Herbig Ae/Be stars and fuors', in L.V.Mirzoyan (ed.). Flare stars and Related Objects. Armenian Academy of Sciences, Yerevan, pp.230-244.
17. Parsamian, E.S. and Gasparian L.G.(1987) 'On fuorlike variations of the Orion association stars',Astrofizika, 27, 447-458.

A NEW FUOR ?

H.M.TOVMASSIAN, R.KH.HOVHANNESSIAN, R.A.EPREMIAN
Byurakan Astrophysical Observatory
Armenian Academy of Sciences, USSR

During observations made in August and November 1987 with the ultraviolet telescope "Glasar" [1] installed at the module "Kwant" of the "Mir" Space Station a very blue star HD 269665 = Sk $-68^{\circ}110$ in the region of the LMC was detected.

The data obtained by ground based observations [2] have showed that this star might belong to spectral types O – B2. In case it is a main sequence star its color m(1640)-V should be about -4^m. On Fig.1 two photographs of the observed region taken by us (right one) and by Sanduleak [3] in photographic rays are presented.

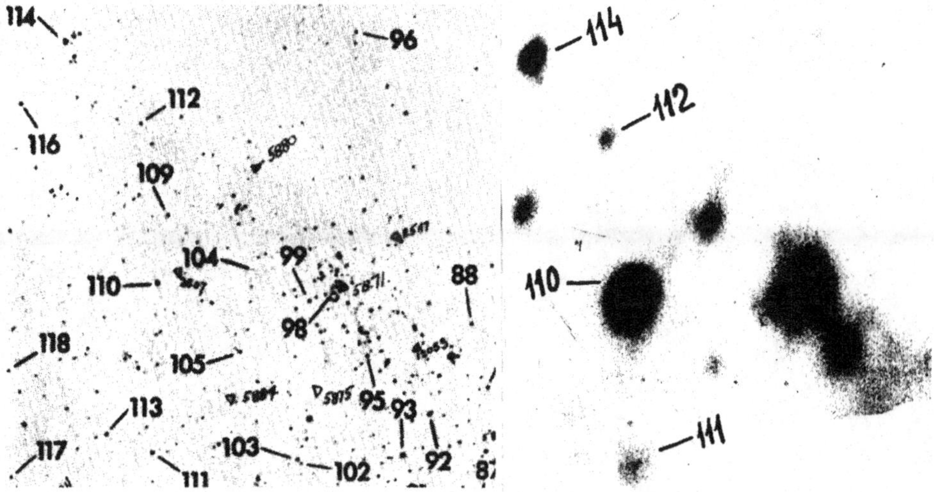

Figure 1.

L. V. Mirzoyan et al. (eds.), Flare Stars in Star Clusters, Associations and the Solar Vicinity, 261–262.

The photometric data on a few in the field stars measured by TD-1 sattellite [4] were used for determination of the stellar magnitude of HD 269665. It is equal to $5^m.0 \pm 0^m.1$. The V magnitude of this star was $\sim 11^m.2$ [2]. Thus, its colour was anusually blue. $m(1640)-V=-6^m.2$. Therefore we suggest that the star was in an enhanced state at the time of our observations, and thus may be a fuor. It is remarkable that this star was not recorded by observations with TD-1 in 1972-1974 though much fainter stars up to $9^m.0$ at 1565 A have been recorded.

REFERENCES

1. Tovmassian, H.M., Khodjayantz, Yu.M., Krmoyan, M.N., Kashin, A.L., Zakharian, A.Z., Hovhanessian, R.Kh., Mkrtchian, M.A., Tovmassian, G.H Huguenin, D., Bootov, V.V., Romanenko, Yu.V., Laveikin, A.J. and Alexandrov, A.P. (1988), '"Glazar" - the orbital ultraviolet telescope', Pisma v A.J., 14, 291-295.
2. Shobbrook, R.R., and Visvanathan, N. (1987), 'The distance to the LMC from uvbyβ photometry of B stars' Monthly Notices Roy. Astron. Soc., 225, 947-960.
3. Sanduleak, N. (1969) 'A deep objective - prism survey for LMC member'. Cerro-Tololo Interamerican Observatory, Contr. No.89.
4. Thompson, G.J., Nandy, K., Jamar, C., Monfiles, A., Houziaux, L., Carnochan, D.J. and Wilson, R. (1978) 'Catalogue of stellar ultraviolet fluxes', The Science Research Council pp. 1-449.

CORRELATION BETWEEN FLARE STARS AND OTHER POPULATIONS IN YOUNG CLUSTERS AND STAR FORMING REGIONS

V. S. SHEVCHENKO, S. D. YAKUBOV.
Astronomical Institut
Uzbek Academy of Sciences
USSR

ABSTRACT. Ratio between number OB-stars and that of comparatively bright flare stars in star forming reions are discussed.

This short note we present in addition to prof. L.V. Mirzoyan review (this issue). We want to draw attention to a very simple ratio between number of OB-stars and that of comparatively bright flare stars(f.s.) in clusters and Star Forming Regions (SFR).

The two following well-known arguments should be reminded.

1) Observed and calculated numbers of f.s.and various relations strongly depend on observational selection.

2) The only reliable fact is that the luminosity of bright f.s. in Orion cluster is much higher than that in Pleiades and older clusters.

The table displays an attempt to find such rations between f.s. and other population in SFR, which should depend from the selection in the least way.

The table consists of two parts. The first one contains some SFR properties. Characteristics of SFR are derived from our results (Shevchenko, 1979, 1989). The distance module and Av are obtained from our photoelectric five colour photometry made on mt. Maidanak. When estimating the masses and sizes of SFR we took into account the observations in CO-line and other data. For comparative analysis we used only f.s. grouping reliable connected whit SFR. Here we do not discuss the more older Hyades and Coma clusters.

Ten years ago we suggested Pleiades and TDC be considered as the united complex (Stalbovski, Shevchenko, 1980). The table also contains the data for Pleiades cluster separately. The Orion f.s. grouping is studied in detail, while the data for NGC 7000/IC 5070 and RSF 2 Cyg due to the large distance and high Av are not available. The data for other regions are too poor.

N OB is the number of OB-stars (spectral type to B9).

N AG is the number of stars with Mv from +1 to +5 (spectral type A, F, and early G).

N Ae/Be is the number of Ae/Be Herbig stars.

L. V. Mirzoyan et al. (eds.), Flare Stars in Star Clusters, Associations and the Solar Vicinity, 263–265.

TABLE. FLARE STARS IN CLUSTERS AND STAR FORMING REGIONS (RSF)

RSF NAME / DATA	Pleiad. Cluster	1 Tau Pleiad. + TDC	1 Ori (M42)	3 Mon NGC 2264	1 Oph	2 Cyg NGC 6910	4 Cyg NGC 7000	1 Cep NGC 7023
STAR FORMING REGIONS PROPERTIES								
Distance d(pc)	135	130	430	730	160	1100	675	300
Mass ($M_\odot$)	4.10^2	3.10^4	2.10^5	5.10^4 ?	10^4	6.10^5	2.10^5	10^3
Size (pc)	10	10X30	20X60	10X20	10	50	40	2
Av(mag)	0.1	1.5	0.5	0.3	1.5	2.0	2.0	2.0
N OB(Mv<+1)	15	18	88	23	16	40?	>25	1
N AG (+1<Mv<+5)	154	171	210	>150	>80			6
N Ae/Be	-	2	4	2	1	7?	20?	1
N ea		120	540	200	100:	>40	210	2?
ASSOCIATIONS AND CLUSTERS FLARE STARS PROPERTIES — THE OBSERVATIONAL DATA								
Mv f.s. for lim 17.5V	+11.3 M4V	+10.0 M2V	+8.1 K7V	+7.4 K5V	+9.0 M1V	+5.0 G4V	+5.9 K0V	+7.6 K6V
Monitor. time	3175	4112	1406	105	43	324	938	?
Sp of brightest f.s.	K2		K0	K0	K			
Mv of br. f.s.	+6.4	+6.4	<4.5	+5.1	+6.8?			+6.6?
N f.s.(total)	546	648	491	42	4	16	67	10
N f.s.Mv<+7.5	32	35	180	40	1?		55	2
CALCULATIONS AND RATIONS								
Total N f.s.	994	1526	1471	442		129	403	
N f.s. 7.5/ /N OB	2.1	1.9	2.1	1.7			<2.3	2
N f.s. 7.5/ /N ea	0.3	0.3	0.2			0.2		

N ea is number of emission stars including T-Tau stars.

An information on f.s. is collected in the second part of the Table First of all we emphasize that it was to be expected the observational selection strongly influence all the f.s. data. Mv for limit 17.5V is the calculated meaning Mv for limited value V=17.5. Sp lim is the corresponding spectral type for Gamma Cyg=NGC 6910 region is G4V. F.s. of such early spectral type are unknown. All discovered f.s. of this region probably belong to the solar visinity.

Sp of brightest f.s., Mv of brightest f.s., N f.s.- are the observational data.

N f.s. (Mv<+7.5) is the total number of observed f.s. more bright than 7.5 Mv.

This number is more or less free from the observational selection for all regions exepting two SFR in Cygnus. Total N f.s. is the calculated total number of f.s. by Ambartsumian method.

The mean value of (N f.s. 7.5/N OB) is aproximately 2. This ratio is the most important as it is connected with luminosity function of young aggregates and shows the fundamental properties of all low mass stars in early stage of stellar evolution.

Besides, we note the following.

1) By increasing of observational limiting magnitude to 23U the discovery of large number new f.s. in region Gamma Cyg and NGC 7000 may be expected.

2) It is interesting to continue the f.s. observations in Rho Oph Dark Cloud region.

3) The f.s. are forming not only in large aggregates but in small SFR like NGC 7023 region. There are a lot of samples of similar compact SFR. After molecular cloud disintegration in small aggregates f.s. becomes a typical solar vicinity f.s.

REFERENCES

Shevchenko V.S. (1979), Astronomicheski Jurnal 57, p.1162.

Shevchenko V.S. (1989), Ae/Be Herbig Stars, FAN, Tashkent.

Stalbovski O.I.,and Shevchenko V.S. (1980), Flare stars, fuors, and Herbig-Haro objects, ed. L.V.Mirzoyan, Academy Sci. Arm. SSR, Yerevan, p.116.

THE RESULTS OF SPECTRAL OBSERVATIONS OF COLLIMATED OUTFLOWS

T.Yu.MAGAKYAN, T.A.MOVSESSIAN
Byurakan Astrophysical Observatory
Armenian Academy of Sciences, USSR

ABSTRACT. The results of spectral investigations of collimated outflows from young stellar objects, including HL/XZ Tau, L723, Bernes 48, RNO 43N, CoKu Tau/1, obtained on 6-m telescope BTA with two-dimensional photon counting system in 1986-1988 are presented.

1. INTRODUCTION

The discovery of collimated optical outflows from young stars [1] and their subsequent investigations (see review [2]) can be considered as a major step in understanding of early evolution of low and intermediate mass stars. The studies of this phenomenon are in continuation.

In this paper the main results of observations of collimated outflows on 6-m telescope BTA of the SAO USSR in 1986-1988 are presented. All spectra were obtained with UAGS spectrograph in prime focus and two - dimensional TV photon counting system "KVANT". For 15 objects (mostly for the first time) the radial velocity fields were obtained and different parameters estimated. Among observed objects are RNO 43N, 1548C27,HL/XZ Tau and HH30 region, L723, Bernes 48, CoKu Tau/1, GGD 34, HH105 and others.

2. RESULTS AND DISCUSSION

2.1. HL/XZ Tau

Even first images of optical outflows in this region [2] revealed rather complicated picture. Subsequent spectral observations confirmed this impression. We obtained the spectra of north-western and southern jets. On the maps of their radial velocity field the regions with splitted spectral lines, i. e. with two radial velocity components were

L. V. Mirzoyan et al. (eds.), Flare Stars in Star Clusters, Associations and the Solar Vicinity, 267–270.

found. This was interpreted as projection of two flows on the line of sight [3]. Recent data [4] confirmed this suggestion.

2.2. L723

The presence of optical flow in this cloud was suspected in [5]. Our spectrum of nebulous object in cloud shows, that it is really the collimated outflow from non-observable in optics source, which can be IRAS PSC 19155+1906. Large angular distance between them indicates, that outflow, practically without expansion, goes through whole dark cloud and only after leaving it becomes visible [6].

2.3. Bernes 48

The presence of collimated outflow in this star with cometary nebula was suspected in [7]. On our new spectrum this outflow is visible directly. Furthermore, the new Herbig-Haro object, consisted of several condensations with different radial velocities, was found nearby (Fig.1). The unusual feature is that HH object lies not on the axis of directed outflow and

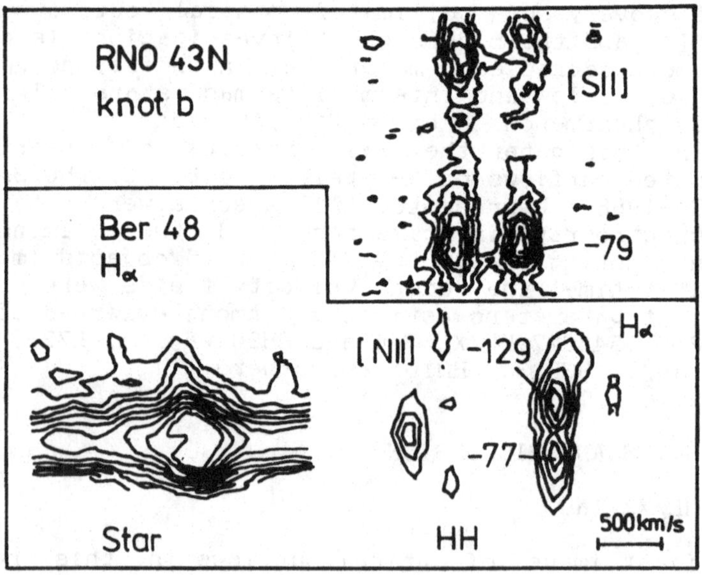

Figure 1. The contours of emission lines in RNO 43N and Bernes 48 (star and HH-object) spectra

cometary nebula, as in most cases. But no other source capable to excite HH object except Bernes 48 star is known in this region.

2.4. RNO 43

In this complex of HH-objects [8] we succeeded to obtain spectra of several knots in the region known as RNO 43N. The detailed study of this field is in progress. But it is worth to mention the interesting "triangular" shape of Hα - line contours of "d" knot. It is in complete agreement with theoretical models for bow-shock [9]. The direction of shock and the object's proper motions are opposite.

2.5. CoKu Tau/1

This rather faint star in CZ/DD Tau field [10] also appeared to be a source of collimated outflow. Most interesting circumstance is the very low luminosity of central star ($L < 0.001L_{\odot}$) . Probably with some other objects it belongs to a separate subgroup [11].

3. CONCLUSION

From the results of observations of collimated outflows and Herbig-Haro objects it may be noted, first of all, that phenomenon of anisotropic outflow from young stars possesses the large diversity of observational effects and physical characteristics. It is nessesary to mention the great capabilities of 6-m telescope for such studies. We are planning to continue this observational program on BTA with modern equipment.

REFERENCES

1. Mundt, R.(1983) 'Jets from PMS stars', Astrophys.J., 265, L71-L75
2. Mundt,R., Brugel,E.W., and Buhrke,T.(1987) 'Jets from young stars', Astrophys.J., 319, 275-303
3. Magakyan,T.Yu., Movsessian,T.A., Afanass'ev,V.L. and Burenkov,A.N.(1989) 'The spectral investigation of collimated flows in HL/XZ Tau region', Sov.Astr.Lett., Vol.15,2.124-130
4. Mundt,R.(1987) 'Flows and jets from young stars', Mitt. Astron. Ges., 70, 100-115
5. Vrba,F.J., Luginbuhl,C.B., Strom,S.E., Strom,K.M. and Heyer,M.H.(1986) 'An optical imaging and polarimetric study of the Lynds and Barnard 335 molecular outflow regions', Astron.J., 92, 633-636

6. Movsessian,T.A.(1989) 'Jet in the darc cloud L723', Sov. Astr.Lett., Vol.15, 2, 131-134
7. Magakyan,T.Yu., Khachikian,E.Ye.(1988) 'Bernes 48 - the case of optically observed anisotropic outflow', Astrofizika,Vol.28,139-147
8. Ray,T.P.(1987)'CCD observations of jets from young stars' Astron.Astrophys., 171, 145-151
9. Raga,A.C., Bohm,K.-H.(1985)'Predicted long-slit, high resolution emission-line profils from interstellar bow shocks', Astrophys. J. Suppl., 58, 201-224
10. Movsessian,T.A., Magakyan,T.Yu.(1989) 'CoKu Tau/1 - a new object with optical bipolar flow', Astrofizika,in press
11. Movsessian,T.A,(1989) 'The statistical analysis of young star optical and infrared luminosities', in this volume

SZECSENYI-NAGY: We were shown a lot of detailed spectral intensity maps Are you able to calibrate these in intensity for real quantitative studies?

MAGAKIAN: Now it is not possible, but we can do it in our future observations with the new multi-lens spectrograph.

H_2O MEGAMASER IN ORION KL

L.I.MATVEYENKO
Space Research Institute
Profsojuznaja 84
Moscow V 485
U.S.S.R.

ABSTRACT. The region of the H_2O megamaser emission in Orion KL had been studied by VLBI method. The structure is a chain of the compact components oriented under the angle $X \cong -80^o$. The size of the components is $\leqslant 0.1$ A.U., the brightness temperature $T_b \cong 10^{17}$ K, the linewidth of each component is $\Delta f \leqslant 7$ kHz. The emission has linear polarization $P \geqslant 80$ % and position angle is changed by 9.2 degree / A.U. The velocity of the components has a gradient 0.41 km/sec A.U. The masers are unsaturated, the kinetic temperature $T_k \leq 120$ K. The structure corresponds to an expanding protoplanet rings, radius of which is equal R = 6 A.U. The rotation and the expanding velocities are equal 5 and 3.8 km/s accordingly. Mass of a protostar is $\sim 0.7\ M_\odot$.

INTRODUCTION. In a number gas-dust complexes of our Galaxy it has been processed a star formation, which accompanied by a hydrocsil and a water vapor maser emission. A H_2O megamaser outbursts observed in a two cases : in a object W 49 and Orion KL,(Matveyenko, (1981)). The active period of the megamaser emission had been observed in Orion KL nebula at 25.09.1979 to end 1987. The maximum flux density was equal to $F = 8 \times 10^6$ Jy and linewidth $\Delta f \cong 40$ kHz (Abraham et al.,(1986), Matveyenko et al., (1988), Garay et al.,(1989)).

OBSERVATIONS AND RESULTS. To study this fenomena one needs measuring a structure of the emission region. But the maser sources are very compact and can be measured by VLBI method only. The H_2O maser outburst in Orion KL observed from a moment, when emission began to grow (Matveyenko , (1981), Matveyenko et al.,(1982)), and continued each year at different interferometers. The Fig 1 shows results of this observations. The megamaser region has a complex st-

L. V. Mirzoyan et al. (eds.), Flare Stars in Star Clusters, Associations and the Solar Vicinity, 271–274.

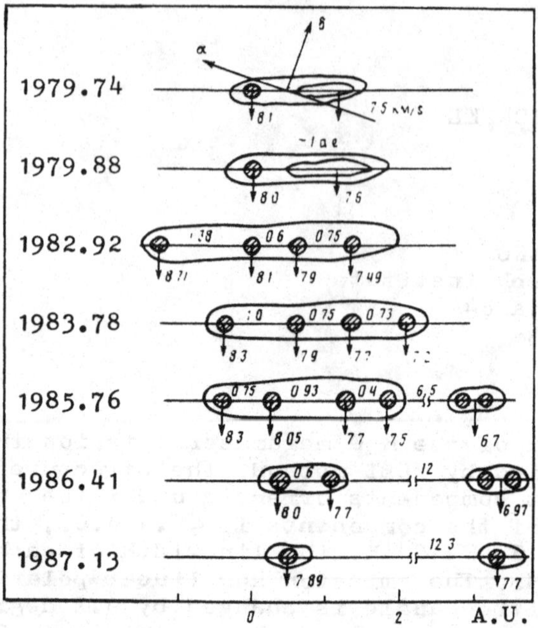

Fig. 1. Relative position of the compact components and velocity.

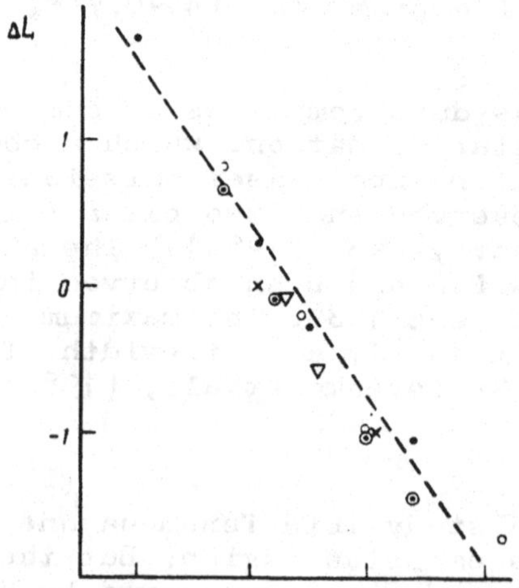

Fig. 2. Reletive position (ΔL, A.U.) plotted against velocity (V^{-2}, $km^{-2}\ s^{2}$).

ructure, which consists of a few compact components distributed along a direction $X = -79^{\circ}$. The size of the components is equal to $\leqslant 0.1$ A.U. A distance between them is equal to 0.4 - 1.4 A.U. One can observe 3 - 4 compact brightness components inside ~ 2.5 A.U.. The velocity of the components are $V = 7.2 - 8.7$ km/s and correlated with their location. The gradient velocity is equal to $dV/dL = 0.41 \pm 0.05$ km/s A.U. . The linewidth of each component is equal to $\Delta f = 10$ kHz, or $\Delta V = 0.135$ km/s. The linewidth corrected by the differential velocity inside the emission region would be $\Delta f = 7$ kHz. A brightness temperature of the compact components is $T_b = 10^{16-17}$ K and changes gradually. Correlation of the T_b with time of the components is not observed. Such a high brightness temperature is defined by a high directivity of emission $\Omega < 10^{-4}$, which is determined by a geometry of the maser source. The shape of source looks like a cylinder by length -l and diameter -d. In this case $\Omega = (d/l)^2$. For $d = 0.1$ A.U. $l \geqslant 10$ A.U.. A chain of the components corresponds to the parallel cylinders, which are observed along its axis, or a protoplanet rings, a plane of which is parallel to a beam of view. A radius of the rings are equal to $R = 6$ A.U.

MODEL OF THE MEGAMASER. The masers are unsaturated or saturated only partialy. Thus, the variability of the flux density would be $I/I_o = e^{-\Delta\tau}$. The optical depth is $\tau = \ln T_b / (T_B + T_S)$, where T_B and T_S correspond to the background and spontaneous emission. $T_B + T_S \cong 100$ K for Orion KL and $\tau = 35$. A few percent variation of τ changes the maser emission strongly.
The dependence of velocity of the components from the location (Fig 2) corresponds to Kepler's low $\Delta L \sim V^{-2}$. In this case $R_K = V\Delta R/2\Delta V$. The velocity of region, where the maser is located is $V = 5.5$ km/s and $R = 3.5$ A.U. A discrepancy R and R_K can be explained by rings expansion and rotation $V_{exp} = 3.8$ km/s and $V_{rot} = 5$ km/s.
The slope of line Fig 2 corresponds to $\alpha \cong MG = 570$ and $M = 0.7\,M_{\odot}$, ($G = 6.67 \times 10^{-8}$ Din$\cdot$cm$^2\cdot$g^{-2}).
The temperature of the rings from the linewidth is equal to $T_k = \Delta f^2 \tau / 2 = 120$ K. The low kinetic temperature proposes IR pumping. The IR source can be a protostar. The IR emission relative to the protoplanet rings would be isotropic and a polarization of the maser emission would be strong linear (Western and Watson (1983)). The emission of the compact components is linear polarized $P > 80$ % (Matveyenko and Romanov (1983)). Position angle of the polarization is changed in dependence with its location. The gradient is equal to $dX/dL = 9.2$ deg./A.U. and corresponds to the expanding rings with a magnetic field, (Matveyenko et al., (1988)).

A non-homogenious destribution of H_2O molecules in the rings would change optical depth in the tangential directions of rings and a maser emission would be variable with rotation. For the change of the emission on the order τ would be change by 2.3 or 6% of τ. This model explaned why correlation of the variability emission of the components is absent. The number of H_2O molecules in colomn is equal to $\sim 10^{18}$ cm^{-2} and the density is $\sim 10^4$ cm^3.

REFERENCES.

Abraham Z., Vilas Boas J.W.S., Giampo del L.F.,(1986),Astron. and Astrophys., V.167, 311.
Matveyenko L.I. (1981), Pisma Astronomich. Zh., V.7, 100.
Matveyenko L.I., Moran J.M., Genzel R.,(1982), Sov.Astron. Lett., V 8, 382.
Matveyenko L.I., Romanov A.M., Kogan L.R., Moiseev I.G., Sorochenko R.L., Timofeev V.V.,(1983), Sov.Astron.Lett, 9, 240.
Matveyenko L.I., Graham D., Diamond Ph., (1988), Pisma Astronom. Zh., V.14, 1101.
Garay Guido, Moran J.M., Haschick A.D., (1989), Astroph. J. V.338, 244.
Western L.R., Watson W.D.,(1983), Astroph.J., V.275, 195.

FLARES OF RADIO LINE EMISSION H_2O IN ORI A AND W49N

I.V.GOSACHINSKIJ, N.A.YUDAEVA
Special Astrophysical Observatory
Academy of Sciences, USSR
R.A.KANDALIAN, F.S.NAZARETIAN and V.A.SANAMIAN
Byurakan Astrophysical Observatory
Armenian Academy of Sciences, USSR

ABSTRACT. The results of observations of H_2O maser sources Ori A and W49N at 1.35cm made the RATAN-600 radio telescope from June 1981 till December 1988 are presented. The light curves of these sources showed several outbursts.Duration of the strongest outbursts for the Ori A and W49N are approximately 1 and 1.5yr respectively. The maximum of the total luminosity (assuming isotropic emission) of these sources for the velocity range ±8km.s^{-1} is egual to 1×10^{47}ph.s^{-1} and 2×10^{49}ph.s^{-1} respectively. The time behaviour of the water maser source in W49N during more than 7yr is monitored for the first time.

1. INTRODUCTION

The flares of maser radio line emission H_2O in Ori A and W49N were discussed in few papers [1-6]. Here we shall discuss the results of our observations carried out from June 1981 till December 1988 in detail.

2. THE RESULTS OF OBSERVATIONS

In Fig.1a the time variations of the peak flux density and line width on half power level are shown for the feature in the Ori A having radial velocity 7.5km.s^{-1}. In Fig.1b the integral intensity variations for the velocity interval from -0.7 till +15 km.s^{-1} are shown. From these Figures one can conclude that:

1.The line flux density variability and its width are not correlate.

2.The integral intensity of Ori A and the peak flux density are changed practically in the same way.However, the integral intensity in the period from June 1985 till April 1986 is approximatly constant, while the peak flux density

L. V. Mirzoyan et al. (eds.), Flare Stars in Star Clusters, Associations and the Solar Vicinity, 275–278.

at that time is decreasing. This difference is due to by the appearing of strong component in the source spectrum with the radial velocity 6.9km.s^{-1}. The dependence of the flux density from line width for the feature 7.5km.s^{-1} is shown in Fig.1c. This dependence does not correspond neither the saturated maser amplification nor the unsaturated case. It should be noted also, that during the period of our observations the drift of the radial velocity 7.5km.s^{-1} was ±0.3km.s^{-1}.

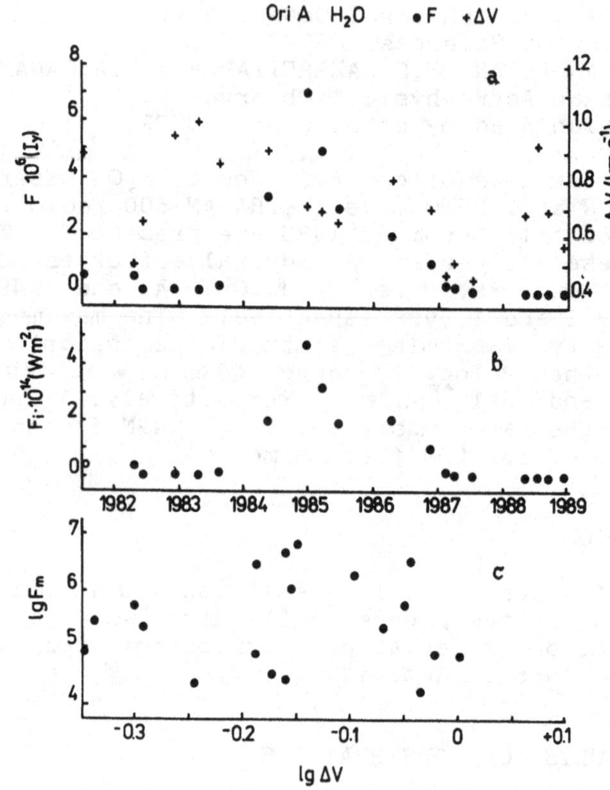

Figure 1. a) The time variation of the peak flux density (the left ordinate) and line width (the right ordinate) of the feature with radial velocity 7.5km.s^{-1}.

b) The integral intensity variation for the velocity interval $-0.7 \div +15$km.s^{-1}.

c) The dependence of the flux density from line width for feature on 7.5km.s^{-1}.

The results of observation of W49N are presented in the

fig.2a,b. Fig.2a is demonstrated the time variations of the peak flux density and line width of the feature on 10.3km.s^{-1} The drift of the radial velocity during our observations was ±0.6km.s^{-1}. The variability of the source integral intensity for the velocity interval from 3 till 23km.s^{-1} is shown in the Fig.2b.

The following characteristic properties we can pick out from observed dependences:

1. The light curve for the radial velocity 10.3km.s^{-1} consists of main maximum (March 1985) around which weaker maximums are observed.
2. The increasing and decreasing time of main maximum are approximately 2yr. If the flare duration is considered to be the time during which the intensity of radiation decreases twice, then it is egual to approximately 1.5yr. The light curve of the integral radiation has almost the same shape. However, it is reached its maximum in the end of June 1985.

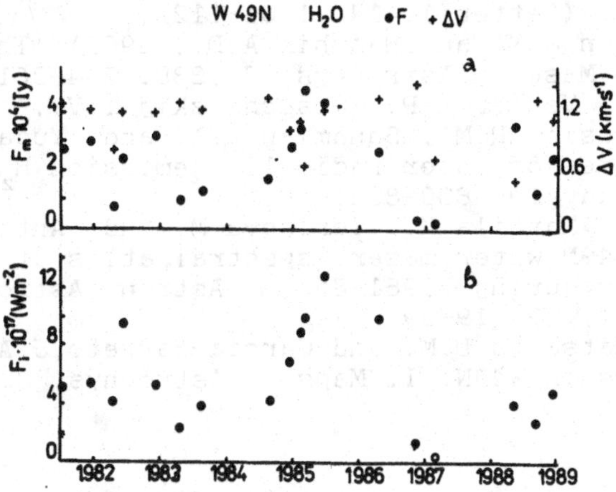

Figure 2. a) Same as in fig. 1a for the W49N on the radial velocity 10.3km.s^{-1}.
b) Same as in fig. 1b for the W49N for the velocity interval +3 ÷ +23km.s^{-1}.

3. CONCLUSIONS

VLBI observations of Ori A [3] and W49N [7] in the H_2O line show that in these objects there are a numberof maser sources clusters which are not resolved by a single radio telescope. Hence the results of our observations undergo of blending effects of the features. But the blending effect can introduce uncertainty when the properties of the concrete feature of the spectrum are discussing. Therfore study of the time variations of the source integral characteristic has certain importance, since it gives the notion about the source as a whole and about the contribution of components to the integral radiation.

REFERENCES

1. Abramian.L.E..Venger.A.P..Gosachinskij,I.V.,Kandalian,R.A Martirosian,R.M., Nazaretian,F.S., Sanamian,V.A. and Yudaeva,N.A.,(1987) 'The variability of H_2O maser emission at 1.35cm wavelenght .I.Observational data', Soviet Astrophys.Investigations, SAO of the USSR Acad. of Sci., 24, 85-92.
2. Abraham Z.,Vilas Boas J.W.S. and del Ciampo L.F.,(1986) 'The time behaviour of the 8km.s^{-1} water source in Orion', Astron. Astrophys, 167,311-314.
3. Matveenko L.I.,Craham D. and Diamond Ph.,(1988).'Region of the outburst of H_2O maser emission in Orion-KL', Soviet Astr. (Letters). 14, 1101-1122.
4. Garay G., Moran J.M. and Haschik A.D.,(1989) 'The Orion-KL Super Water Maser', Astrophys. J.,338, 244-261.
5. Abramian L.E., Venger A.P., Gosachinskij I.V., Kandalian R.A.,Martirosian R.M., Sanamian V.A. and Yudaeva N.A. (1983) 'Flares of maser radio line emission H_2O in W49N ',Astrofyzika, 19, 830-834.
6. Liljestrom T., Mattila K., Toriseva M. and Anttila R., (1989) ' W49N water maser: spectral atlas of time variability during 1981-85 ', Astron. Astrophys. Suppl. Ser., 79, 19-39.
7. Walker R.C.,Matsakis D.M. and Garcia-Barreto J.A.,(1982), ' H_2O masers in W49N. I. Maps ', Astrophys. J., 255, 128-142.

ON THE CONNECTION BETWEEN THE IRAS POINT SOURCES AND GALACTIC NONSTABLE OBJECTS

A.L.GYULBUDAGHIAN
Byurakan Astrophysical Observatory,
Armenian Academy of Sciences, USSR
R.D.SCHWARTZ
University of Missouri, St. Louis, USA
L.F.RODRIGUEZ
Instituto de Astronomia, UNAM, Mexico

ABSTRACT. The connection between the IRAS point sources catalog and following objects is studied: 1. H_2O masers, 2. Herbig - Haro objects, 3. Cometary nebulae, 4. Compact nebulae,situated in the large nebulae, and 5.Dark globules. It is shown, that more than 50% of the 1,2,3,and 4 type objects are connected with the IRAS sources. More than 10 new H_2O masers, connected with IRAS sources are found, one of them having very unusual properties. More than 10 dark globules are found in the Cep OB2 association, which are connected with IRAS sources. One of these sources consists of 4 components, forming a Trapezium-like configuration.

1. INTRODUCTION

It is well known, that many IRAS point sources [1] are connected with star-formation regions. We were interested to reveal the connection between IRAS point sources [2] and unstable young objects of our Galaxy (mainly situaited near star-formation regions): 1. HH objects, 2. cometary nebulae, 3. Trapezium-type tight systems, 4. compact nebulae,situated in the large nebulae, 5. dark globules.

We were looking for IRAS point sources in the circle with radius 2', centered on the object. The IRAS sources, satisfied two following restrictions, were choosed: 1.$F_{100} >$ 10 Jn, 2. the flux for IRAS source is given not only for 100 μm,but also for at least one of other wavelengths (12,25 or 60 μm).

2. UNSTABLE OBJECTS

In Table 1 the results of this search are given. In the first column the types of objects are given. HHL are the

L. V. Mirzoyan et al. (eds.), Flare Stars in Star Clusters, Associations and the Solar Vicinity, 279–282.

objects, found in Byurakan, CLN - all known so far cometary and cometary-like nebulae, Tr- Trapezium-type tight systems, consisted mainly of T Tau type stars [3], c.n. - compact nebulae, embedded in the large nebulae [1]. In the second column the number of objects is given; in the third column - the number of objects, taking into account their doubling (if two objects accidentally fall in the same circle with 2' radius, we count them as a single object). Corresponding to [4], the IR colours of IRAS sources can be used for the determination of the type of objects, connected with IRAS sources. In dependence of the values of three quantities

$$R_{12} = \lg\frac{F_{25}\cdot 12}{F_{12}\cdot 25}, \quad R_{23} = \lg\frac{F_{60}\cdot 25}{F_{25}\cdot 60}, \quad R_{34} = \lg\frac{F_{100}\cdot 60}{F_{60}\cdot 100}$$

the IRAS sources in [4] are devided into three types: 1. probably connected with H_2O masers, 2. probably connected with T Tau type stars, 3. probably connected with embedded not evolved objects. In the fourth column of Table 1 is the percentage of IRAS sources of type 1, in the fifth- of type 2, and in the sixth - of type 3, in the seventh column - the percentage of sources, not involved in this classification, in the eight - the number of objects, connected with IRAS sources are given.

TABLE 1. Nonstable objects and IRAS sources.

1	2	3	4	5	6	7	8
HHL	79	70	59	10	14	17	32
CLN	176	165	49	23	14	14	82
Tr	12	12	37	0	36	27	11
c.n.	10	9	71	0	0	29	9

2.1 HHL OBJECTS

The percentage of type 1 sources is rather high. It is known, that 13 HHL objects are connected with H_2O masers, and 10 of these objects are also connected with IRAS sources (all these sources are of type 1) [2]. Hence we can conclude, that the percentage of HHL objects connected with IRAS sources is much higher among these 13 objects, than among all HHL objects.

2.2. CLN OBJECTS

These objects have the highest percentage of type 2 sources (many CLN objects are connected with T Tau type stars).

2.3.TR AND C.N.OBJECTS

From the Table 1 we can conclude, that the percentage of connected with IRAS sources Trapezium-like tight systems, as well as compact nebulae is very high, and it is remarkable, that all the classified IRAS sources, connected with the compact nebulae, are of type 1.

3. IRAS SOURCES IN THE ASSOCIATION CEP OB2

We choosed IRAS sources, which are projected on this association and are of type 1. There are 21 such sources. The search of H_2O masers, connected with these sources, revealed 11 such masers. The percentage of masers is several times higher, than among the field IRAS sources. Especially interesting is the maser, connected with IRAS 21144+5430, this maser is unusual, because the ratio $F_{H_2O}/F_{100\mu m}$ is at least 10 times larger than for already known objects.

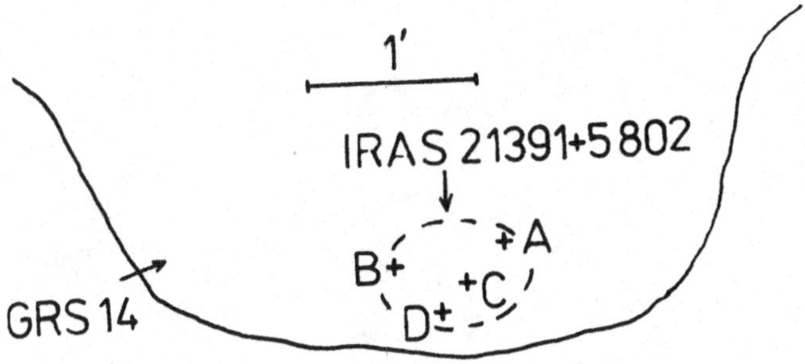

Figure 1. IRAS source with 4 components in GRS 14.

There are at least three radial systems of dark globules in this association. They contain at least 36 globules , and 11 of them are connected with IRAS sources. Especially interesting is the globule GRS 14. The IR observations of IRAS 21391+5802, connected with this globule, revealed an unusual situation. This source consists of 4 components, which form the Trapezium-type configuration (Fig. 1). The near-IR colours of these components correspond to A-F class stars. Hence the birth of the stars in Trapezium-type groups can take place not only among the massive early type stars (as in Kleinmann-Low nebula), but also among the not so bright and not so massive stars. In this globule a CO molecular outflow and water vapor maser are also revealed.

REFERENCES

1. IRAS Point Source catalog (1985), Washington, D.C.
2. Gyulbudaghian,A.L. and Schwartz,R.D. (1989) 'On the connection between infrared sources with the objects situated in the star-formation regions', Soobsh. Byur. Obs., in press.
3. Gyulbudaghian,A.L.(1986), 'Trapezium type close systems containing T Tauri type stars'.in L.V.Mirzoyan (ed.), Flare Stars and Related Objects, Publishing House of the Academy of Sciences ,Yerevan, 250-254.
4. Wouterloot,J. and Walmsley,C.(1986), 'H_2O masers associated with IRAS sources in region of star formation' Astron. Astrophys., 168, 237-247.

THE STATISTICAL ANALYSIS OF THE OPTICAL AND INFRARED LUMINOSITIES OF YOUNG STARS

T.A.MOVSESSIAN
Byurakan Astrophysical Observatory
Armenian Academy of Sciences, USSR

ABSTRACT. The infrared IRAS data for supposed sources of optical outflows and young stars in Taurus-Auriga complex are studied. It is shown that there exists a group of stars with extremely low optical and infrared luminosities. It's suggested, that they are the stars with anisotropic sourface activity.

1. INTRODUCTION

The study of non-stable phenomena, connected with young stellar objects, always attracted attention of astronomers. This interest was further increased after discovery of high-velocity outflows from these objects [1]. Herbig-Haro-objects, collimated jets, cometary nebulae and molecular flows, all of which being the consequences of anisotropic outflows from young stars are good indicators of violent non-stable processes in star formation regions.

During the investigation of optical collimated outflows our attention was attracted by sources with very low optical luminosity ($L<0.001L_{\odot}$)[1-3]. It was usually suggested, that extremely low luminocity of these objects is due to the fact that outflow is perpendicular to the line of sight and radiation of the source is absorbed in the plane of circumstellar disk. The model with thick dust disk is universally adopted now. According to this, such objects must have comparatively strong emission in far infrared.

In this paper a new approach to the problem of anisotropic outflows from young stars based on the analysis of the their optical and IRAS luminosities is used.

L. V. Mirzoyan et al. (eds.), Flare Stars in Star Clusters, Associations and the Solar Vicinity, 283–286.

2. RESULTS

In this study optical and IRAS data for 23 suggested excitation sources of HH objects from Cohen and Schwartz list [4] and 33 T Tau stars from Taurus-Auriga complex were included.

Factor analysis of luminosities in different spectral regions confirmed, that continuous emission of young stars can be interpreted as a sum of photospheric and envelope (T ~ 50K) emissions.

IRAS colours for anisotropic outflow sources and for T Tau stars show the marked differences between them. This is in good agreement with the data of other authors [5] and can be conditioned by different evolutionary stages of these objects.

From the analysis of IRAS luminosities was found that objects with low optical luminosities have also rather low luminosities in far infrared region.

On Fig.1 for the sources of optical outflows the IRAS - luminosities and shortwave ($\lambda < 7\mu$) luminosities diagram is presented. A group of low luminosity sources, which

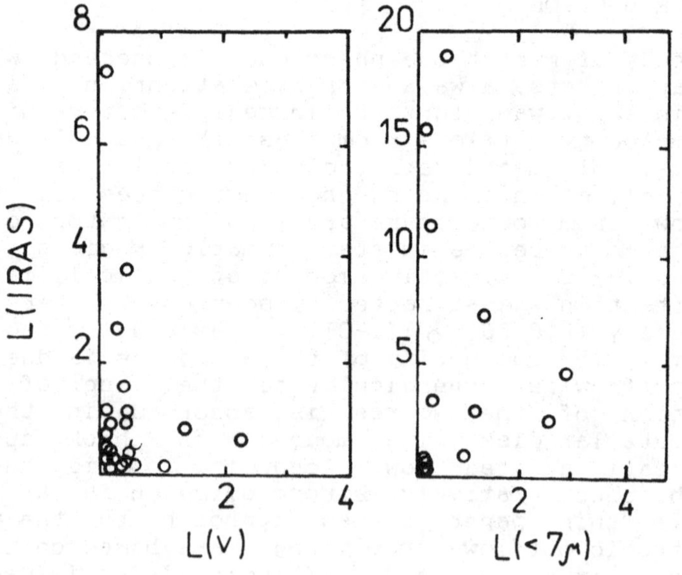

Figure 1. The sources of optical outflows (left), T Tau stars (right).

stands out from general tendency of increase of far infrared emission as shortwave emission decreases, can be easily distinguished. On Fig.1 the dependence of L(IRAS) from L(V) for T Tau stars is presented. The same picture can by seen here, but in this case the group of faint stars isn't separated so clearly, and there are some intermediate luminosity stars.

3. CONCLUSION

Thus a conclusion can be made that there exist a group of objects with extremely low shortwave luminosities, which cannot be explained by the presence of thick dust disks around stars. One can think that these stars are deeply embedded in dark cloud, but this assumption cannot be true for the sources of jets because jets have very low surface brightness and can't be seen through the cloud. May be the lack of so sharply detached group for the T Tau stars can be due to their different immersion in clouds.

It is worth to note that whole detached group on Fig.1 consists from sources, immediately connected with optical jets (HH30, DG Tau, FS TauB, Th28). And these outflows are rather strong: for example, HH30 outflow stretches on great distance (0.1pc)[6], and the Th28 outflow velocity exceeds 300km/s [7].

The existence of such a group of objects, at our opinion, is an evidence in favour of the idea that collimated outflows are not the consequence of outer anisotropy, connected with the presence of dust disks, but are the consequence of anisotropic activity of stars itself on early evolutionary stages.

REFERENCES

1. Mundt,R., Brugel,E.W., and Buhrke,T.(1987) 'Jets from young stars', Astrophys.J., 319, 275-303
2. Movsessian,T.A., Magakyan,T.Yu.(1989) 'CoKu Tau/1 - a new object with optical bipolar flow',Astrofizika, in press
3. Magakyan,T.Yu., Movsessian,T.A., Afanass'ev,V.L. and Burenkov, A.N. (1989) 'The spectral investigation of collimated flows in HL/XZ Tau region', Sov.Astr.Lett., Vol.15, 124-130
4. Cohen,M. and Schwartz,R.D.(1987) 'IRAS observations of the exciting stars of Herbig-Haro objects', Astrophys.J., 316,311-322
5. Berrilli,F., Ceccarelli,C., Liseau,R., Saraceno,P. and Spinoglio,L.(1989) 'The evolutionary status of young stellar mass loss driving sources as derived from IRAS

observations',Mon.Not.R.Astr.Soc., 237, 1-15
6. Mundt,R.(1987) 'Flows and jets from young stars', Mitt.Astron.Ges., 70, 100-115
7. Krautter,J.(1986) 'Th28: a new bipolar Herbig-aro jet', Astron.Astrophys., 161, 195-200

FAR INFRARED EMISSION FROM POST-T-TAURI STARS

Vassiliki Tsikoudi
University of Ioannina
Physics Department
45110 Ioannina
Greece

ABSTRACT. We have investigated a large sample of faint Post-T-Tauri stars located far from dark cloud complexes, for far-infrared emission and infrared excess, using the IRAS data-base. ∿12% of the stars studied here were detected by IRAS at 12 μm. Their 12 μm luminosities L_{12} are in the range of $4x10^{30}$ - $6x10^{33}$ erg/sec and correlate well with the stars' bolometric luminosities (L_{bol}).

IRAS has detected emission from the photospheres of the stars studied here and for a few of them it may have observed far-IR excess.

1. INTRODUCTION

Post-T-Tauri stars are young stars, <150 million years of age, in the pre-main sequence phase of their evolution; they are more evolved than T-Tauri stars. Many P-T-Tauri stars show CaII and Hα emission lines as well as Li absorption lines, in their spectra; these features appear weaker here than in T-Tauri stars.

A very extensive and detailed study of eighty such stars was done by Lindroos (1986), in the optical wavelength region. Here we investigate the far irfrared emission and possible IR excess of these 80 stars, using the IRAS data-base.

T-Tauri stars are strong sources of infrared emission and show IR excess, which is attributed to cold dust surrounding the star. P-T-Tauri stars, on the other hand, do not exhibit appreciable IR excess, as the scanty observations at 1 μm, show. The stars studied here are mainly F, G and K spectral type and are physical companions to B and A primaries, with which they form wide pairs. About thirty such systems were detected by IRAS; here we consider only the detections (eleven of them) for which the infrared fluxes can be attributed to the late-type secondary, i.e. to the P-T-Tauri star.

2. OBSERVATIONS

The Infrared Astronomical Satellite (IRAS) scanned the entire sky (98%)

L. V. Mirzoyan et al. (eds.), Flare Stars in Star Clusters, Associations and the Solar Vicinity, 287–292.

between 8 and 120 μm, with peaks at 12, 25, 60 and 100 μm. The IRAS survey of the IR sky was extremely reliable and very sensitive. Point sources detected during the survey mode are given in the Point Source Catalogue (PSC 2; 1987). In addition to the sky survey, the infrared satellite performed a series of pointed observations over selected areas of the sky, which were three to five times more sensitive than the survey mode and resulted in extending the IRAS detection threshold for point sources. Sources detected during pointed observations appear in the Serendipitous Survey Catalogue (SSC; 1986).

We searched the PSC 2 and the SSC at the positions of the eighty P-T-Tauri stars, for positional coincidences with the IRAS-detected sources. For stars with positive detections (in the PSC 2) we proceeded with coadding the multiple survey scans in order to enhance flux sensitivity. The fluxes given here are 10% better than those in the PSC 2; The flux errors are 0.03 - 0.05 Jy.

3. RESULTS AND DISCUSSION

IRAS has detected eleven P-T-Tauri stars at 12 μm and a few at 25 μm. For these stars it was possible to obtain 12 μm fluxes (f_{12}) and in turn estimate their 12 μm magnitutes (m_{12}) through the expression, $m_{12} = 4.03 - 2.5 \log f_{12}$ (IRAS Explanatory Supplement, 1985). Table I gives the IRAS-detected sources (columns 1 and 2), together with their far IR fluxes (columns 3 and 4), magnitudes (column 5) and luminosities (column 6). In column 7 the ratio of 12 μm luminosity (L_{12}) to the star's total luminosity (L_{bol}) is shown. In column 8 a measure of the IR excess is estimated.

The L_{12} of the detected stars is in the range of 4×10^{30} - 6×10^{33} erg/sec and comprise only 0.01-1% of the star's bolometric luminosity. A good correlation is found to exist between L_{12} and L_{bol}, as seen in Fig. 1. The two stars, SAO 196352 and 218788, deviate considerably from the mean relationship; they are not included in the least squares fit, which is indicated by the line drawn through the data points. The correlation coefficient of this relationship is 0.96.

In order to investigate whether the detected far IR emission of the P-T-Tauri stars is solely due to the stellar photosphere, we compare it to that of normal stars (stars with no IR excess), of the same spectral type. A large number of normal stars has been studied extensively by Waters et al. (1987), at the far infrared. They found a very tight color-color relationship to exist between B-V and V-12 of all main-sequence stars (from O to K spectral type) with no IR excess or other peculiarities. In terms of an equation this relationship is best described by:

$$V-12 = 0.05 + 3.13\,(B-V) - 1.26\,(B-V)^2 + 0.29\,(B-V)^3 + 0.16\,(B-V)^4$$

Making use of the above expression the predicted V-12 and m_{12} were obtained for the detected P-T-Tauri stars. The predicted values were in turn compared with the corresponding observed quantities. The last column in Table I gives the deviation of the observed 12 μm magnitude from that expected of a normal photosphere. As it can be seen, there are several stars for which the $m_{12}(obs) - m_{12}(pred)$ is considerably larger than the errors associated with the estimated quantities; such large deviations would tend to indicate the presence of excess emission. The stars that deviate the most from the normal stellar situation are SAO 196352, 218788 and 207208 with $\Delta m(12\mu m)>2.0$ mag. The first two stars were also seen (on Fig.1) to emit more radiation at 12 μm than their bolometric luminosity would tend to indicate.

It is concluded that for the P-T-Tauri stars studied here, the far IR emission observed by IRAS is photospheric in origin with the exception of a few stars for which there is an apparent excess emission. This IR excess tends to indicate the presence of an additional emitting component which could be material surrounding the star.

The IRAS data were obtained and partly processed at Rutherford Appleton Laboratory (RAL), England. I wish to thank the scientific and technical staff at RAL, and especially Drs.G.Bromage and B.Stewart, for their assistance and useful discussions during my visit there.

REFERENCES

Beichman, C.A., Neugebauer, G., Habing, H.J., Clegg, P.E. and Chester, T.J., editors of the IRAS Catalogs and Atlases, Explanatory Supplement 1985, JPL Publ. No. D-1855, Internal Document.

IRAS Point Source Catalog - 2, 1987; Joint IRAS Science Working Group (Washington D.C., U.S. Goverment Printing Office).

IRAS Sevendipitous Survey Catalog - Explanatory Supplement 1986, Kleinmann, S.G. Cutri, R.M., Young, E.T., Low., F.J. and Gillett, F.C.

Lindroos, K.P., 1986, Astron. Astroph. 156, 223.

Waters, L.B.F., Coté, J. and Aumann, H.H., 1987, Astron. Astroph. 172,225

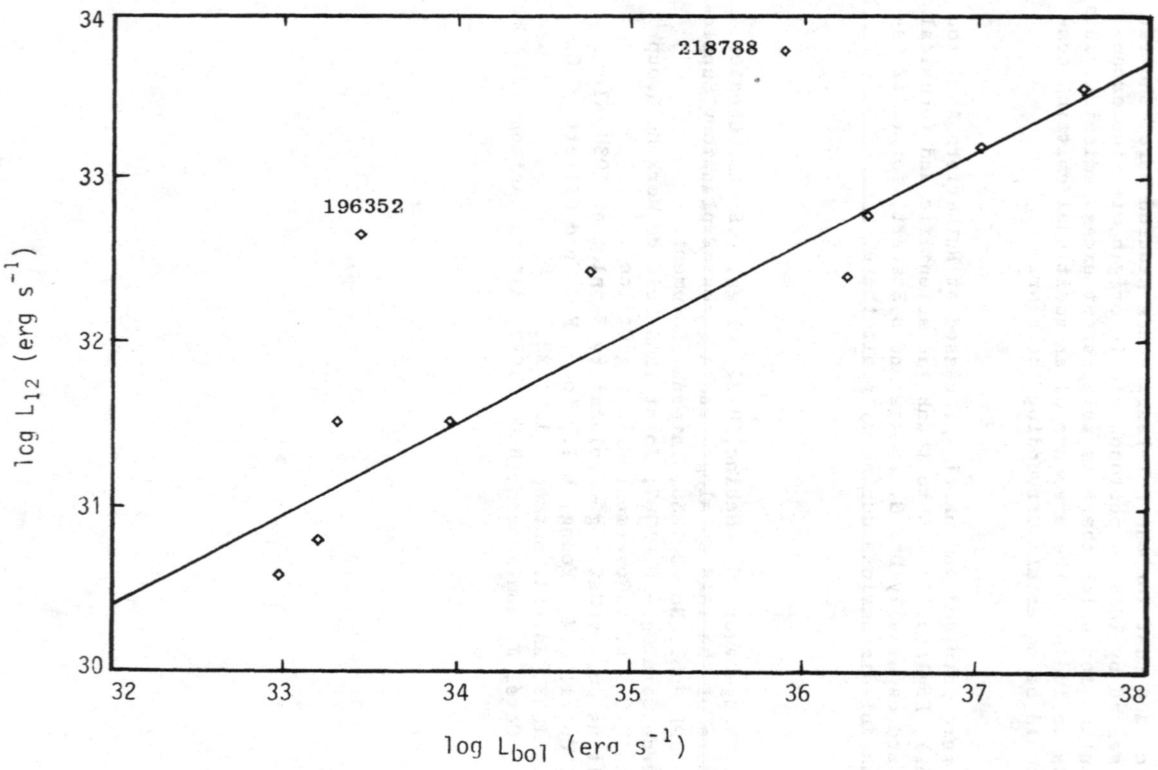

Fig. 1: Plot of 12 μm luminosity L_{12} vs bolometric luminosity L_{bol} for IRAS-detected Post-T-Tauri stars. The straight line shows the best least squares fit, not including SAO 196352 and 218788.

TABLE I

Far-infrared properties of IRAS-detected P-T-Tauri stars

IRAS source name	SAO name	f_{12} Jy	f_{25} Jy	m_{12} mag	$\log L_{12}$ [erg s^{-1}]	L_{12}/L_{bol} x10^{-2}	Δm (12 μm) $m_{obs}-m_{pred}$
00075+1052	91750	0.17	-	5.95	31.52	1.6	-1.62
04195+2530	76573	0.18	-	5.89	31.52	0.36	-0.15
05325-0602	132301	0.44	-	4.92	33.56	0.01	-1.02
05557-3517	196352	0.17	-	5.95	32.66	17.0	-4.74
07314-4905	218788	0.31	-	5.30	33.78	0.77	-2.74
07367-2641	174199	0.52	0.36	4.74	32.40	0.01	0.23
10056+1212	98967	0.28	-	5.53	30.81	0.41	0.13
12272-1614	157323	0.25	-	5.54	30.59	0.42	-1.00
S12516-5653	240367	1.00	0.64	4.03	32.78	0.02	-0.40
15568-3815	207208	0.37	-	5.10	32.43	0.47	-2.10
16025-1940	159682	1.75	0.39	3.42	33.20	0.02	-1.24

Notes: Columns 1 and 2 give the IRAS-source name and the name of the associated star as it appears in the SAO catalogue. Columns 3 and 4 give the 12 and 25 μm fluxes in Jy. Columns 5 and 6 show the estimated (using the fluxes in column 3) 12 μm magnitudes and luminosities. Column 7 gives the ratio of the 12 μm luminosity to the star's total luminosity. Column 8 shows the difference between the IRAS (observed) 12 μm magnitude and that of a stellar photosphere (predicted).

MIRZOYAN: How can a B star be a post-T Tauri star?

TSIKOUDI: The stars are above the main sequence. The early type stars could be called post-Herbig Ae,Be stars.

MIRZOYAN: How is the age determined?

TSIKOUDI: From Stromgren photometry of the primary main sequence component and calibration over standard main sequence models.

GAHM: In the survey the separation between the components were not very large. Are the IRAS sources separated?

RESIDUAL EXTINCTION EFFECTS IN SPECTRA OF NEWLY FORMED STARS

Jacek Krełowski[1], Jacek Papaj[1], Walter Wegner[2]
[1]Institute of Astronomy, N. Copernicus University
ul. Chopina 12/18, Pl-87-100 Toruń, Poland
[2]Institute of Mathematics, Pedagogical University,
ul. Chodkiewicza 30, Pl-85-064 Bydgoszcz, Poland

ABSTRACT. Extinction laws observed in spectra of B_e stars are shown to be evidently different from the mean extinction curve. They do not contain in many cases the prominent extinction feature - the 2200A bump. The results are derived both from multicolour (ranging from the far-IR until the ultraviolet ANS) photometry and from TD-1 UV spectra. It is strongly suggested that the observed extinction phenomena are originated in circumstellar, disk-shaped shells as the shape of resultant extinction curve suggests both the presence of big, core-mantle grains and continuous infrared emission of circumstellar origin.

1. INTRODUCTION

The variability of extinction law (Massa and Savage 1989) is certainly caused by differences in physical properties of interstellar grains. Dust particles may grow in the dark, cold clouds and thus their sizes, shapes and chemical composition may vary as well as their crystalline structure from one cloud to another. Evolutionary tracks of interstellar grains have been recently discussed by Greenberg (1984).

The growth of grains may be especially fast in dense and cold clouds - the expected birth-places of new stars. Thus big, core-mantle grains are to be expected in globules or elephant trunks observed in vicinities of newly formed objects as well as in circumstellar shells, possibly disk-shaped. The remnants of the parent clouds are strongly irradiated with the stellar radiation in close vicinities of young stars. It seems thus to be of basic importance to check whether such grains produce the extinction obeying or not the mean law observed in the truly interstellar medium. Similar extinction effects are to be expected in vicinities of all young stars.

2. DESCRIPTION OF THE METHOD

B_e stars are well known objects closely related to some circumstellar matter - possibly remnants of their parent clouds. Very peculiar extinction curves derived from spectra of at least some of them (Sitko

L. V. Mirzoyan et al. (eds.), Flare Stars in Star Clusters, Associations and the Solar Vicinity, 293–297.

et al. 1981) make a more systematic investigation really attractive. Let's select a sample of B_e stars of nearly the same spectral type and luminosity class. A difference in the extinction law should be most easy to detect in the UV, particularly around the famous 2200A bump. Thus an analysis of possible peculiarities of extinction in B_e stars should start from a consideration of this prominent feature.

The spectral type and luminosity class most frequently observed from astronomical satellites is B2V. Thus we have selected the sample of these stars from the ANS Catalogue (Wesselius et al. 1982). The colour index between the ANS 2200A band and the V band was correlated with the B-V in the selected set of normal B and B_e stars. The result is shown in Fig.1.

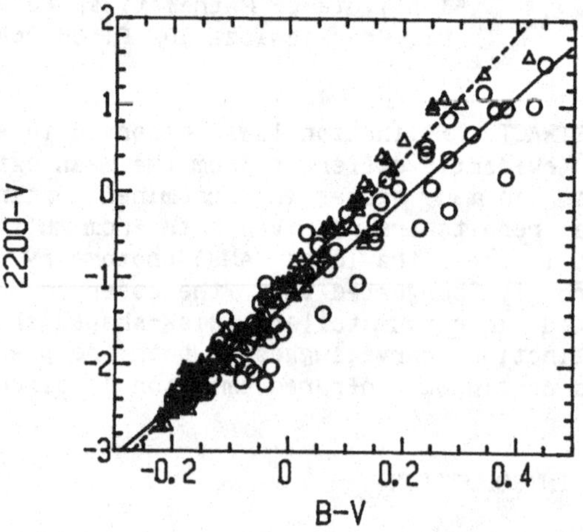

Fig.1 *Two-colour diagram relating the colour index between the 2200A ANS band and V band to B-V. Normal B2V stars - triangles; B2Ve - open circles. Note the different slopes of the mean relations - they intersect close to the assumed $(B-V)_0$ indicating that stars of both samples are intrinsically identical.*

The observed variety of colour indices among the sample of the same Sp/L should result only from different reddenings. The observed different slopes of the two-colour relations for normal B2V and B2Ve stars, as shown by Krełowski and Strobel (1987), indicate for different extinction laws towards the stars of both samples. The 2200A extinction bump is apparently stronger in relation to E_{B-V} when observed in spectra of normal B2V stars. Let's emphasize that the scatter observed in Fig.1 is also greater in the case of B_e stars. Thus the latter sample is probably much less homogeneous - reflecting different stages of the evolution of circumstellar grains or different geometry of intervening clouds (disks?).

The most interesting cases are, however, these of very low reddenings. In such cases we may expect that extinction effects are caused by single clouds or solely by circumstellar shells in the case of B_e stars. Thus a comparison of extinction curves derived from spectra of slightly reddened B_e stars with the mean extinction curve may be very interesting. In spectra of heavily reddened B_e stars the truly interstellar effects become more and more important and thus their extinction curves get more and more similar to "normal" ones.

The stars selected for this purpose are: HD 20336, 32343, 44458, 57150, 60606, 65875, 88661, 202904, 59878, and 63462. They are mostly

B2V stars; the criterion of selecting the above sample was the availability of stellar photometric data from far infrared (Gezari et al. 1984) through the UBV up to the ANS UV bands as well as of the TD-1 spectra (Jamar et al. 1976, Macau-Hercot et al. 1978). All these data allow to construct extinction curves in a very broad wavelength range.

The individual extinction curves of the chosen stars have been calculated using the standards recommended in our recent paper (Papaj, Wegner and Krełowski 1989). After deriving all these curves with the aid of the standard pair method the results have been averaged over the whole sample. The comparison of the B_e extinction curve with the mean galactic one (Savage and Mathis 1979) is shown in Fig.2 together with that for the B_e stars (HD10516, 68980, 41335, 45910, 59878, 63462, 200775, and 153261) for which only photometric data are available. The error bars show possible uncertainties.

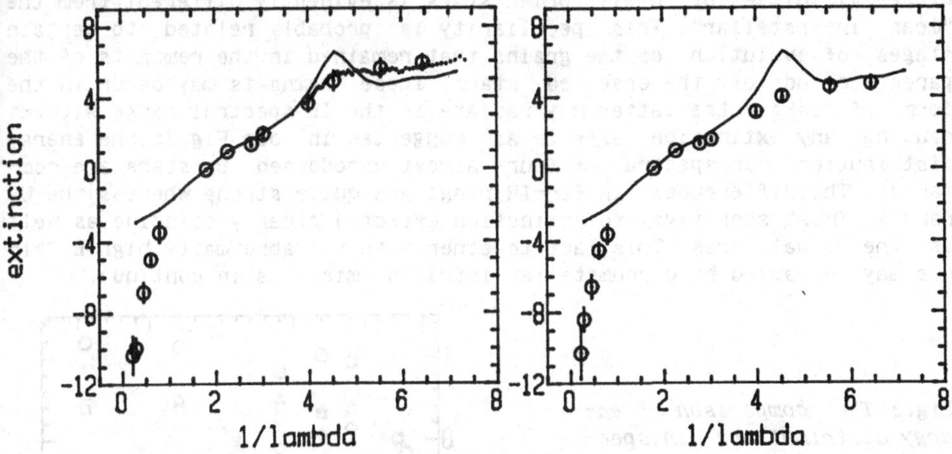

Fig.2 *The comparison of the extinction curves (normalized to E_{B-V}) averaged over the B_e samples including TD-1 spectra marked as dotted line (left) and based on photometric data only (right) with the mean extinction curve of Savage and Mathis (1979) - full line.*

Two very important phenomena are observed in the extinction curves derived from B_e stars. One of them is the lack of the 2200A bump. This astonishing fact is observed in the TD-1 as well as in the ANS data and thus it is proved beyond a doubt. It is interesting that Be and mean curve differ only shortward of 2200A in far-UV. In cases of heavily reddened B_e stars the dominating obscuration of interstellar clouds produces the bump, but it is usually weaker than in normal B stars (Fig.1) because of the circumstellar contribution. Another important phenomenon is the far-IR behaviour of the curve. The data seem to indicate for the enormously high value of the R constant (the total-to-selective extinction ratio) which should be situated at the point where the extinction curve intersects the ordinate axis. This phenomenon may be, however, at least partly due to emissions of relatively hot dust particles situated in close vicinities of the stars under consideration. But the R value caused by very big particles, suggested by the flat far-UV segment of the extinction curve, may be higher than the usually accepted value ≈3, due to saturation effects between B and V bands. Let's emphasize: for a great majority of B_e stars neither photometric nor spectral data exist in such a wide wavelength range. Two colour plots (like that in Fig.1) coincide in the case of very small reddenings indicating that B_e and normal B stars are characterized by the same intrinsic colours. This fact makes the results completely reliable.

3. DISCUSSION AND CONCLUSIONS

The results shown in Sec.2 show that the extinction originating in close vicinities of newly formed stars is evidently different from the "mean interstellar". This peculiarity is probably related to certain stages of evolution of the grains that remained in the remnants of the parent clouds of the observed stars. These remnants may occur in the form of disks - the latter may radiate in the IR spectral range without causing any extinction effects as suggested in our Fig.3: the energy distributions in spectra of four almost unreddened B_e stars are compared. The differences in far-IR range are quite strong whereas the UV ranges (most sensitive to extinction effects) clearly coincide as well as the visual ones. This fact together with the abnormally high R values may be caused by circumstellar infrared emissions in continua.

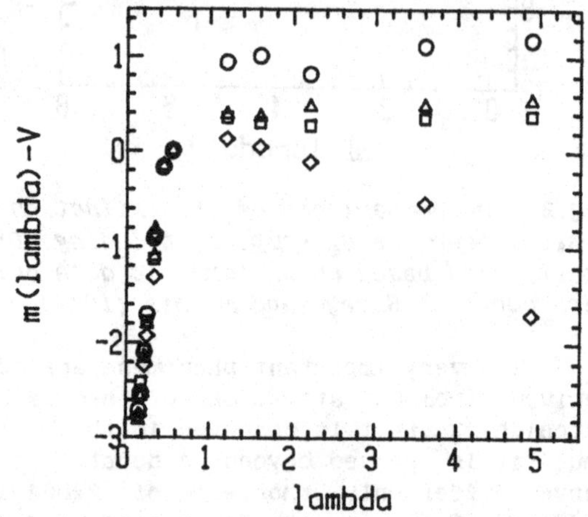

Fig.3 *The comparison of energy distributions in spectra of four unreddened B2Ve stars. HD56139 (squares), HD57219 (triangles), HD120324 (open circles) and HD158427 (diamonds). Note the great scatter in far infrared suggesting circumstellar emissions. Lambda in microns.*

We conclude that "normal" B stars and those with emission lines are intrinsically identical at least in the sense they have the same intrinsic colour indices. The extinction (together with the infrared emission) modifies however their spectra in quite different ways. The above presented "peculiar" extinction curve is probably a rather typical product of the grain evolution in vicinity of a newly formed star. This result stresses once again the fact of basic importance: every "mean interstellar extinction curve" is probably a mixture of individual contributions of clouds differing strongly in many cases. Such an "average" can hardly be used to remove small reddening effects (=single cloud effects) from high quality spectra.

ACKNOWLEDGEMENTS. This project has been supported partially by the Polish Academy of Sciences under the grant RPB-R 1.11.

REFERENCES

Gezari, D.Y., Schmitz, M., and Mead J.M. 1984, *Catalog of Infrared Observations*, NASA Ref. Publ. 1118.

Greenberg, J.M. 1984, in *Laboratory & Observational Infrared Spectra of Interstellar Dust*, eds. R.D. Wolstencroft and J.M. Greenberg, (Edinburgh: Royal Observatory Edinburgh), p. 1.

van de Hulst, H.C. 1986, in *Light on dark matter*, ed. F.P. Israel, (Dordrecht: Reidel), p. 161.

Jamar, C., Macau-Hercot, D., Monfils, A., Thompson, G.I., Houziaux, L. and Wilson, R. 1976, *Ultraviolet Bright Star Spectrophotometric Catalogue*, ESA SR-27.

Krełowski, J., and Strobel, A. 1987, *Astr. Ap.*, **175**, 186.

Macau-Hercot, D., Jamar, C., Monfils, A., Thompson, G.I., Houziaux, L. and Wilson, R. 1978, *Supplement to the Ultraviolet Bright Star Spectrophotometric Catalogue*, ESA SR-28.

Massa, D., Savage, B.D. 1989, in *Interstellar Dust*, eds. L.J. Allamandola and A.G.G.M. Tielens, (Dordrecht: Kluwer), p. 3.

Papaj J., Wegner W., and Krełowski J. 1989, *M. N. R. A. S.* (submitted).

Savage, B.D. and Mathis J.S. 1979, *Ann. Rev. Astr. Ap.*, **17**, 73.

Sitko, M.L., Savage B.D., and Meade M.R. 1981, *Ap. J.*, **246**, 161.

Wegner, W., and Krełowski, J. 1989, *Astr. Nachr.*, (in press).

Wesselius, P.R., van Duinen, R.J., de Jonge A.R.W., Aalders, J.W.G., Luinge, W., and Wildeman, K.J. 1982, *Astr. Ap. Suppl.*, **49**, 427.

REVIEW OF THE THEORETICAL MODELS OF FLARES OF THE UV CETI-TYPE STARS

V.P.GRININ
Crimean Astrophysical Observatory
Crimea, 334413, P.O. Nauchny, USSR

ABSTRACT. The observations of stellar flares in optics, radio and X-ray show the general similarity of the flares on UV Ceti-type stars and on the Sun. At the same time there exist a lot of data showing that the analogy between these processes is not full. From this point of view the analysis of the theoretical models of stellar flares is given.

1. INTRODUCTION.

The first flares on UV Ceti-type stars were observed about 50 years ago and we known now from statistical investigations initiated by professor Ambartzumyan (1988), that all (or almost all) dMe stars are the flare stars. During this time many different ideas about the nature of the flares were suggested (Gershberg, 1978; Gurzadjan, 1980), but only one of them - based on the community of the physics of the stellar and solar flares - was confirmed by the followed observations. The logical design of this idea was the IAU Colloquium N 104 "Solar and Stellar Flares" at the Stanford University where these two astrophysical topics were presented for the first time as a part of a single whole.

In this review we will discuss from this point of view the problems of the physical modelling of stellar flares with the accent on the models provided by the optical emission. Some questions such as the physics of the primary energy output, dynamics of flare loops and some others are not considered here since they are discussed in the recent reviews by Mullan (1989).

2. THE MAIN STAGES OF THE INVESTIGATIONS OF STELLAR FLARES.

The main characteristics of stellar flares are very different from the solar ones. The energy of strong stellar flares exceeds by the 2 - 3 orders the energy of the strongest events on the Sun. The development of stellar flares is more rapid: at the flares with well-pronounced impulsive phase the brightness of the star increases in tens times on the time scale of about one minute. That is why the analogy with the Sun was not obvious initially and the large efforts of the observers and theorists were needed for its evidence.

L. V. Mirzoyan et al. (eds.), Flare Stars in Star Clusters, Associations and the Solar Vicinity, 299–312.

2.1 OPTICAL EMISSION OF THE FLARES.

The first spectra of stellar flares with the moderate time resolution were observed in the Crimea by Gershberg and Chugainov (1967) and at the McDonald Observatory by Kunkel (1970). They showed, that during the flare the optical continuum is accompanied by the intensification of the Balmer lines, H and K Ca II and neutral helium lines. It meant that the main component of optical flares is the thermal radiation of the gas heated up to the temperature $T \approx 10000$ K.

The obtained by Kunkel Balmer decrements of the flares turned out to be unusual: instead of "normal" intensities ratio: $H\alpha > H\beta > H\gamma > \ldots$. the inverse ones has been observed: $H\beta < H\gamma < H\delta < \ldots$. at maximum of light. Such decrements are typical for optically thick almost fully thermalized gas.Using their Kunkel estimated the electron number density in the flares: $N_e \approx 10^{13} - 10^{14}$ cm^{-3}. It was the first direct evidence that the flare emission lines are formed at the chromospheric level.

The next important result has been obtained by Bopp and Moffatt (1973) and by Pettersen (1983). They showed that:

1. the development of emission lines in the flares is more extended in time compared to the continuum (see Figure 1), that is typical also for solar flares;
2. the contribution of the emission lines to the total radiation of the optical flares at maxima of light is rather small: 5 - 11% in B - band and hence UBV - observations near the maxima of the flares characteri--zed the properties of its continuum in main.

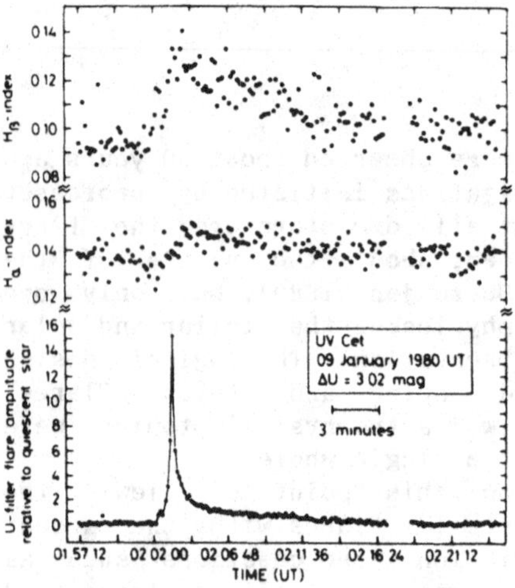

Figure 1. The flare of UV Ceti in $H\alpha$, $H\beta$ and U - band (Pettersen, 1983).

The interpretation of continuum emission especially in the peaks of the flares is extremely important for the understanding of the flare process and has been considered by many authors (see the review of Kodaira, 1983). The initial suggestion, according to which the continuum emission has a purely recombinational origin and is formed in the optically thin gas faced with two problems: a) the observations showed that the Balmer jump at the maxima of light is rather small and systematically lower than the values observed by Kunkel. b) the colour indexes at the peaks of the flares appeared to be in disagreement with the theory. The cloud of the observational points on the two-colour diagram is markedly concentrated near of values: $U - B \approx -1^m.0$ and $B - V \approx 0^m.2$

These important observational facts were interpreted as the direct evidence for bremstrahlung radiation of a hot gas with the temperature $T \approx 10^6 - 10^7$ K as the main component of the optical continuum of flares at the maxima of light. Such interpretation explained the mentioned above properties of continuum emission but was in conflict with the "low-temperature" composition of emission lines in the optical spectra of the flares at the maxima of light.

Another interpretation was proposed by Grinin and Sobolev (1977): we argued that the main part of the optical continuum at the flare maxima is formed in the deeper layers of the stellar atmosphere where the number density of atoms is about $10^{15} - 10^{17}$ cm^{-3}. At these conditions two additional factors effected the energy distribution of the flares: at low temperature (T < 8000 K) the H^-- emission is added to the recombinations of hydrogen atoms; at larger T the gas becomes optically thick beyond the Balmer jump and its radiation is quasi-black-body. The last conclusion has been confirmed by the followed observations of Mochnacki and Zirin (1980), Kaler et al. (1982), Chugainov (1987), de Jager et al. (1989).

Such approach permitted us to draw a direct analogy between stellar flares and solar white flares, whose optical continuum arises also in transition region between photosphere and chromosphere (Neidig, 1989). We estimated also the area of continuum formation region of strong stellar flares (at the maxima of light): $S_{con} \approx 1 - 2\%$ of stellar disk area.

Summarizing the data on the optical spectra of the flares let us note briefly some properties of the emission lines profiles. The observations show that near the maxima of strong flares the hydrogen lines are broadening by several angstroms as a rule. On the spectra with high signal to noise ratio the broad wings are well pronounced in the Balmer lines, that might be due to a classical Stark- effect or plasma turbulence broadening. In a set of the flares a redward asymmetry of the line profiles was observed near the maximum of light. On the spectrogrammes obtained with high spectral resolution Shneeberger et al. (1979) observed the absence of central reabsorption in Hα at the flare of EV Lac. It means that the excitation temperature of the atoms and consequently the electron temperature in the flare increases outward.

2.2. THE ULTRAVIOLET SPECTRA.

The existence of hot gas in upper layers of the flares follows from IUE spectra (see review by Giampapa, 1983). They showed the intensive emission lines of C II, Si II, C IV, Si IV and some others. For the ionization of latter the electron temperature of about 10^5 K are needed. However the ultraviolet continuum of flares may be connected with the radiation of deeper layers heated ut to $(1 - 2)*10^4$ K as was suggested by Baljunas and Rymond (1984) and Butler et al. (1981). It is noteworthy that the exposure time of the flare spectra obtained with IUE was usually about several dozens of minutes . That is why they can be hardly used for quantitative analysis.

Two short-living flare bursts of EV Lac were registered on board the ASTRON station with high time resolution (0.6 s) and a pass-band $\Delta\lambda$ = 28 Å (Gershberg and Petrov, 1986; Burnashova et. al 1989).One burst was observed in the spectral region centered at λ = 2430 Å, free from the strong emission lines, the other - at the wavelength λ = 1550 Å of resonance transition of C IV. In both cases the UV bursts were accompanied by the flares in longward wavelengths. These data are analyzed in the framework of the hydrodynamical flare model (Katsova and Lifshitz, 1989; see Section 3).

2.3 RADIO OBSERVATIONS.

At the beginning from the work of Lovell et al.(1963) radio observations of stellar flares were carried out by many authors at the meter-waves and microwaves. Their properties are summarized in the reviews by Gibson (1983), Kuijpers (1989) and Lang (this volume). The main difference from solar microwave flares is the absence or poor correlation with optical ones. The observations show that the stellar radio bursts follow usually the optical events (Figure 2.), that resemble the development of solar radio bursts of IV - type. It is possible that better correlation will be observe in shorter radio wavelengths (see Rodono et al. (1989).

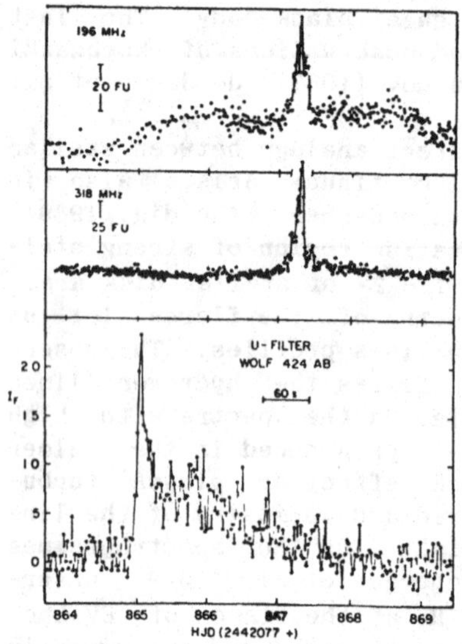

Figure 2. The example of the flare of Wolf 424 in radio and optics from Spangler and Moffett (1976).

The important properties of the stellar radio flares are the high brightness temperature of radiation (up to 10^{15} K) and the high circular polarization, reaching sometimes almost 100%. Both properties are typical for the coherent plasma process. According to Melrose and Dulk (1982) the most probable is an electron-cyclotron maser but other coherent emission mechanisms are not fully excluded (Kuijpers, 1989).

2.4. THE X-RAY OBSERVATIONS.

The first soft X-ray burst was registered by Heise et al. (1975) on red dwarf star YZ CMi on board ANS satellite. Since then the X-ray observations of flare stars have been carried out in a number of specialized space projects. Most of them were observed on EINSTEIN and EXOSAT satellites in the energy ranges 0.2 - 4 keV and 0.1 - 10 keV, respectively. The results of these investigations are summarized in the reviews by Haish (1983), Ambruster et al. (1987) and Pallavicini et al.(1989) The main conclusions are as follows:

The total energies of X-ray bursts on the flare stars are: $E_x \approx 3*10^{33} - 10^{35}$ ergs that is two or three orders higher than on the Sun.

The temperatures of the X-ray flares in both cases are about the same: $T \approx (1 - 3) * 10^7$ K. The hardness of X-ray emission is maximal usually at the flare maximum. The electron number density of a hot plasma found from the models of the flare loops is of the order of $10^{11} - 10^{13}$ cm^{-3}.

According to Pallavicini et al. (1989) the stellar X-ray flares can be classified into two main groups: a) the impulsive flares with the dumping time of the order of a few minutes and b) the slow flares with the time decay of about an hour. Their analogies on the Sun are so-called compact and two-ribbon flares.

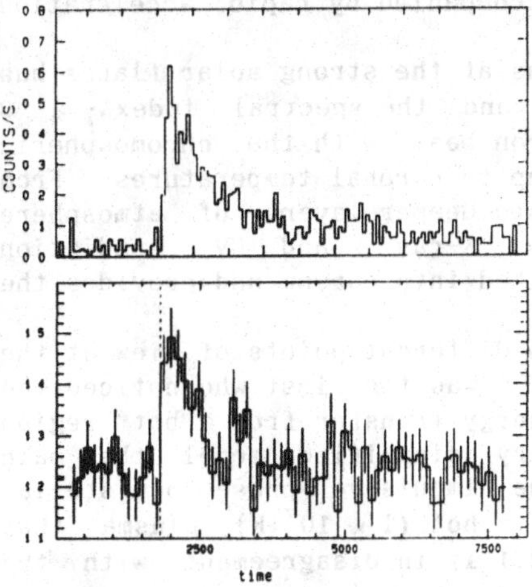

More informative are the coordinate observations of the flares in X-ray and optics. Unfortunately however only small part of X-ray observations were supported by simultaneous optical photometry and we do not known at present what are typical ratios of the luminosities L_x/L_{opt} at the maxima of light and the total energies E_x, E_{opt} of stellar flares (Byrne, 1989) Some of the optical flares had the X-ray counterparts. The best example of such event is the strong flare of UV Ceti (de Jager et al. 1989) which was observed synchronously on the LE and ME detectors of EXOSAT (see Figure 3). Some of flares were observed without X-ray bursts (Karpen et al. 1977; Doyle et al. 1988)

Figure 3. The strong flare of UV Ceti in X-ray (from de Jager et al. 1989). Top: the count rates in the Low Energy detector (40 - 200 Å); Bottom: the count rates in the Medium Energy detector (1 - 6 keV). The abscissa is the time in seconds. The vertical broken line in the lower diagram marks the time of maximum of the optical flare.

Therefore the radio and X-ray observations show the existence of a hot phase of stellar flares. At the same time they demonstrate one important difference between solar and stellar flares: at the flares on the Sun the strong correlation between optical, microwave and soft X-ray events exists.

In conclusion of this Section let us note two important results connected with the solar - stellar analogy. These are: a) the discovery by Saar et al. (1987) of the strong magnetic fields (4 - 6 kgs) on the flare stars, and b) the evidence of the similarity of the energetic spectra of stellar and solar flares (Gershberg, 1989; Schakhovskaja, 1989). According to Pustylnik (1988) this similarity may be due to a turbulent character of the magnetic energy output on the surface of the stars by the convective elements of different scales. It means, that starting from the microflares (Beskin et al. 1988) up to giant flares with the total energy of the order of 10^{35} ergs, we are dealing with the physical phenomena of the same nature.

3. THE THEORETICAL MODELS

The general similarity between solar and stellar flares means, that the processes of the primary energy release in both cases are qualitatively the same. As in the solar case we will distinguish two main phases of stellar flares: impulsive and gradual. The impulsive phase is undoupted-ly the key for the understanding of the flare mechanism. At the solar flares the primary process is realized in upper part of the magnetic loops and is due to magnetic reconnection *). Many important details of this process are unknown at present. But it is well known from the hard X-ray observations that this phase is accompanied by rapid acceleration of the charged particles.

The energy spectrum of fast electrons at the strong solar flares has the cut-off energy $E_1 \approx 10 - 20$ keV and the spectral index $\gamma \approx 3$ (Lin, 1974). The interaction of electron beam with the chromospheric plasma provides the heating of the gas up to coronal temperatures. From this hot region the energy penetrates into deeper layers of atmosphere by the conductivity, the shock front or X-ray and UV radiation. Simultaneously the hot plasma is evaporated into corona and provides the gradual phase of the flares.

These processes were considered from different points of view at the modelling of stellar flares. Mullan (1977) was the first who noticed the important role of conductivity in the energy transfer from a hot region of stellar flares. According to his very simplified model the main component of the optical continuum in the impulsive phase of stellar flares is the bremstrahlung radiation of a hot ($T \approx 10^7$ K) plasma that as has been mentioned in the Section 2.1 is in disagreement with the properties of the flare spectra.

3.1. THE HYDRODYNAMIC MODEL OF THE FLARES

Another flare model was considered by Katsova et al. (1981). They calculated the hydrodynamical response of the chromosphere of the red dwarf on the impulsive heating by the electron beam (taken into account the energy transfer by the conductivity). The parameters of the beam in their model was adopted as for strong solar flares: $E_1 = 15$ keV, $\gamma = 3$ and the initial energy flux $F = 10^{12}$ erg/s cm^{-2}. The duration of impulsive heating was equal to 10 seconds.

According to this model the optical flare arises in the dense layer beyond the shock front (see Figure 4). This layer is moving down into denser part of stellar atmosphere with the initial velocity of about 100 km/s which is in agreement with the observed red asymmetry of the emission line profiles in flares spectra.The parameters of optical flare in this model correspond to the parameters of the weak stellar flares.

---

*) In principle another mechanism of the primary energy output at the stellar flares based on the idea of Z - pinch was suggested recently by Hayrapetyan et al. (1988) (see Hayrapetyan, this volume).

The further investigations of this type models are needed in the connection with the recent observations of ultra-short flare events (see Tovmasyan and Zalinyan, 1988).

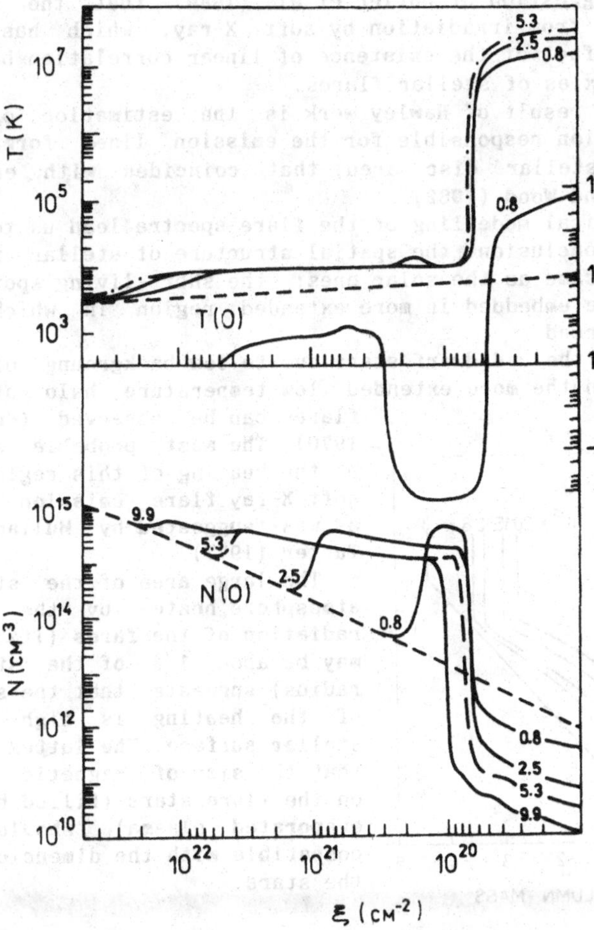

Figure 4. The hydrodynamical response of the atmosphere of a flare star on the impulsive heating by electron beam (from Katsova et al. 1981). (See text for details)

3.2. THE X-RAY HETING MODELS

The soft X-ray emission of the flares is another important mechanism of the heating of the stellar atmosphere. In the solar flare models it was considered by Somov (1975). Recently Hawley (1989) made a detailed investigation of this source of the heating taking into account the accurate non-LTE radiative cooling functions on hydrogen atoms and ions of Ca II and Mg II (see Figure 5). According to her calculations the soft X-ray emission with the parameters close to the observed ones provide the gas heating which explains fully the observational fluxies in the main emission lines of stellar flares. This conclusion is in agreement with the suggestion of Butler et al. (1989) that the Balmer emission lines result from irradiation by soft X-ray, which has been established from the fact of the existence of linear correlation between Hγ and soft X-ray fluxies of stellar flares.

Another important result of Hawley work is the estimation of the area of the flare region responsible for the emission lines formation: $S_{em\ l} \approx 10\%$ of the stellar disc area, that coincides with earlier estimations of Cram and Wood (1982).

Thus, the theoretical modelling of the flare spectra lead us to the following important conclusion: the spatial structure of stellar flares is qualitatively the same as the solar ones: the short living spots of continuum emission are embedded in more extended region in which the emission lines are formed.

Unlike the Sun on the red dwarf stars due to low backgroung of the photospheric radiation the more extended low-temperature halo of the flares can be observed (Kunkel, 1970). The most probable source of the heating of this region is soft X-ray flare emission also, as was suggested by Mullan and Turter (1977).

The large area of the stellar atmosphere heated by the X-ray radiation of the fares (its size may be about 1/3 of the stellar radius) suggests, that the source of the heating is high over stellar surface. The latter means that the size of magnetic loops on the flare stars (filled by the evaporated plasma) should be compatible with the dimensions of the stars.

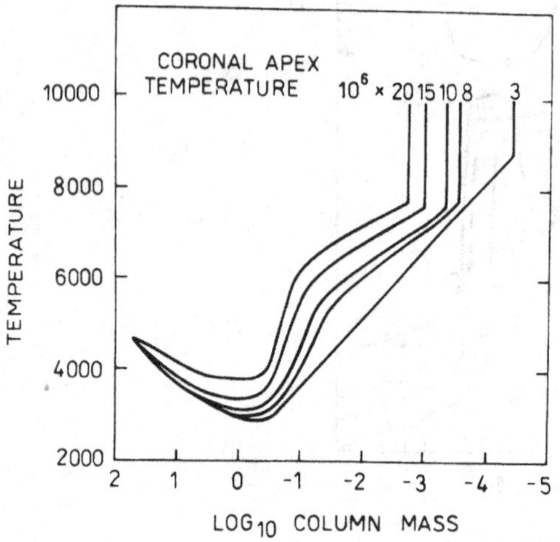

Figure 5. The temperature profiles in the flare model with the X-ray heating from Hawley (1989).

Thus the soft X-ray emission of the flares is an important (and probably the main) source of the gas heating at the gradual phase. It should be note however, that the role of this mechanism cannot be decisive at the impulsive phase of the stellar flares. This follows for example from the lack of the X-ray counterpart at the set of optical flares (see Section 2.4).

3.3 THE MODEL OF THE IMPULSIVE PHASE OF THE STELLAR FLARES

In the case of solar flares the bulk of accelerated electrons has a non-relativistic energies (tens keV). Such electrons are stopping in the chromospheric layers and heat them up to $T \approx 10^7$ K. Only a small part of the accelerated electrons penetrate into deeper layers of the solar atmosphere and their energy deposition is insufficient to produce the optical emission of white flares (Canfield et. al. 1986), (see, however Aboudarham and Henoux, 1986). There is a strong argument in favour of the fact that in case of stellar flares the situation is opposite and the direct heating of the low-temperature region of the flares by charged particles play more important role.

As was mentioned in the Section 2.1, the gas emitting in the optical continuum at the maxima of strong stellar flares is optically thick beyond Balmer jump. At the same time even the strongest white flares on the Sun are optically thin (Machado and Rust, 1974). Since the electron temperatures in both cases are about the same ($T \approx 10^4$ K), it means that the primary heating agent penetrates into deeper layers of stellar atmosphere (in the units of the column density), that at solar flares.

This idea has been used recently at the modelling of the impulsive phase of stellar flares (Grinin and Sobolev, 1989 a,b). We investigated the direct heating of the low-temperature region of the stellar flares by the beam of accelerated particles taking into account the radiative transfer.

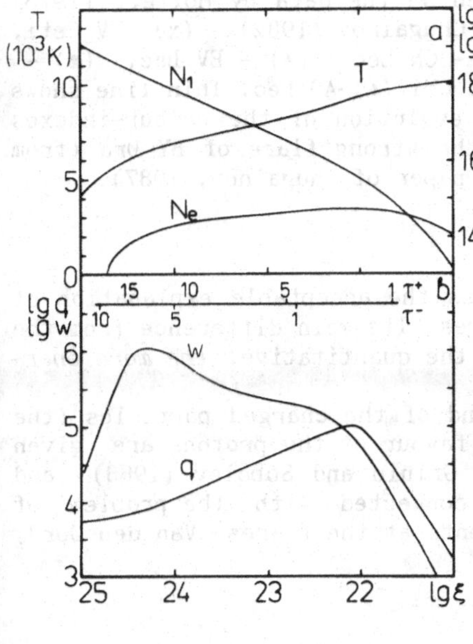

Figure 6 illustrates the parameters of the flare calculated in a thick target approximation at the following assumptions: the primary heating is due to the high energy proton beam with a power-low energetic spectrum. The parameters of the beam: the spectral index $\gamma = 3$, the cut-off energy $E_1 = 5$ MeV, the the initial energy flux $F = 5*10^{11}$ erg/cm^2 s. The free-free and free-bound transitions of hydrogen atoms and H^- in the LTE approach were taken into account at the calculations of the radiative cooling functions.

Figure 6. The model of the low-temperature flare region in the impulsive phase (see text for details). Top: the temperature, electron (N_e) and neutral atom (N_1) number density as a function of the column density ξ. Below: the energy deposition by charged particles (q) and by radiation of the flare itself (w) in erg/cm^3 s. The optical depths scale beyond (τ^+) and before (τ^-) Balmer jump.

Our main conclusions are the following:

a) The proton beam with the cut-off energy $E_1 \approx 3 - 10$ MeV and the initial flux $F \approx 10^{11} - 10^{12}$ erg/cm^{2}s provides the gas heating, which explains the observational properties of the optical continuum of the stellar flares at the maxima of light. The same results give the heating by the electron beam with the cut-off energy in $(m_p/m_e)^{1/2} \approx 50$ times less and the same spectral index.
b) In both cases the energy deposition by charged particles is maximal near the value of $\xi \approx 10^{22}$ cm^{-2} and the radiation of the flare heats deeper layers of the atmosphere ($\xi \approx 10^{24}$ cm^{-2}) . The role of the radiative transfer is especially important at the energy fluxies in the beam: $F > 3*10^{11}$ erg/cm^2 s, when the flare region is optically thick beyond Balmer jump. The radiation of the flare is quasi-black-body at these conditions in the agreement with the observations (see Figure 7).

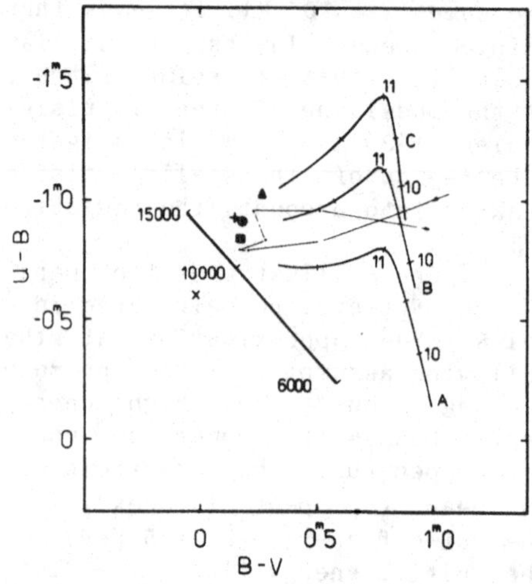

Figure 7. The theoretical two-colour diagram for the set of the models: A, B, C - corresponds to the values of the cut-off energy E_1 = 10, 5, 3 MeV. The values of Log F are given along theoretical lines. The mean values of the colour-indexes of the flares near maxima of light are given on the data by Moffett (1973) and Chugainov (1982): (x) -UV Ceti, (■) - CN Leo, (+) - EV Lac, (●) - - YZ CMi,(▲)-AD Leo. Thin line shows the evolution of the colour-indexes of the strong flare of BY Dra (from the paper of Chugainov, 1987).

Thus, the discussed above model gives the acceptable explanation of the most important phase of stellar flares. Its main difference from the classical models of the solar flares is the quantitative: *the more energetic accelerated particles in the beam.*

This model is insensitive to the kind of the charged particles (the protons or electrons). The arguments in favour of the protons are given in the papers by Van den Oord (1988), Grinin and Sobolev (1988) and Simnett (1989). The main argument is connected with the problem of the neutralization of the electric currents at the flares (Van den Oord, Simnett).

4. CONCLUSION

In the conclusion of this review let us return again to the problem of the solar-stellar analogy. It is well known from the solar physics that the energy spectra of the charged particles generated at the solar flares depend from the power of the events (Lin, 1974): most purely electron events are observed at the weak solar flares (importance 1 and subflares). The proton events are observed mainly at the strong flares (importance 2 and 3). The electron energy spectrum during proton events tends to be harder than for purely electron ones. The simple physical extrapolation of this tendency to the flare stars supports the main conclusion of the previous Section that the more energetic charged particles are generated in the impulsive phase of stellar flares. ***This conclusion is not unexpected from the point of view of the solar-stellar analogy.***

REFERENCES

Aboudarham, J. and Henoux, J.C. (1986), Astron. Astrophys. 156, 73.

Ambartzumyan, V.A. (1988), "The Scientific Transections", Vol.3, Erevan, Publ. Acad. Sci. Armenian SSR.

Ambruster, C.W., Sciortino, S. and Golub, L. (1987), Astrophys. J., 65, 273.

Baljunas, S.L. and Rymond, J.C. (1984), Astrophys. J., 282, 728.

Beskin, G.M., Gershberg, R.E, Neizvestnyi, S.I., et al. (1988). Izv. Krymsk. Astrofiz. Obs., 79, 71.

Bopp, B.W. and Moffett, T.J. (1973), Astrophys. J., 185, 239.

Burnashova, B.A., Gershberg, R.E., Zvereva, A,M., et al. (1989), in B.M.Haish and M.Rodono (eds.), "Solar and Stellar Flares", Poster Papers, Catania Astrophys. Obs., Special Publication, p. 76.

Butler, C.J., Byrne, P.B., Andrews, A.D. and Doyle, J.G. (1981). M.N.R. A.S., 197, 815.

Butler, C.J., Rodono, M. and Foing, B.H. (1989), in B.M.Haish and M.Rodono (eds), "Solar and Stellar Flares", Poster Papers, Catania Astrophys. Obs. Special Publication, p. 67.

Byrne, C.J, Rodono, M and Foing, B.H. (1989), ibid, p. 21.

Canfield, R.C., Bely-Dubau, F., Broun, J.C., et al. (1986), in M.Kundu and B.Woodgate (eds.), "Energetic Phenomena on the Sun", NASA CP 2439, Chapter 3.

Chugainov, P.F. (1982), Izv. Krymsk. Astrofiz. Obs., 65, 155.

Chugainov, P.F. (1987), Izv. Krymsk. Astrofiz. Obs., 76, 54.

Cram, L.E. and Wood, D.T. (1982), Astrophys. J., 257, 269.

De Jager, C., Heise, J., Van Genderen, A.M., et al. (1989), Astron. Astrophys., 211, 157.

Gershberg, R.E. (1978), "Flare Stars of the Small Masses", Nauka, Moskow

Gershberg, R.E. (1989), Memorie Soc. Astron. Ital., 60, 263.

Gershberg, R.E. and Chugainov, P.F. (1967), Astron. Zh., 44, 260.

Gershberg, R.E. and Petrov, P.P. (1986), in L.V.Mirzoyan (ed.), "Flare Stars and Related Objects", Erevan, Publ. Acad. Sci. Armenia, 38

Giampapa, M.S. (1983), in P.B.Byrne and M.Rodono (eds.), "Activity in Red-Dwarf Stars", Reidel, Dordrecht, p. 223.

Gibson, D.M. (1983), ibid, p. 273.

Grinin, V.P. and Sobolev, V.V. (1977), Astrofizika, 13, 586.

Grinin, V.P. and Sobolev, V.V. (1988), Astrofizika, 28, 355.

Grinin, V.P. and Sobolev, V.V. (1989 a), in B.M.Haish and M.Rodono (eds.) "Solar and Stellar Flares", Poster Papers, Catania Astrophys. Obs., Special Publication, p. 297.
Grinin, V.P. and Sobolev, V.V. (1989 b), Astrofizika, 31, N 3, in press.
Gurzadyan, G.A. (1980), "Flare Stars", Publ. Pergamon Press, Oxford.
Haish, B.M. (1983), in P.B.Byrne and M.Rodono (eds.), "Activity in Red--Dwarf Stars", Reidel, Dordrecht, p. 255.
Hawley, S.L. (1989), in B.M.Haish and M.Rodono (eds.), "Solar and Stellar Flares", Poster Papers, Catania Astrophys. Obs., Special Publication, p. 49.
Hayrapetyan, V.S., Nikhogossian, A.G. and Vikhrev, V.V. (1988), in Proc. Int. Workshop on Recconnection in Space Plasma, Potsdam, GDR, (ESA SP - 285, Vol. 11), p 163.
Heise, J., Brinkman, A.C., Shrijver, J., et al. (1975), Astrophys. J. (Letters), 202, L73.
Kahler, S., et al. (1982), Astrophys. J., 243, 234.
Karpen, J.T., et al. (1977), Astrophys. J., 216, 479.
Katsova, M.M, Kosovichev, A.G. and Livshits M.A. (1981), Astrofizika, 17, 285.
Katsova, M.M. and Livshits M.A. (1989), in B.M.Haish and M.Rodono (eds) "Solar and Stellar Flares", Poster Papers, Catania Astrophys. Obs., Special Publication, p. 71.
Kodaira, K. (1983), in P.B.Byrne and M.Rodono, (eds.), Activity in Red--Dvarf Stars", Reidel. Dordrecht, p. 561.
Kuijpers, J. (1989), in B.M.Haish and M.Rodono, (eds.), "Solar and Stel--lar Flares", Kluwer Acad. Publ., Dordrecht, p. 163.
Kunkel, W.E. (1970), Astrophys. J. 161, 503.
Lin, R.P. (1974), Space Sci. Rev. 16, 189.
Lovell, B., Whipple, F.L. and Solomon, L.H. (1963), Nature, 198, 228.
Machado, M.E. and Rust, D.M. (1974), Solar Phys., 38, 499.
Melrose, D.J. and Dulk, G.A. (1982), Astrophys. J., 259, 844.
Mochnacki, S.V. and Zirin, H. (1980), Astrophys. J., 239, L27.
Moffett, T.J. (1974), Astrophys. J. Suppl. S., 29, 1.
Mullan, D.J. (1977), Astrophys. J., 212, 171.
Mullan, D.J. (1989), in B.M.Haish and M.Rodono (eds.), "Solar and Stel--lar Flares", Kluwer Acad. Publ., Dordrecht, p. 239.
Mullan, D.J. and Tarter, C.B. (1977), Astrophys. J., 212, 179.
Neidig, D.F. (1989), in B.M.Haish and M. Rodono (eds.), "Solar and Stellar Flares", Kluwer Acad. Publ., Dordrecht, p. 261.
Pallavicini, R., Stella, L. and Tagliaferri, G.(1989), Astron. Astrophys. in press.
Pettersen, B.R.(1983), in P.B.Byrne and M.Rodono (eds.) "Activity in Red--Dwarf-Stars", Reidel, Dordrecht p. 239.
Pustylnik, L.A. (1988), Letters in Astron. J., 14, 940.
Rodono, M., Houdebine, E., Catalano, S., et al. (1989), in B.M.Haish and M.Rodono (eds.), "Solar and Stellar Flares", Poster Papers, Catania Astrophys. Obs., Special Publication, p. 53.
Saar, S., et al. (1987), in 27th Liege Int. Astrophys. Colloquium.
Shakhovskaya, N.I. (1989), in B.M.Haish and M.Rodono (eds.), "Solar and Stellar Flares", Kluwer Acad. Publ., Dordrecht, p. 375.

Shneeberger, T.J., Linsky, J.L., McClintock, W. and Worden, S.P. (1979), Astrophys. J., **231**, 148.

Simnett, G.M. (1989), in B.M.Haish and M.Rodono (eds.), "Solar and Stellar Flares", Poster Papers, Catania Astrophys. Obs., Special Publication, p. 357.

Somov, B.V. (1975), Solar Phys., **42**, 235.

Spangler, S.R. and Moffett, T.J. (1976), Astrophys. J., **203**, 497.

Tovmasyan, G.M. and Zalinyan, V.P. (1988), Astrofizika, **28**, 131.

Van den Oord, G.H.J. (1988), Astron. Astrophys., **207**, 101.

PETTERSEN: Since Suzanne Hawley could not be present at this symposium, I would like to briefly comment on her interesting results, just discussed by Grinin. Her model calculations of the time evolution of hydrogen Balmer lines are in agreement with observations during flares. Energy is transported from coronal to chromospheric levels by radiation and thermal conduction in her model, and she is not able to obtain continuum results of the kind that we observe during flares. For that it is probably necessary to include accelerated particles, electrons and/or protons, directed towards the photosphere.

Suzanne Hawley is now preparing her results for journal publication, and I refer you to this.

RODONO: Your prediction of more energetic electron spectrum for stellar flares does it imply a larger than solar hard-X-ray flux from stellar flares? If so, the detection of hard X-ray emission from stellar flares would become possible.

GRININ: From an extrapolation of gamma given above we may expect a larger ratio of hard X-ray/soft X-ray emission in stellar flares.

MONTMERLE: Your Lx versus electron spectral index extrapolation would indicate delta=2 for flare stars. Could this index be determined directly by radio observations (gyrosynchrotron mechanism)?

GRININ: I believe this would be problematic because of the high sensitivity of the radio maser mechanism to different parameters of the plasma.

Quiescent chromospheric response to the (E)UV/Optical flare radiation field on dMe stars

E.R. Houdebine[1] and C.J. Butler[2]

[1] *Institut d'Astrophysique Spatiale, Univ. d'Orsay, bt. 120, 91400 Orsay, France.*
[2] *Armagh Observatory, Armagh, BT61 9DG, N. Ireland.*

Abstract: We examine the response of the quiescent chromosphere to the large (E)UV and optical continuum and spectral line radiation field arising from a flare. We show that during a UV Ceti type flare, which displays a large U-band enhancement, a major part of the Balmer line flux may arise from the "quiescent chromosphere", rather than the heated flare plasma itself. This leads us to distinguish two main phases in the Balmer lines, as first proposed by Houdebine et al. (1989): an early, mainly impulsive, phase, driven by radiative pumping of the quiescent chromosphere, which is mainly correlated with continuum variations, and a later one, related to the thermal flare phase, which arises from the cooling of the flare plasma itself. The effect of the radiative pumping is much larger for stellar (dMe) than for solar flares, due to substantial differences in the flare, relative to the quiescent, level and the quiescent chromospheric density and temperature.

.1 Introduction

Several detailed studies have been carried out on the soft X-ray backwarming from flares but, so far, none dealing with the chromospheric changes due to optical and (E)UV enhancements. The observed U-band variations - up to 5 magnitudes during a stellar flare - drastically changes the ionisation equilibrium of Hydrogen and other elements.

.2 Lyman, Balmer and Paschen photo-ionisation and photo-excitation

In order to estimate, qualitatively, the effect of the flare radiation field on the quiescent chromosphere, we compare the additional net radiative rates produced by the flare radiation to the net rates for the quiet state. The quiescent chromosphere from which we computed the rates has been presented by Houdebine and Panagi (1989) (16 levels in NLTE, 70 lines and 9 continua). Then, to compare their effects, we considered a local increase of ten times the quiescent level, (equivalent to a ∼1.4 mag. U-band flare), in the Lyman, Balmer and Paschen continua. We computed the net photoionisation rates produced by such a radiation field at several levels in the atmosphere: at the temperature minimum, and at the Balmer and Lyman line formation regions. These rates are given in Figures 1a,b for two regions, as the *b- flare* rates, while the atomic net rates in the quiet state are labelled as *b-b* and *b-f*.

L. V. Mirzoyan et al. (eds.), Flare Stars in Star Clusters, Associations and the Solar Vicinity, 313–316.

From this diagram, it is striking that even small enhancements of the Lyman continuum will have a large impact on the Hydrogen ionisation equilibrium, at any atmospheric level, as the flare rate is at least 20 times larger than any other net rate.

Conversely, the same relative increase in the Paschen continuum (8210 Å, between the Johnson R and I bands) creates a rather small rate, such that the relatively weak enhancement here during flares has little effect on the quiescent chromosphere.

On the other hand, such an increase of the Balmer continuum introduces a rate of the order of, or larger than, the rates in the quiescent state. Furthermore, the local Balmer continuum strength depends on the distance from the flare site, and for a 5 magnitude U-band flare, it will, in the vicinity of the flare, be ~2000 times the quiescent level, (i.e. 200 times that above). Such large variations drastically and instantaneously change the Hydrogen ionisation equilibrium throughout the surrounding quiescent chromosphere with the amplitude of the effect related to the magnitude and duration of the flare. An increase of the Lyman continuum flux has the most important effect on the atomic equilibrium, but its penetration in the quiescent chromosphere is limited by the large opacity of the chromospheric plasma. Conversely, although its effects are weaker, the Balmer continuum penetration is much larger. The main results are an increase of the Hydrogen ionisation fraction and local electron density, simultaneously with the depletion of the level 1 and 2 populations. This implies increased photo-recombination rates, decreased opacities of the medium in the Lyman and Balmer series, and thus *an increase in the Balmer line emission from the surrounding "quiescent" chromosphere.*

Therefore, we are led to distinguish two main phases for the evolution of Balmer line fluxes with time, (Houdebine et al. 1989); a phase driven by radiative phenomena and most often impulsive, and a second gradual one related to the thermal phase of the flare plasma. We note that no restriction is proposed for the relative strength or timing of these two phases; they can be totally blended or one of them can be so weak that it becomes negligible. The only constraint is that a large U-band increase should be acompanied by a subsequent Balmer line increase.

An important point that one has to take into consideration, is the region where the flare radiant energy originates. If the flare continuum radiation arises from the photosphere, as proposed by de Jager et al (1989), the effects of the Lyman continuum will probably be confined within a very small volume around the flare site, due to the large chromospheric opacity. This also applies to the Balmer continuum, but to a lesser extent. Therefore, we can state that in that case, the flare will affect the upper chromosphere down to deeper chromospheric layers, but with a limited lateral extent. Conversely, if the radiation arises mostly from the low corona or high chromospheric levels, as in the solar two-ribbon white light flares, a surprisingly large chromospheric region is subjected to substantial changes, particularly the external layers where the Lyman and Balmer lines are formed.

.3 Correlation between U-band enhancements and Balmer line profiles and fluxes

Although no observations have been made of the Lyman continuum during stellar flares, the solar analogy as well as IUE stellar flare observations (Butler et al. 1981, Bromage et al. 1986) lead us to expect large and impulsive variations.

We saw that the enhancements in the Balmer and Lyman continua should increase the quiescent chromospheric emission in the Balmer lines. Thus, the rather good correlation between the U-band and Balmer lines enhancements, (see Butler, 1988), may be due to Balmer and Lyman continuum variations.

To illustrate this point, we report to Houdebine et al. (1990) and Rodonò et al. (1989) who give respectively the continuum and normalised line flux time-profiles, and the U-band variations, during the AD Leo flare on the 28 March 1989 at 3:21 UT (see also Houdebine et al. 1989). The continuum and U-band impulsive changes are followed by a similar behaviour in the Balmer lines, with the time delay observed in the lines in agreement with the mean recombination time of electrons at chromospheric densities. Similarly, Doyle et al. (1988) found a good time correlation during flares on YZ CMi.

The flare on AD Leo displays two rather distinct Balmer line phases; the former more impulsive than the latter. During the former phase, the line profiles of the Balmer lines show two very different components, (Figure 2). A two Gaussian fit gives FWHM values of $\sim$4.3 and $\sim$15.3 Å respectively, with fluxes of $\sim$2700 and $\sim$ 11000 10^{-16} ergs/cm^2/s. We interpret these interesting features, as being respectively the signature of a highly turbulent flare plasma, and the quiescent chromospheric reponse to the large flare radiation field.

In order to confirm the previous suggestions, we computed the effect of a Balmer continuum backwarming, arising from the top of the chromosphere, with a modified version of the Carlsson and Sharmer NLTE-Radiative-Transfer code (see Scharmer and Carlsson 1985 for the original version). A few runs have been performed that confirm the previous discussion. We show in Figure 3 the H_γ profiles of the quiescent chromosphere, and of the quiescent chromosphere irradiated by an increased Balmer continuum. These computations show that the flux has notably increased for all the Balmer series. Detailed results will be published later.

References

Bromage et al., *MNRAS*, **220**, 1021.

Butler, C.J., Byrne, P.B., Andrews, A.D., Doyle, J.G.: MNRAS, **197**, 815 (1981)

Butler, C.J., Rodono, M., Foing, B.H., Haisch, B.M.: *Nature*, **321**, 679 (1986)

Butler, C.J.: *IrishAstr.J.* **18**, 198 (1988)

De Jager et al.: *Astron.* and *Astrophys.*, **211**, 157 (1989)

Doyle, J.G., Butler, C.J., Byrne, P.B., Van den Oord, G.H.J.: *Astron. Astrophys.*, **193**, 229 (1988)

Hawley, L.S.: IAU Coll. on *Solar and Stellar Flares*, Poster Papers, Catania Astrophys. Obs. Special Volume, Eds. Haisch and Rodono (1989)

Houdebine, E.R., Butler, C.J., Panagi, P.M., Rodono, M., Foing, B.H.: IAU coll. 104 on *Solar and Stellar Flares*, Poster Papers, Eds. Haisch and Rodono, (1989)

Houdebine, E.R., Butler, C.J., Panagi, P.M., Rodono, M., Foing, B.H.: Submitted to *Astron. Astrophys.*, (1990)

Houdebine, E.R., Panagi, P.M.: To appear in *Astron. Astrophys.*, (1989)

Houdebine, E.R., Butler, C.J., Panagi, P.M., Rodono, M., Foing, B.H.: submitted to *Astron.* and *Astrophys.*, (1989)

Mullan, D.J., Tarter, C.B.: *Ap. J.*, **212**, 179 (1977)

Rodono, M., Houdebine, E.R., Catalano, S., Foing, B.H., Butler, C.J., Scaltriti, F., Cutispoto, G., Gary, D.E., Gibson, D.M., and Haisch, B.M.: IAU coll. 104 on *Solar and Stellar Flares* Poster Papers, Eds. Haisch and Rodono, (1989)

Scharmer, G.B, Carlsson, M.: *J. Comp. Phys.*, **59**, 56 (1985)

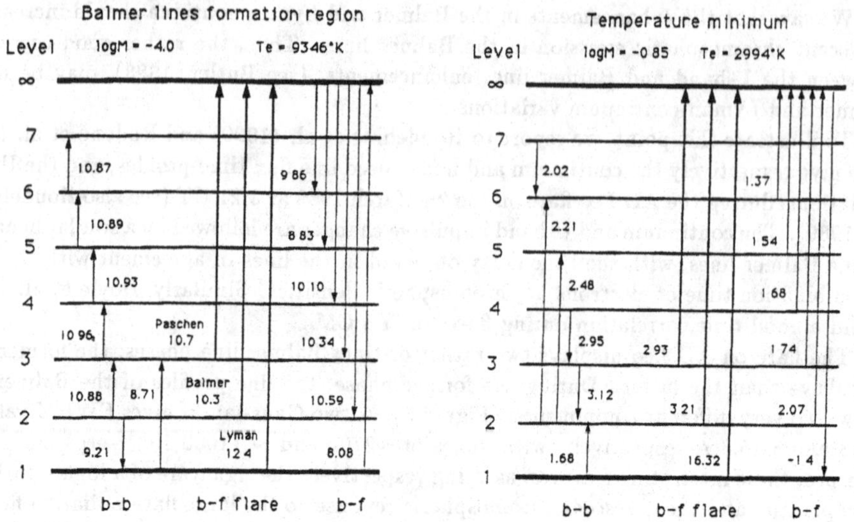

Fig. 1a, 1b: The logarithm of the main rates between the Hydrogen levels are given on the right and left sides (b-f and f-f). The values given as *b-f flare*, are the rates that would be created by an increase of the local photoionisation continua equal to, ten times the stellar quiescent flux. The amazingly high amplitudes of such flare induced rates imply a large atomic equilibrium re-arrangement in the quiescent chromosphere (see text).

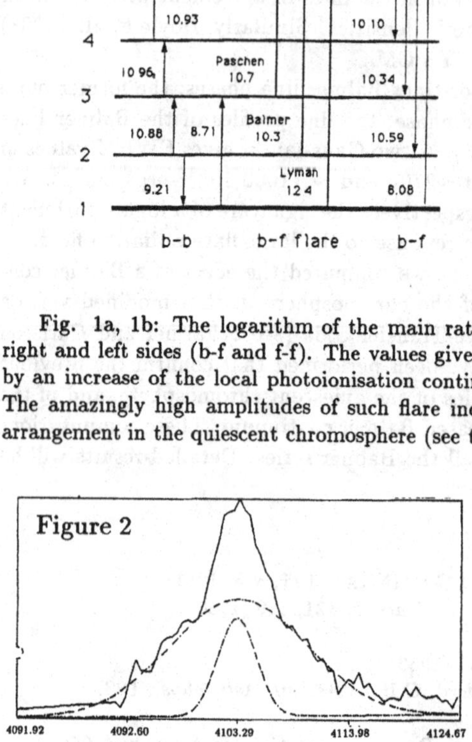

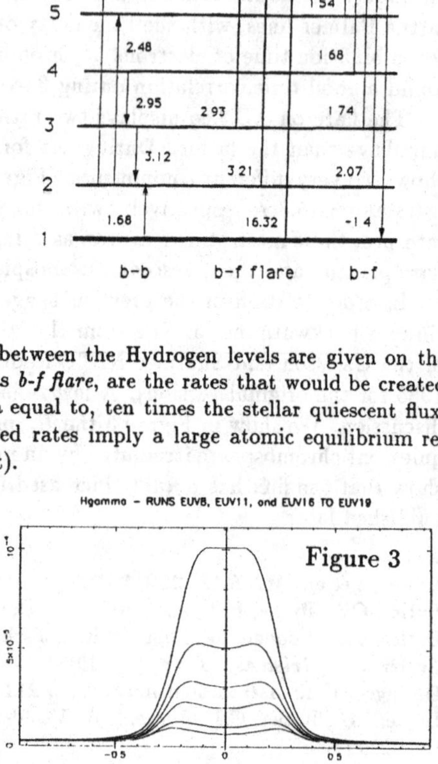

Wavelength (A)

Fig. 2: The H_γ line profile during the first spectrum of the AD Leo flare (28 March 1984, 3:21 UT). The profile displays two main components that here are fitted with two Gaussians. The narrower component is interpreted as being the signature of the quiescent chromospheric response to the flare radiation field.

Fig. 3: H_γ profiles, for the quiescent chromosphere, and the quiescent chromosphere under a simulated flare radiation field (only the Balmer continuum). The enhancement in the Balmer continuum (~U-band) relative to the quiescent stellar level was respectively, 1.0, 1.95, 5.13, 8.91, 16.2 and 38.0, with the lower profile the quiescent emission.

THE HYDROGEN ATOM KINETICS IN FLARE STAR CHROMOSPHERES

E. A. BRUEVICH, M. A. LIVSHITS
Institute of Terrestrial Magnetism, Ionosphere and Radio Wave Propagation USSR Academy of Sciences, Troitsk, Moscow Region, 142092, USSR

ABSTRACT. The solution of steady-state equations for hydrogen is carried out. The diffuse radiation field was taken into account in these equations the value β≡NRB - the escape probability of photons from the medium. This value is calculated for a layer of motionless plasma with finite optical depth. Our results are compared with analogical ones using Sobolev's method.

1. INTRODUCTION

The spectra of red dwarf stars with high temporal resolution were recently obtained. These data show very peculiar behaviour of the Balmer decrement during the flares and out-of-flare for several stars. The theoretical analysis of this problem was possible if one uses calculations by Gershberg and Schnoll, 1974 (further GS). However, these authors employed Sobolev's method for consideration of the radiative transfer problems. In our case, the large gradients of velocity in red dwarf atmospheres are absent and increase of transparence of medium at the Lyman and Balmer lines is not connected with motions. We have carried out the consideration of hydrogen atom kinetics for the case of motionless medium.

2. THE BASIS OF METHOD FOR OPTICALLY THICK STATIONARY MEDIUM

Here, as in GS, we have been examined the statistical equilibrium equations for hydrogen plasma without account of the outer radiation field:

$$n_i\Big[\sum_{k=1}^{i-1}(A_{ik}+B_{ik}J_{ki})+\sum_{\ell=i+1}^{\infty}B_{i\ell}J_{i\ell}+n_e\Big(\sum_{k=1}^{i-1}q_{ik}+\sum_{\ell=i+1}^{\infty}q_{i\ell}+q_{ic}\Big)\Big]=$$
$$=\sum_{\ell=i+1}^{\infty}n_\ell(A_{\ell i}+B_{\ell i}J_{i\ell})+\sum_{k=1}^{i-1}n_k B_{ki}J_{ki}+n_e\Big(\sum_{\ell=i+1}^{\infty}n_\ell q_{\ell i}+\sum_{k=1}^{i-1}n_k q_{ki}\Big)+ \qquad (1)$$
$$+\,n_e n_p C_i + n_e^2 n_p Q_{ci}$$

where n_i, n_k, n_l are populations of relevant quantum levels of neutral

L. V. Mirzoyan et al. (eds.), Flare Stars in Star Clusters, Associations and the Solar Vicinity, 317–319.

hydrogen atoms, n_e, n_p are electron and proton densities, A and B are the Einstein probabilities, J is mean intensity $\int I_\nu d\omega/4\pi$ at the frequences of lines, q is the rate of excitation by electron collisions, C_i is the coefficient of recombination, Q_{ci} is the coefficient of triple recombination to the level i.

We transform system (1), as GS by introducing the Mensel's parameter b_i

$$\frac{n_i}{n_e n_p} = b_i \frac{i^2 h^3}{(2\pi m k T_e)^{3/2}} \exp \chi_i \qquad (2)$$

where $\chi_i = h\nu/kT_e$ is the ionization potential of relevant quantum state in units kT_e, n_i is the population of the same level. Taking into account the previous results, in particular, the conclusion about the equilibrium populations of the upper level, i.e. $b_i=1$ for $i \geq 9$ in all the range of physical conditions discussed, we are confined to consideration of the hydrogen atom model "8 levels plus continuum" only.

The influence of radiative transfer in lines on the hydrogen atom kinetics will take into account by introduction in (1) of the escape probability of a photon from a layer β_{ik}=NRB for each of the lines. Otherwise, the diffuse radiation field J_{ki} for each of i→k transitions is excluded from (1) as follows

$$n_i(A_{ik} + B_{ik}J_{ki}) - n_k B_{ki} J_{ki} = n_i A_{ik} \beta_{ik} \qquad (3)$$

In our case the value β_{ik} is determined by escape of photons at frequencies of the line wings. Taking into account a weak dependence of solution of system (1) on β_{ik}, i.e. the solutions of radiative transfer equations, we shall use in (3) the value β_{ik} averaged over the layer.

We compute the value β_{ik} for layer with fixed optical depth at $L\alpha$ line. Assume for simplicity that primary sources of photons are distributed uniformly over a layer: S^*=const. Then using an expression for the mean number of scatterings $\bar{N}_0$ (Ivanov, 1973) and Ivanov's, 1972 approximation for the function $X(\infty, \tau)$, we can obtain

$$\bar{N}_0 = \frac{1}{\tau_0}\int_0^{\tau_0} X^2(\infty,\tau)\,d\tau = \frac{1}{\tau_0}\int_0^{\tau_0} \frac{d\tau}{1-\lambda+\lambda L(\tau)}$$

or

$$\bar{\beta}_{ik} = \beta = 1-(1-\lambda)\bar{N}_0 = 1 - \frac{1-\lambda}{\tau_0}\int_0^{\tau_0} \frac{d\tau}{1-\lambda+\lambda L(\tau)} \qquad (4)$$

Here λ is the probability of photon surviving by single scattireng for each line, $L(\tau)$ is the function, which determines the escape probability of a photon from a medium without any scattering:

$$L(\tau) = 2\int_0^\infty \alpha(x)\,Ei_2\left(\frac{\alpha(x)}{\alpha(0)}\tau\right) dx$$

where $\alpha(x)$ is absorption coefficient from Vidal et al., 1973, x is the distance from the line centre. This function for $n_e=10^{13}$ and 10^{14} cm^{-3} and $T=10^4$ K was given earlier (Bruevich et al., 1989).

3. RESULTS AND DISCUSSION

We have calculated $L(\tau)$ and β for each set of physical parameters ne, T, $\tau(L\alpha)$ and all the lines. For solution of steady-state equations (1), the

method NONLINEAR was used, which leads (1) to the system of weak nonlinear equations.
First of all, we repeated the GS'computation of n_i for $n_e=10^{12}-10^{14}$ cm^{-3} with calculation of β after the formulae given there. Taking these n_i values as a first approximation, we find the optical depths τ_{ki} for all the lines as

$$\frac{\tau_{ki}}{\tau_{12}} = \frac{\alpha_{ki}(0)}{\alpha_{12}(0)} \cdot \frac{n_k}{n_1}$$

Then we calculate β after (4) and carry out the solution of system (1). Then, finding once again the relative scale of optical depths, we determined once more τ_{ki} and repeated the solution of system (1) with recalculated value of β. This is enough to obtain the final result, given in Table for $T=10^4$ K.
Thus, our consideration supports the GS's results and shows that reference to velocity field is not necessary. If we put GS's parameter β_{21} equal to $1/\tau(L\alpha)$, their solution is described well by b_i, n_i-values, especially in the case of $n_e=10^{14}$ cm^{-3}; also the general conclusions of GS's consideration hold true. For $n<10^{14}$ cm^{-3} some numerical discrepancies appear, see Table.
The solution of steady-state equations depends on the value β_{ik} weakly. The precision in computation of β from (4) is enough to determine of values n_i with sufficient accuracy.

REFERENCES

Bruevich E.A., Katsova M.M., Livshits M.A.:1989, Preprint IZMIRAN No 44; Astron.J. of USSR, in press
Gershberg R.E., Shnoll E.E., 1974, Izv.Krimsk.Astrophys.Obs., V.50, P.122 (GS)
Ivanov V.V.:1972, Astron. J. of USSR, V.49, P.115
Ivanov V.V.:1973, Transfer of Radiation in Spectral Lines. Washington U.S.Cov.Print.Office
Vidal C.J., Cooper J., Smith E.W.: 1973, Astrophys.J.Suppl.Ser.V.25,P.37

TABLE

T=1E4	τ(Lα)=1E8		ne=1E14		ne=1E13 τ(Lα)=1E8		Method GS β_{21}=1E-8 ne=1E13	
i	β_{i1}	β_{i2}	b_i	n_i	b_i	n_i	b_i	n_i
1	-	-	4.96	1.5E14	231.8	6.9E13	27.52	8.2E12
2	1.0E-3	-	1.51	1.3E9	7.44	6.4E7	4.87	4.2E7
3	4.0E-3	4.32E-3	1.12	2.4E8	2.31	5.0E6	1.6	3.5E6
4	1.5E-2	2.48E-2	1.02	1.8E8	1.19	2.1E6	1.13	2.0E6
5	3.4E-2	5.18E-2	1.00	2.0E8	1.04	2.0E6	1.04	2.0E6
6	7.8E-2	8.55E-2	1.00	2.3E8	1.01	2.3E6	1.01	2.3E6
7	0.126	0.123	1.00	2.8E8	1.00	2.8E6	1.00	2.8E6
8	0.226	0.175	1.00	3.4E8	1.00	3.4E6	1.00	3.4E6
9	0.3	0.225	1.00	-	1.00	-	1.00	-

4.32E-3 means 4.32 10^{-3}

THE BALMER DECREMENT IN RED DWARFS SPECTRA DURING THE FLARES AND QUIESCENT STATE

M.M.KATSOVA, Sternberg State Astronomical Institute,
Moscow State University, 119899 Moscow V-234, U.S.S.R.

ABSTRACT. The Balmer decrements observed from flare stars during the flares and out-of-flare are analyzed on the basis of our new consideration of hydrogen atom kinetics. The steep Balmer decrements correspond to the emission source with electron density $n_e \leqslant 10^{13}$ cm^{-3} optical depth at the Lα line centre $\tau(L\alpha) \leqslant 10^7$ for $T=10^4$. The low-sloping Balmer decrement, being typical for flare maximum of red dwarfs as well as for several flare stars with measurable X-ray emission out-of-flare, are described by $n_e = 3\cdot10^{13}-10^{14}$ cm^{-3} and $\tau(L\alpha) = 10^7 - 10^8$ for $T=(1-1,5)\cdot10^4$ K.

1. INTRODUCTION. THE THEORETICAL BALMER DECREMENT.

The problem of interpretation of the Balmer decrements of red dwarfs out-of-flare and especially during the flares is far from being solved. Recent observations with high spectral and temporal resolution are called for new consideration of this problem. First of all, the hydrogen atom kinetics under conditions of late-type star chromosphere is studied. The solution of statistical equilibrium equations is carried out with approximate consideration of radiative transfer problem for motionless medium (Bruevich, Livshits, this proceeding).

Let us define the Balmer decrement (B.d.) as the ratio of the flux in a given line to the Hγ line flux:

$$\frac{F(Hn)}{F(H\gamma)} = \frac{n_1}{n_5}\cdot\frac{A_{12}}{A_{52}}\cdot\frac{h\nu_{12}}{h\nu_{52}}\cdot\frac{\beta_{12}}{\beta_{52}}$$

where n_k - the k-level population, A_{k2} - the probability of spontaneous transitions and β_{k2} - the averaged over the layer escape probability of a photon. Taking the escape probabilities β and level populations from the theoretical calculations, we found the theoretical B.d. When $\tau<10$ at the line centre considered, so approximate expression for β is not valid. In such a case, the profiles of the higher Balmer line were calculated.

The theoretical B.d., for instance, for $T=10^4$ K and electron densities $n_e=10^{13}$ and 10^{14} cm^{-3} are given at Fig.1. By each of curves theoretical depth at the Lα line centre $\tau(L\alpha)$ is written. Firstly, it

L. V. Mirzoyan et al. (eds.), Flare Stars in Star Clusters, Associations and the Solar Vicinity, 321–323.

is easy to see that the relative Hα- and Hβ-fluxes are most sensitive to the $\tau(L\alpha)$ variations. The same conclusion was made earlier by Grinin, 1969. Secondly, there is a certain value of $\tau(L\alpha)$ for each fixed value of T and ne in a layer, exceeding of which leads to a change of the B.d.slope: instead of being steep, decrement becomes low-sloping one. This transition occurs when $\tau=10$ at the Hδ line centre; that corresponds to $\tau(L\alpha)=3\cdot10^7$ for $n_e=10^{14}$ cm^{-3} and $T=10^4$ K. Otherwise, when the Balmer lines considered become optically thick, the slope decreases noticeably and it can become inverse for some range of $\tau(L\alpha)$: for instance, the Hδ-line fluxe can be larger than the Hγ-line flux. It should be noted that the curve for $\tau(L\alpha)<10^6$ on Fig.1 does not vary practically for $n_e<10^{13}$ cm^{-3} and does not depend on concrete value of $\tau(L\alpha)$. For $T>10^4$ K and the same densities $n_e=10^{13}$-10^{14} cm^{-3}, B.d. can be more sloping one at lower $\tau(L\alpha)$ by several orders of magnitude.

2. INTERPRETATION OF THE BALMER DECREMENT OBSERVED.

a)UV Ceti-stars out-of-flares Full set of the observational B.d. of quiescent flare stars was published by Gershberg, 1974a and Pettersen, Hawley, 1988.

Comparison of these data with the above mentioned B.d. theory (Fig.2, 3) shown that the observations are described quite well by the theoretical curve with $T=10^4$ K, $n_e=10^{13}$ cm^{-3} and $\tau(L\alpha)=10^8$. 1988 data allow to separate two types of behaviour of B.d., which conform to two sets of parameters at $T=10^4$ K: 1) $n_e=10^{13}$ cm^{-3}, $\tau(L\alpha)\leqslant10^8$; and 2) $n_e\leqslant10^{13}$ cm^{-3}, $\tau(L\alpha)<10^6$. This distinction is due to, more probably, decrease of electron density in chromospheres of these stars. An analysis shows that these two types of stars are distinguished by the X-ray levels: for stars with more low-sloping Balmer decrements Lx is higher by 1-2 orders of magnitude.

b)stellar flares Analysis of flare spectra, obtained with temporal resolution of a few minutes and discussed by Gershberg, 1974b, shows that in the first approximation they are described by theoretical B.d. with $T=10^4$ K, $n_e=10^{14}$ cm^{-3} and $\tau(L\alpha)>3\cdot10^7$ (Fig.4). For several spectra near the flare maximum, characteristic are somehow higher temperature and/or larger optical depth at the Lα line centre. Even for the flare 11 Dec.1965 on EV Lac we were able to trace the evolution of B.d.: its slope becomes steeper as the flare decays during 15 minutes (observations were made with 3-min temporal resolution).

Modern observations allow to give a more detailed analysis of B.d. evolution. Fig.5 shows comparison of the observations of 4 March 1985 flare on YZ CMi (Doyle et.al,1987) with the results of theoretical calculations. It is seen, that B.d. is much more low-sloping at the emission lines maximum than out-of-flare. This B.d. is consistent with the temperature $T=10^4$ K, electron density $n_e=10^{14}$ cm^{-3} and $\tau(L\alpha)=3\ 10^7$. Observations at the maxima of several other flares show that B.d. can be low-sloping and even inverse, that corresponds to some higher temperatures ($T\leqslant2\cdot10^4$ K). Subsequently, decrease of plasma opacity at all the lines and/or some decrease of electron density occur.

Parameters of low-temperature, dense emission source, forming after the U-band flare maximum, are good agree with our gas-dynamic model of stellar flare (Katsova and Livshits, 1988).

REFERENCES

Bruevich E.A., Livshits M.A. this proceeding

Doyle J.G., Butler C.J., Byrne P.B., van den Oord G.H.J.: 1988, Astron. Astrophys. V.193, P.229

Gershberg R.E.: 1974a. Astron. Zhurn. of USSR, V.51, P.552

Gershberg R.E.: 1974b. Izv. Krymsk. Astrophys. Obs., V.51, P.117

Grinin V.P.: 1969, Astrofizika, V.5, P.213

Katsova M.M., Livshits M.A.: 1988, "Activity in Cool Star Envelopes" Eds. Havnes O. et al. Dordrecht: Kluwer D., P.143

Pettersen B.R., Hawley S.L.: 1988, Prepr. Ser. Inst. of Theor. Astrophys. Univ. of Oslo, No 005

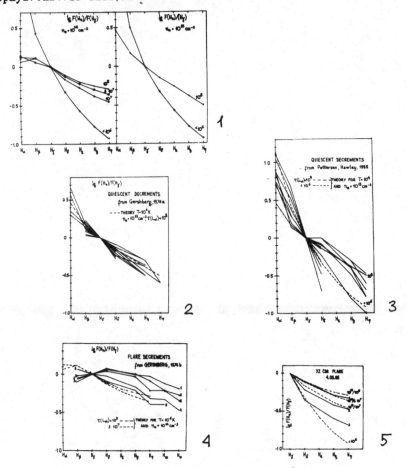

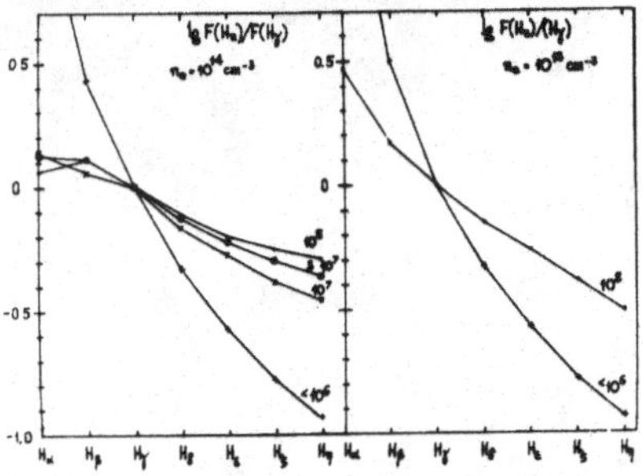

1

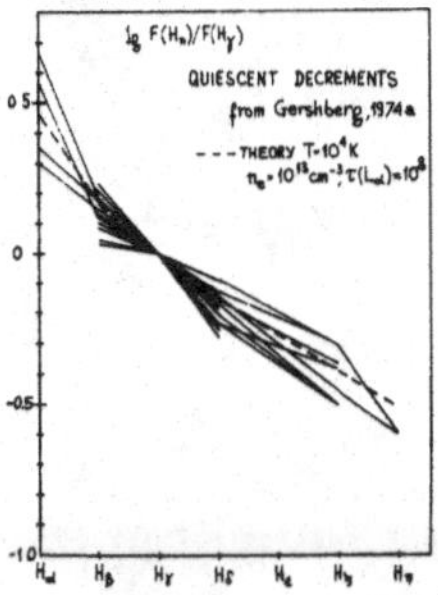

2

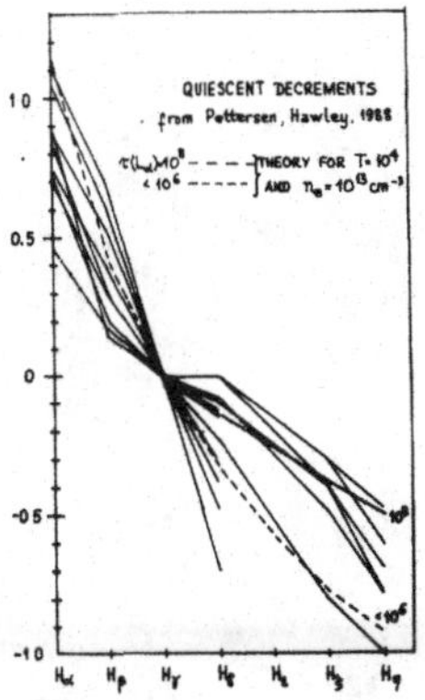

3

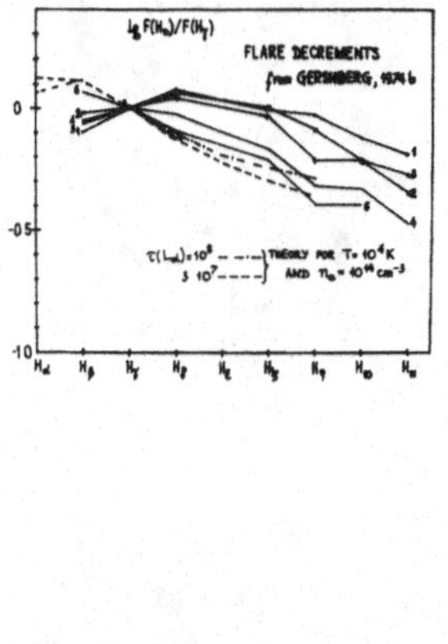

4

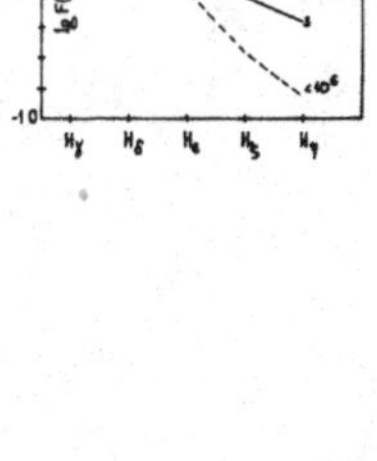

5

[illegible]
The [illegible] of [illegible] temperature, dense emission sources, forming [illegible] the [illegible]-band flare maximum, are [illegible] with our gas-dynamic [illegible] flare [illegible]

REFERENCES

[illegible] this proceeding.
Doyle J.G., [illegible] (Dynamics of [illegible] Solar Flare [illegible] 1988 Astron. Astrophys. [illegible]
[illegible] 1976 Astroph. Journ. [illegible] P. 553.
[illegible] S.[illegible] 19[illegible] [illegible] Astrophys. Obs. [illegible] P. 117
[illegible] V.R. 1969, Astrofizika [illegible]
[illegible] M. [illegible] 1986 [illegible] Catalogue [illegible]
[illegible] Dordrecht [illegible] P. 149
Patterson B. [illegible] Hawley S. [illegible] 1983 [illegible] Instit. Theor. Astrophys. [illegible]

A model for the observed periodicity in the Flaring Rate on YY Gem

J.G. Doyle[1] G.H.J. van den Oord[2] C.J. Butler[1] T. Kiang[3]

[1]Armagh Observatory, Armagh, N. Ireland
[2]Observatoire de Paris, Meudon, France
[3]Dunsink Observatory, Dublin 15, Ireland

Summary: Four flares were observed on the late-type binary YY Gem in March 1988 during a total monitoring time of 408 min. The flares were unusual in that there is a periodicity in their occurrence, being separated by 48 ± 3 min. Considering the flares to be formed as a stochastic process, we find that the probability of these events occurring by chance is 0.5%. Modelling indicates that for quite reasonable input parameters (e.g. a spot field strength of 1000 G and a filament with mass per unit length of $10^6\,g\,cm^{-1}$), the flare periodicity can be explained in terms of filament oscillations. The only requirement is that there should be a filament at these heights where the magnetic field drops inversely proportional to the height.

1 Introduction

Here, we report on observations of flares on the eclipsing binary YY Gem (Castor C = Gl 278C), which has a photometric period of 0.8142822 days (van Gent 1931). Both components are classified as dMe1 (Joy and Abt 1974), where the 'e' refers to the Balmer lines being in emission in the 'quiescent' state. Flare activity on this star was first reported by Moffett and Bopp (1971). The flares reported here are unusual in that there seems to be a periodicity in their occurrence. Below we discuss an interpretation of the flare data in terms of filament oscillations.

2 Observations and data reduction

The data were taken with a photometer on the University of Hawaii 60cm telescope on Mauna Kea from 2 to 7 March 1988. The integration time in the Johnson U-band was 5 secs. The figure below shows a section of U-band data (sky subtracted) in cts/s versus time in U.T. obtained on 6 March 1988. The first event at 7:30 UT is just detectable above the noise, with the last event at 9:59 UT being the largest. The 10 - 15 min. gaps in the data were times where we were either taking measurements of the sky or a comparison star.

L. V. Mirzoyan et al. (eds.), Flare Stars in Star Clusters, Associations and the Solar Vicinity, 325–328.

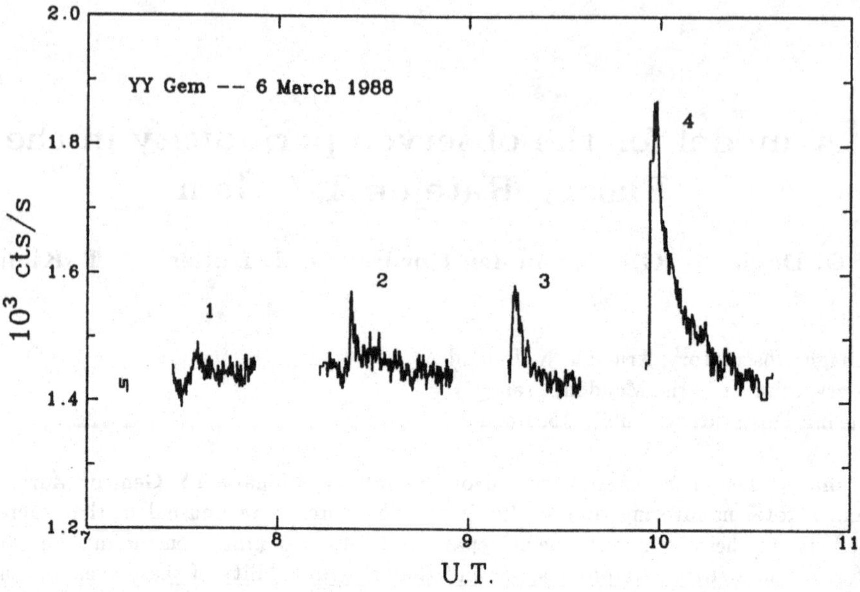

3 Flare statistics

The most interesting aspect of the flare data is the time-spacing between the peaks of the four events on 6 March 1988, i.e. 48 ± 3 mins. Previous authors have looked for such a periodicity in flare occurrence, e.g. Lukatskaya (1976) analysed 782 U-band flares from UV Ceti but found no evidence for a periodicity. Similar, Pettersen (1983) also attempted a correlation between the time-spacing of flares on V780 Tau and their energies, but also obtained a negative result. The reason for such interest is the idea of a "flare reservoir", i.e. an active region has a certain amount of stored energy which is then released either as a single flare or as a multiple of smaller events. Pazzani and Rodono (1981) in a statistical analysis of flare data on three stars, UV Ceti, EQ Peg and YZ Cmi showed that there was a significant departure from Poissonian behaviour in the distribution of time interval between successive flares and the number of events observed within a fixed time interval (these events were described as precursors). The four flares of 6 March 1987 on YY Gem occurred at intervals of 49, 51 and 44 min. The question one has to ask, is there a significant periodicity?, or rather assuming flaring to be a Poisson process, what is the probability of four flares occurring with more regular spacings than the above? Applying a detailed statistical analysis we find that the probability of the four flares arising out of a chance process is rather small ($\leq 0.5\%$), and so we conclude that there is a significant periodicity in their occurrence (a complete description of the statistical analysis is given in a paper to be published in *Astron. Astrophys.*)

4 A model

The observed periodicity of 48 minutes is long compared to MHD time-scales, e.g. for a loop of length 10^{10} cm and Alfven speed of 300 $km\,s^{-1}$, the Alfven travel time is only 6 minutes. It is therefore unlikely that the flares are triggered by an MHD process inside a coronal loop. We therefore consider the trigger mechanism to be related to filament oscillations.

The net force F on a filament in vertical direction and per unit length of the filament is given by

$$\lambda \frac{d^2 h}{dt^2} = F(h) \tag{1}$$

where h is the height of the filament above the photosphere and λ the mass of the filament per unit length. If the filament is in equilibrium at position h_o, we have $F(h_o) = 0$. Suppose that the filament is displaced over a small distance ϵ, e.g. by a passing shock wave, then we can write $h = h_o + \epsilon$. Substituting this in equ. (1) and making a Taylor expansion for ϵ gives

$$\lambda \frac{d^2 \epsilon}{dt^2} = \epsilon \frac{dF}{dh}\Big|_{h_o} \tag{2}$$

This can be rewritten as

$$\frac{d^2 \epsilon}{dt^2} + \omega^2 \epsilon = 0 \tag{3}$$

with

$$\omega^2 = -\frac{1}{\lambda}\frac{dF}{dh}\Big|_{h_o} \tag{4}$$

Under the condition $(dF/dh)_{h_o} < 0$, we see that the filament behaves as a harmonic oscillator with period $P = 2\pi/\omega$. The net force F on a filament is given by van Tend and Kuperus (1978) as

$$F(h) = \left(\frac{I}{c}\right)^2 \frac{1}{h} - \left(\frac{I}{c}\right) B(h) - \lambda g \tag{5}$$

with I the current in the filament (in statamps), $B(h)$ the active region potential field strength (in Gauss) as a function of height and g the gravitational acceleration. This expression is valid for heights h smaller than a stellar radius. The first term on the right-hand side accounts for the Lorentz force between the filament current and the currents induced in the photosphere by the magnetic field of the filament (see van Tend and Kuperus 1978 for further details). The terms are the Lorentz force on the filament current by the active region magnetic field and gravitational force on the filament.

In order to calculate the oscillation period of the filament at a specific height h in the corona we need to evaluate dF/dh for $F(h) = 0$. From equ. (5) we conclude that the current I required at height h for the filament to be in equilibrium (i.e. $F(h) = 0$) is

$$\left(\frac{I}{c}\right) = \frac{1}{2} h B(h) + \left[\left(\frac{1}{2} h B(h)\right)^2 + \lambda g h\right]^{1/2} \tag{6}$$

The gradient of the force is then given by

$$\frac{dF}{dh} = -\left(\frac{I}{c}\right)^2 \frac{1}{h^2} - \left(\frac{I}{c}\right) \frac{dB}{dh} \tag{7}$$

This then allows us to determine the oscillation period of the system as a function of the equilibrium position. We assume that the magnetic field strength in the stellar spots is $B_o \approx 1000$ Gauss and the mass per unit length of the filament $\lambda = 10^6\, gram\, cm^{-1}$ (see van Tend and Kuperus). Taking the masses of the components to be $0.6 M_\odot$ and the radii $0.6 R_\odot$, this results in a gravitational acceleration of $g_\odot/0.6$. This gives for $D/2 < h < D$ (where D is the active region size)

$$\omega^2 = \frac{g}{h} = \frac{2g}{D}(\frac{D/2}{h}) \tag{8}$$

For heights corresponding to $D/2 < h < D$, the period of the oscillations is given by

$$P = \frac{2\pi}{\omega} = 2\pi(\frac{h}{g})^{1/2} = 70.8(\frac{D}{R_*})^{1/2}(\frac{2h}{D})^{1/2}\, minutes \tag{9}$$

This time scale is indeed comparable to the observed period. Taking P = 48 minutes, we find for the ratio of the size of the active region to stellar radius

$$(\frac{D}{R_*}) = 0.46(\frac{D}{2h}) \tag{10}$$

For $D/2 < h < D$ we find $0.23 < (D/R_*) < 0.46$ or $9.6\, 10^9\, cm < D < 19.2\, 10^9\, cm$.

For $h < D/2$, the oscillation frequency is $\omega \approx 1 s^{-1}$. The reason for this extremely short frequency is that close to spots, gravity in unimportant and stability is maintained by the Lorentz force of the photospheric surface currents and the Lorentz force of the active region.

These results indicate that for quite reasonable input parameters ($B_o \approx 1\, kG$, $\lambda = 10^6\, g\, cm^{-1}$), filament oscillations can be expected. The only requirement being that there should be a filament at these heights where the active region varies like $1/h$.

References

Joy, A.H., Abt, H.A.: 1974, *Astrophys. J. Supp.* **29**,1

Lukatskaya, F.L.: 1976, *Soviet Astron. Lett.* **2**,61

Moffett, T.J., Bopp, B.W.: 1971, *Astrophys. J. Lett.* **168**,L117

Pazzani, V., Rodono, M.: 1981, *Astrophys. & Space Sci.* **77**,347

Pettersen, B.R.: 1983, *Astron. Astrophys.* **120**,192

van Gent, H.: 1931, *B.A.N.* **6**,99

van Tend, W., Kuperus, M.: 1978, *Solar Phys.* **59**,115

PINCH-MODEL OF FLARES AND ITS OBSERVATIONAL CONSEQUENCES

V.S.HAYRAPETYAN, A.G.NIKOGHOSSIAN
Byurakan Asrophysical Observatory
Armenian Academy of Sciences, USSR

ABSTRACT The paper presents the basic ideas of pinch-model for UV Cet star flares. The observational consequences and predictions are discussed.

1. INTRODUCTION

Cooperative observations of UV Cet star flares carried on recent years revealed the important features of stellar activity such as the noticeable local surface magnetic fields, the spot structure of the surface, the fine temporal structure of the light curves in optics and X-rays. The interpretation of flare phenomenon is based on sunlike MHD models. However this approach, as was shown in [1], encounters a number of significant difficulties, primarily, in predicting the energy of powerful stellar flares ($\sim 10^{35}$ ergs) and explaining the fine structure of light curves. To overcome in some extent the disagreements with observational data, we developed the pinch-model of flares. The model is based on the qualitative and quantitative analysis of observational data, the properties of the laboratory pinch-discharge as well as the numerical modelling of the pinch processes in stellar atmospheres. The comprehensive description of the model is given in [2].

This paper reports the main ideas and observational consequences of the pinch-model.

2. THE MAIN IDEAS OF THE PINCH-MODEL

According to the model, the flare process can be represented as follows. The formation of the closed magnetic configuration with low value of β can be due to floating (or stimulated floating, see below) of the magnetic fluxes from the convective zone of a star with origination of an arch

L. V. Mirzoyan et al. (eds.), Flare Stars in Star Clusters, Associations and the Solar Vicinity, 329–332.

structure of local magnetic field and it's "cleaning" during floating. If the force tubes in the lower parts of an arch are reconnected, such configuration will compress with the containing plasma owing to magnetic field tension. This process can lead to origination of solenoidal configurations analogous to that formed in laboratory pinch-discharge. The numerical modelling of magnetic thor evolution has been performed by present authors within the framework of ideal one-liquid two-dimensional MHD. The calculations show the evolution process to consist of two stages of compression. At the initial stage, thor collapses to the symmetry axis Z, stretching along that. As a result, both the external and internal boundaries become plane. The final stage is characterized by pinch-column and "hot-spots"(the regions with a high-temperature dense plasma) formation. According to model, the column radiation has a form of an impulsive optical flare, whereas the "hot-spot" radiation is manifested as the X-rays flare.

3. THE OBSERVATIONAL CONSEQUENCIES OF THE PINCH-MODEL

3.1. Time-scale and energetics of the flare process

The duration of the first stage is due to sonic speed of plasma compression within a thor and is of the order 300 s (if the following values for parameters of initial configuration are adopted: radius ~ 10^{8} cm, temperature kT~ 10 kev, density ~ 10^{-12} cm^{-3}). The final stage is caused by stretchnings compression with Alfven speed during 1-10 s. The stretchnings have a fluctuative nature with lifetime of the order to 1-10 s. Further evolution of the process lead to an increase in stretchnings temperature and density. Under some conditions (see [3]), the thermonuclear channel of energy release is switched. If the initial energy of the magnetic field is > 10^{33} erg, under above mentioned initial conditions, the thermonuclear release is dominant and reaches the values of the order 10^{35}-10^{36} erg. Thus the energy of powerful flares in considering model can be of thermonuclear nature.

3.2. Flare maxima in optics and X-rays.

The simultaneous observations of solar flares in optics and soft X-rays point to the delay (of about 2 min) of the maximum in soft X-rays relative to that in optics. Kodaira [7] also arrived at the same conclusion. when analysing the stellar flares. But, in this case, the delay is somewhat greater and amounts to 5 min. The interpretation of this fact encounters the difficulties within the framework of traditional flare models. At the same time, according to the developed model thesoft X-ray radiation is due to impulsive compression of stretchnings, therefore the difference between Alfven and sonic time-scales leads to agreement with observed time interval between maxima in optics and soft X-rays. However this problem requires the comprehensive statistical investigations.

The radiation in optical range results, as was said, from high-temperature ($\sim 10^5$ K), dense ($\sim 10^{15}$ cm^{-3}) and opaque hydrogen gas-pinch-column. According to [4], it is such values of plasma parameters are necessary to explain the wide-range-variations of flare colours (U-B, B-V) and Balmer jump in flare maximum.

3.3. Energy spectrum of particles and X-ray spectrum of non-thermal phase.

The preliminary research on soft X-ray energy spectra of the solar an stellar flares shows that the particles spectra for stellar flares are rather sloping ($\gamma \sim$ 1.5-2.0). In the pinch model, the particles gather their energy by consecutive collisions with converging current shell of the pinch-column. This mechanism also predicts the sloping energy spectra.

3.4. Asymmetry and central depression of emission lines.

The asymmetry and central depression of emission lines in flares spectra were detected quite recently with appearing of high-resolution spectra. This effect is also peculiar to solar flares. It was also shown, that it cannot be accounted for by absorption of infalling matter or by Stark-effect and other models. However in the pinch-model, the compression of plasma and its subsequent extension with a velocity gradient leads to either asymmetry or central depressions of Hα line for both opaque and transparent gaseous shell [5]. The results of this paper are in satisfactory agreement with observational data.

3.5. γ - bursts of powerful stellar flares.

Under some initial conditions the compression of stretchnings generates the thermonuclear wave of deiterium burning, resulting in γ-ray radiation by free-free transitions and lines of heavy elements. The details of the problem are given in [6].

4. ON STIMULATED HYDROMAGNETIC NATURE OF STELLAR FLARES

Solar and stellar flares are characterized by preflare

activity. At the present time, there is not generally accepted model for this stage, but it's clear that it can be considered as a trigger for flare. Our analysis of solar preflares permits to come to conclusion that under-photospheric impulsive energy release is the cause of this stage. The energy release heats plasma ($n \sim 10^{15} \div 10^{16}\ cm^{-3}$, $R < 10^{6}$ cm) up to $T \sim 10^{7}$K. It leads to stimulated floating of spontaneously lifting magnetic fluxes from convective zone via high-tempersture plasma extension. The stimulated floating speed is one order greater than spontaneous lifting speed. The recent data on solar anomal neutrino flux and its connection with flares indicates the under-photospheric nature of preflare.

REFERENCES

1. Gershberg,R.E., Mogilevskii,E.I., Obridko,V.W. (1987) 'Energetics of activity of flare stars and the Sun. A synergetical approach',Fiz. i Kinem.neb.tel,3(5),3-17
2. Hayrapetyan,V.S., Nikoghossian,A.G., Vikhrev,V.V.(1988) 'Pinch-effect and stellar flares physics', in T.D.Guyenne and J.J.Hunt (eds.), Reconnection in Space Plasma ESA SP-285,II,163-167
3. Hayrapetyan,V.S.(1989) 'Gamma-bursts and flares on red dwarfs',Astr.Tsirk.,1337,17-18
4. Buslavskii,V.G., Severny,A.B.(1969)'Asymmetry effect of solar flares emission', in V.V.Sobolev (ed.), Stars, Nebulae, Galaxies, pp.129-134.
5. Grinin,V.P., Sobolev,V.V.(1977)'On the theory of flare stars',Astrofisika,13,587-603
6. Harutyunian,H.A.,Hayrapetyan,V.S.(1989)'On the gamma activity of stellar flares',this volume
7. Kodaira,K.(1983)'Empirical models of stellar flares: constraints on flare theories, in P.B.Byrne and M.Rodono (eds.),Activity in Red Dwarf Stars,pp.561-578.

KONSTANTINOVA-ANTOVA: How can your model explain the very shortlived spike flares?

HAYRAPETYAN: They can be explained in a pinch-model as the "hot spot's" manifestation in the optical part of the spectrum. The characteristic time scales are from 0.1 to 1 second.

RODONO: The optical-X-ray delay of 2 minutes you mentioned is a minimum value or can your interesting model allow also for shorter delays?

HAYRAPETYAN: The pinch-model yields a wide interval of delay values, from some 10 seconds to several minutes. The actual value depends on the characteristic time of the second (Alfven) stage.

ON THE GAMMA-ACTIVITY OF STELLAR FLARES

H.A.HARUTYUNIAN, V.S.HAYRAPETYAN
Byurakan Astrophysical Observatory,
Armenian Academy of Siences, USSR

ABSTRACT. The principal possibility of γ-ray manifestation of stellar flares in the framework of pinch-model is shown. Validity of assumption that γ-bursts are connected with magnetically active dwarfs (MAD) is discussed.

1. INTRODUCTION.

Present-day models of γ-bursts postulate that for interpretation of the emission and absorbtion features of their energetic spectra it is necessary to invoke the neutron stars of various forms. However, all suggested models encounter with difficulties within the used mechanism. In [1] the pinch-model of stellar flares was suggested which predicts that γ-bursts must follow the powerful optical flares. In present paper an analysis of observatinal data on flares and cosmic γ-bursts in the light of pinch-model is given.

2. CONDITIONS FOR THE THERMONUCLEAR BURNING OF DEUTERIUM

The general picture of flare evolution within the pinch-model was discribed in [1,2]. The model supposes a generation of the closed magnetic configuration (magnetic thor) with low plasma β in upper layers of stars which is unstable to system compression relative to the symmetry axis. The numerical modeling shows that configuration compression with plasma within the thor take place. It leads to warm moderate density pinch-column and dense high temperature stretching unstability formation. The plasma column (with parameters $n \propto 10^{15}-10^{16}\,cm^{-3}$, $T \propto 10^{4}K$) is responsible for optical flare and the stretchings ($n \propto 10^{17}-10^{18}\,cm^{-3}$, $T \propto 10^{7}K$) are able to give X-ray radiation.

L. V. Mirzoyan et al. (eds.), Flare Stars in Star Clusters, Associations and the Solar Vicinity, 333–336.

However, a further compression of stretchings is also possible if the magnetic energy density exceeds the density of plasma thermal energy on this stage. Then if this condition is available magnetic compression must be continued up to balance realisation. And when the plasma temperature reachs the value $\propto$ 1 kev and deuterium concetration n_d as well as its life-time τ satisfy the Lawson criterion $n_d \tau > 10^{14}\,cm^{-3}s$, a thermonuclear wave of deuterium burning owing to $d(d,n)^3He$, $T(p,n)^3He$, $n(p,\gamma)d$ reactions can be generated.

The possibility of plasma compression in stretchings up to these conditions depends on plasma energy losses due to heat conductivity, thermal bremsstrahlung and plasma radiation in lines of heavy elements. The characteristic time of bremsstrahlung looses can be estimated by means of $\tau \propto nkT/\varepsilon$ (where ε is thermal bremsstrahlung emission coefficient) and τ happens to be less than the stretching hydrodinamic time or, so called, "flight time". The energetic losses by other ways also occur for a less time. So, the losses of energy lead to plasma cooling and, therefore, the balance between plasma and magnetic pressures will be disturbed. It holds a particular promise for the further compression of stretching plasma (radiational collapce). Just on this stage can be switched the thermonuclear channel of energy release. Every stretching then must show short γ-pulses.

Energy production by the thermonuclear channel is proportional to I^4 where I is the equivalent current intensity according to simple plasma focus model. Depending on the initial density of magnetic field the producted energy in γ-rays can reach values up to 10^{35}-10^{37} ergs. It is easy to obtain the condition for thermonuclear burning initiation in pinch

$$B \geq 5x10^5/\sqrt{\tau_0}$$

where B is the initial magnetic field strength, τ_0 is the burst duration time. One can easily find from the mentioned expressions 100 kGs for the magnetic energy strength. It corresponds to magnetic energy of the order of 10^{34} erg (under column volume $V \propto 10^{25}\,cm^3$). In this case the energy of optical flare is proposed 10^{34} ergs and the γ-burst energy is 10^{36} ergs. Thus the luminosity ratio gamma/optics can have values more than 100. And because of stretchings formation after the column stage a delay of X- and γ-flares to optical flares must take place.

It must be mentioned also that the pinch column stretchings have a fluctuative nature and, therefore, it assums an uncorrelated γ-burst generation.

3. COSMIC γ-BURSTS AND MAD

Let us consider now basic observation data on γ-bursts in detail. Those are the very short duration times (0.01-10 s), fine structure of light curves, irregularities in time, thermal energetic spectrum of 10^{6}-10^{10} K temperature plasma with emission (440 kev) and absorption (30-70 kev) features, existence of γ-"precursors" and etc [3].Thermonuclear models of γ-burst radiation meet with insuperable difficulties under "precursors" and time scale interpretation. Moreover, explanation of emission and absorption features requires the existence of strong magnetic ($\propto 10^{12}$ Gs) and gravitational ($\propto$ 0.1-0.3 c^2) fields. However, it was shown in [4] that these features can be explaned without above mentioned requirements. Emission feature around 400 kev, for example, can be caused not only by red shifted e^+e^- annihilation but also due to $^4He(\alpha,n)^7Be$ reaction.

The theoretical preconditions allow us to interprete the γ-bursts in the pinch-model framework, namely, that MAD with deep covective zones and low absolute magnitudes can be considered as effective γ-sources. First of all, both MAD and γ-bursts optical identification show that stellar magnitudes are limited by $M_v > 14^m$. Secondly, MAD stars and γ-ray bursts are distributed isotropically in the sky. In third place, there exist sufficiently serious indications to optical identification of 34 objects, being flare stars of solar vicinity, RS CVn stars, cataclysmic variables with γ-sources. Moreover,the temporal and positional corellations between X-ray flares from AR Mon and UX Ari and γ-burts are found [4]. It must be noted also that there are preliminary evidences for identification of T Tau stars in ρ Oph cloud with γ-sources [5]. It means, that all magnetically active dwarfs can be treated as γ-sources. Such a treatment seems to be plausible, because all of the mentioned objects possess of deep convective zones, spot structure, noticeable magnetic fields of local structure and,as a consequence, flare activity. In forth place, taking into account, that the distance of MAD are of order 10-30 ps on one hand and stellar pinch-model predicts 10^{35}-10^{37} ergs for γ-bursts energy on other hand, we can obtain γ-fluxes in solar vicinity. which compose 10^{-7}-10^{-5} erg/cm^2 and show a good agreement with observation data [3]. And finally, from our point of view, it is important, that "precursors" are observed for both powerful optical flares and γ-bursts. This is an indirect evidence of analogy of these two phenomena.

Thus, the results of pinch-model and recent observational data analysis speak in favour of flare nature of cosmic γ-bursts. And so, the further cooperative optical and gamma observations, for example, of stellar associations seem to be necessary.

REFERENCES

1. Hayrapetyan,V.S.(1989) 'Gamma-bursts and flares on red dwarfs',Astr.Tsırk.,17-18.
2. Hayrapetyan,V.S.,Nikoghossian,A.G. and Vikhrev,V.V.(1989) 'Pinch-effect and stellar flares physics',in T.D.Guyenne and J.J.Hunt (eds.), Reconnection in Space Plasma, ESA SP-285,II,163-167.
3. Mazets,E.P. and Golenetskii S.V.(1987) 'Cosmic gamma-burst observations',in R.A.Sunyaev (ed.),Itogi nauki i tekhnik ser. Astronomia, Moskva,16-42.
4. Vahia,M.N. and Rao,A.R.(1988) 'Origin of the gamma-ray bursts', Astron.Astrophys.,207,55-69.
5. Montmerle,T.,Koch-Miramond,L.,Falgarone,E. and Grindlay, J.E. (1983),'Einstein observations of the Rho Ophiu-Chi dark cloud: an X-ray Christmas Tree',Astrophys.J.,269, 182-201.

ON A DIFFERENTIAL EQUATION FOR ELECTROMAGNETIC WAVE TRANSMISSION IN FLARE STARS AND THE POSSIBLE EXISTANCE OF COHESIVE WAVE SOLUTIONS

by

CHARLES HERACH PAPAS
California Institute of Technology
Pasadena, California 91125

Abstract

Ambartsumian's celebrated hypothesis that stellar flares and other phenomena of stellar instability are due to a novel source of energy and a novel means of transporting this energy to the outer layers of the star has drawn the attention of electrodynamicists to a number of fundamental problems. One of these problems, namely the energy transport problem, is the subject of this communication. Herein, by assuming that the matter of the star is an isotropic collisionless plasma, from Maxwell's field equations and Newton's equation of motion with nonlinear Lorentz driving force, we have derived a vector differential equation for electromagnetic wave propagation. This equation contains the Debye radius and the plasma frequency as parameters, and reduces to the well-known wave equation when its nonlinear terms are neglected. We have indicated that the nonlinear equation has cohesive (solitary) wave solutions for both the longitudinal and transverse components of the electromagnetic field. Such cohesive waves are appropriate for transporting energy from the prestellar core of the star to its outer layers since they hold their shape, are free from dispersive distortion, and can carry energy in discrete amounts.

L. V. Mirzoyan et al. (eds.), Flare Stars in Star Clusters, Associations and the Solar Vicinity, 337–342.

Introduction

In a previous communication[1] based on Ambartsumian's famous hypothesis on stellar instabilities, we pictured a flare star as a kind of transformer that converts the low-entropy energy of the star's pre-stellar matter into the high-entropy energy of the star's flare radiation, and we reasoned that energy is drawn from the pre-stellar matter and deposited in discrete amounts on the outer layers of the star by means of cohesive waves having the form of solitary waves or solitons.

In the present communication we focus our attention on the cohesive waves and submit that such waves are mathematically possible if the nonlinearity of the plasma comprising the star is taken into account.

We proceed by deriving from Maxwell's equations and Newton's equation of motion a differential equation for the propagation of electromagnetic waves in an isotropic collisionless electronic plasma which we suppose resembles closely the plasma of the star. We find that the resulting equation is a nonlinear vector equation. To handle such an equation, we scalarize it and obtain a nonlinear system of two coupled scalar equations. And it is these coupled equations that we regard as the mathematical starting point of the problem.

Nonlinear Vector Differential Equation

We assume that the star's plasma is an isotropic collisionless electronic plasma; and we recall that for such a plasma the electric field $\mathbf{E}$, in the linear approximation, must satisfy the well-known equation[2]

$$c^2\nabla \times \nabla \times \mathbf{E} - 3\alpha^2\omega_p^2\, \nabla(\nabla \cdot \mathbf{E}) + \frac{\partial^2 \mathbf{E}}{\partial t^2} + \omega_p^2\, \mathbf{E} = 0\ , \tag{1}$$

where c denotes the velocity of light, ω_p denotes the plasma frequency, and α denotes the Debye radius. We also recall

$$m\alpha^2\omega_p^2 = \kappa T\ , \tag{2}$$

where κ is Boltzmann's constant, T is the temperature, and m is the electronic mass. The vector $\mathbf{E}$ can be expressed as the sum of a transverse field $\mathbf{E}^T$ and a longitudinal field $\mathbf{E}^L$. That is,

$$\mathbf{E} = \mathbf{E}^T + \mathbf{E}^L\ , \tag{3}$$

where, by definition,

$$\nabla \cdot \mathbf{E}^T = 0 \quad \text{and} \quad \nabla \times \mathbf{E}^L = 0\,. \tag{4}$$

Accordingly, from equation (1) it follows that for transverse waves

$$\nabla^2 \mathbf{E}^T - \frac{1}{c^2}\frac{\partial^2 \mathbf{E}^T}{\partial t^2} - \frac{\omega_p^2}{c^2}\mathbf{E}^T = 0\,, \tag{5}$$

and for longitudinal waves

$$3\alpha^2\omega_p^2\, \nabla^2 \mathbf{E}^L - \frac{\partial^2 \mathbf{E}^L}{\partial t^2} - \omega_p^2 \mathbf{E}^L = 0\,. \tag{6}$$

Since equation (5) does not involve $\mathbf{E}^L$ and since equation (6) does not involve $\mathbf{E}^T$ we see that in the linear approximation there is no interaction between the longitudinal and transverse waves. In other words, if a wave is initially transverse, it remains transverse, and if a wave is initially longitudinal it remains longitudinal. This is true in the linear approximation but not in the nonlinear case.

Taking into account the nonlinearity of the plasma that is quadratic with respect to the electric field $\mathbf{E}$ one can show that $\mathbf{E}$ must now satisfy the equation[(3)]

$$c^2 \nabla \times \nabla \times \mathbf{E} - 3\alpha^2\omega_p^2 \nabla(\nabla \cdot \mathbf{E}) + \frac{\partial^2 \mathbf{E}}{\partial t^2} + \omega_p^2\, \mathbf{E} = \frac{\omega_p^2\, e}{2m}\frac{\partial}{\partial t}(\mathbf{Z} + 2\mathbf{\Psi}\nabla \cdot \mathbf{\Phi})\,, \tag{7}$$

where e denotes the electronic charge and where the vectors $\mathbf{Z}$, $\mathbf{\Psi}$, $\mathbf{\Phi}$ satisfy

$$\frac{\partial \mathbf{Z}}{\partial t} = \nabla(\mathbf{\Psi} \cdot \mathbf{\Psi}), \quad \frac{\partial \mathbf{\Psi}}{\partial t} = \mathbf{E}, \quad \frac{\partial^2 \mathbf{\Phi}}{\partial t^2} = \mathbf{E}\,. \tag{8}$$

The right side of equation (7) expresses the quadratic nonlinearity of the plasma.

The derivation of equation (7) is based on the Maxwell field equations and on the Newton equation of motion for the electrons of the plasma. From Maxwell's equations we have

$$\nabla \times \nabla \times \mathbf{E} + \frac{1}{c^2}\frac{\partial^2 \mathbf{E}}{\partial t^2} = -\frac{4\pi}{c^2}\frac{\partial \mathbf{J}}{\partial t}\,. \tag{9}$$

The current density $\mathbf{J}$ is given by

$$\mathbf{J} = ne\mathbf{v}\,, \tag{10}$$

where n is the electron density and $\mathbf{v}$ is the electron velocity, and the conservation of charge is given by

$$\frac{\partial n}{\partial t} + \nabla \cdot (n\mathbf{v}) = 0 \,. \tag{11}$$

From Newton's equation of motion we have

$$\frac{\partial \mathbf{v}}{\partial t} + (\mathbf{v} \cdot \nabla)\mathbf{v} = \frac{e}{m}(\mathbf{E} + \frac{1}{c}\mathbf{v} \times \mathbf{H}) - \frac{1}{mn}\nabla p \,. \tag{12}$$

The left side of this equation is the convective derivative of the velocity, and the first term on the right side is the Lorentz force, and the second term on the right side is the force due to electron pressure. With the aid of equations (10), (11), and (12) the current $\mathbf{J}$ can be expressed in terms of $\mathbf{E}$, and by substituting the resulting equation into the right side of equation (9) we can obtain equation (7).

When we neglect the nonlinear terms $(\mathbf{v} \cdot \nabla)\mathbf{v}$ and $\mathbf{v} \times \mathbf{H}$ in the equation of motion we obtain equation (1), but if take these nonlinear terms into account we obtain equation (7). The second term on the left side of equation (7) comes from the electron pressure term ∇p of equation (12), and the right side of equation (7) comes from the nonlinear terms.

The nonlinear vector differential equation (7) is the equation we must solve to see whether or not cohesive (solitary) wave solutions are possible. To make this equation mathematically tractable it must be scalarized.

Coupled Scalar Equations

To reduce the vector wave equation to scalar form we orient the Cartesian coordinates x, y, z so that the x-axis becomes the longitudinal direction (the direction of propagation) and the z-axis becomes the transverse direction. From equation (8) we see that in component form $\mathbf{E}$ is given by

$$\mathbf{E} = \left[\frac{\partial^2 \Phi^L}{\partial t^2},\ 0,\ \frac{\partial^2 \Phi^T}{\partial t^2} \right] , \tag{13}$$

or by

$$\mathbf{E} = \left[\frac{\partial \Psi^L}{\partial t},\ 0,\ \frac{\partial \Psi^T}{\partial t} \right] . \tag{14}$$

Here Φ^L and Φ^T are the longitudinal and transverse components of the vector $\mathbf{\Phi}$, Ψ^L and Ψ^T are the longitudinal and transverse components of the vector $\mathbf{\Psi}$, and all four scalars Φ^L, Φ^T, Ψ^L, Ψ^T are functions of only x and t.

In view of representation (13) we can write equation (7) as two coupled scalar equations:

$$\mathcal{L}_1\,\Phi^L = \frac{e}{m}\omega_p^2\,(2\Phi^T_{xt}\Phi^L + \Phi^L_{tt}\Phi^L_x + \Phi^T_{tx}\Phi^T_t)\,, \tag{15}$$

$$\mathcal{L}_2\,\Phi^T = \frac{e}{m}\omega_p^2\,(\Phi^T_{tt}\Phi^L_x + \Phi^L_{xt}\Phi^T_t)\,, \tag{16}$$

where the subscripts x and t denote partial differentiation with respect to x and t, and where

$$\mathcal{L}_1 = \frac{\partial^4}{\partial t^4} - 3\alpha^2\omega_p^2\,\frac{\partial^4}{\partial x^2\partial t^2} + \omega_p^2\,\frac{\partial^2}{\partial t^2}\,, \tag{17}$$

$$\mathcal{L}_2 = \frac{\partial^4}{\partial t^4} - c^2\,\frac{\partial^4}{\partial x^2\partial t^2} + \omega_p^2\,\frac{\partial^2}{\partial t^2}\,, \tag{18}$$

are linear operators.

Since we are interested in a wave profile that is moving at a constant speed U, the x and t derivatives are related to each other linearly, i.e.

$$\frac{\partial}{\partial t} = -U\frac{\partial}{\partial x}\,. \tag{19}$$

For waves that satisfy relation (19) we can rewrite equation (15) and (16) in terms of Ψ^L and Ψ^T. That is we can write

$$\mathcal{M}_1\,\Psi^L = \frac{e}{m}\omega_p^2\,(3\Psi^L_x\Psi^L + \Psi^T_x\Psi^T)\,, \tag{20}$$

$$\mathcal{M}_2\,\Psi^T = \frac{e}{m}\omega_p^2\,(\Psi^T\Psi^L)_x\,, \tag{21}$$

where the linear operators $\mathcal{M}_1$ and $\mathcal{M}_2$ are given by

$$\mathcal{M}_1 = \frac{\partial^3}{\partial t^3} - 3\alpha^2\omega_p^2\,\frac{\partial^3}{\partial x^2\partial t} + \omega_p^2\,\frac{\partial}{\partial t}\,, \tag{22}$$

$$\mathcal{M}_2 = \frac{\partial^3}{\partial t^3} - c^2\,\frac{\partial^3}{\partial x^2\partial t} + \omega_p^2\,\frac{\partial}{\partial t}\,. \tag{23}$$

The coupled scalar equations (20) and (21) comprise the mathematical starting point of the problem. By inspection of these equations one can see that $\Psi^L = 0$ implies that $\Psi^T = 0$ but $\Psi^T = 0$ does not imply $\Psi^L = 0$; that is, purely longitudinal waves may exists whereas purely transverse waves may not. When transverse waves exist they are accompanied by longitudinal waves; and the interaction between transverse and longitudinal waves is due to the inclusion of the nonlinearity of the plasma.

Cohesive Wave Solutions

Using the nonlinear coupled equations (20) and (21) as a point of departure, we have shown elsewhere[3] that cohesive (solitary) vector wave solutions are possible. We do not reproduce the calculations here because they are tedious and irrelevant to the matter at hand. The point of importance is that cohesive wave solutions can exist. For such waves the distortions produced by plasma dispersion are cancelled by distortions produced by plasma nonlinearity and the waves can travel through the plasma at a constant velocity and without any change of profile shape.

References

(1) C. H. Papas," On the electrodynamical implications of flare stars", in: L. V. Mirzoyan, Ed., *Flare Stars*, Academy of Sciences, Yerevan (1977) 175-180.

(2) V. L. Ginzburg, *Propagation of Electromagnetic Waves in Plasmas*, Pergamon Press, Oxford (1970); V. N. Tsytovich, *Nonlinear Effects in Plasmas*, Plenum, New York (1970).

(3) J. Z. Tatoian and C. H. Papas, " On solitary waves in plasma", Wave Motion, **8** (1986) 415-438.

Acknowledgment

The author wishes to thank Dr. S. Bassiri for his assistance and technical advice.

THE STOCHASTICAL APPROACH TO THE MODELLING OF THE LINE PROFILES IN T TAURI STELLAR WINDS.

V.P.GRININ, A.S.MITSKEVICH
Crimean Astrophysical Observatory.
P.O.Nauchny, Crimea, 334413
U.S.S.R.

ABSTRACT. Line profiles calculations were carried out for spherically-symmetrical stellar wind models with discrete structure. It is shown the possibility to explain two-component emission profiles typical for T Tauri-type stars in the framework of such stochastical approach.

1.INTRODUCTION.

Intensive outflow of the matter is one of the most important property of T Tauri - type stars (TTS). There is a set of theoretical models in which this phenomenon is considered from different positions. But any of them can explain the emission line profiles which are observing in the spectra of TTS: two-component emission with a blue-shifted absorption. This defect, as we believe, is providing by the using of continuum medium assumption.

Here we present a new approach to the modelling of the emission line profiles in the framework of discret stellar wind with stochastical approach

2.CALCULATIONS.

The calculations are organized by the following way:

1. Our initial models (the radial-symmetric continuum outflow with a giving velocity field and $\dot{M}$) assumed to be isothermal.
2. Non-LTE multilevel problems for H I, Ca I-III, Mg I-III and calculation of the source functions for the main emission lines observed in the spectra of TTS ($H\alpha$, K CaII, k MgII and some others) were carried out with the using of Sobolev escape probability method.
3. The discretization of the envelope with the help of random number code was final (and main) step. This step simulate the erruptive nature of the stellar wind, consisting from the large number (up to 10^4) of blobs.

Such models are spherically-symmetric, but not selfconsistent fully because of the using of the source functions from continuum wind models.

L. V. Mirzoyan et al. (eds.), Flare Stars in Star Clusters, Associations and the Solar Vicinity, 343–346.

3. RESULTS.

Examples of the main emission line profiles (Hα, Hβ, K CaII and k Mg II) calculated for isothermal wind are presented on Fig. 1-4. They are obtained for the following model parameters: electron temperature T_e = 7500 K , temperature of the star T_* = 5000 K (stellar radiation was approximated by Plank function), stellar radius $R_* = 3*R_\odot$.

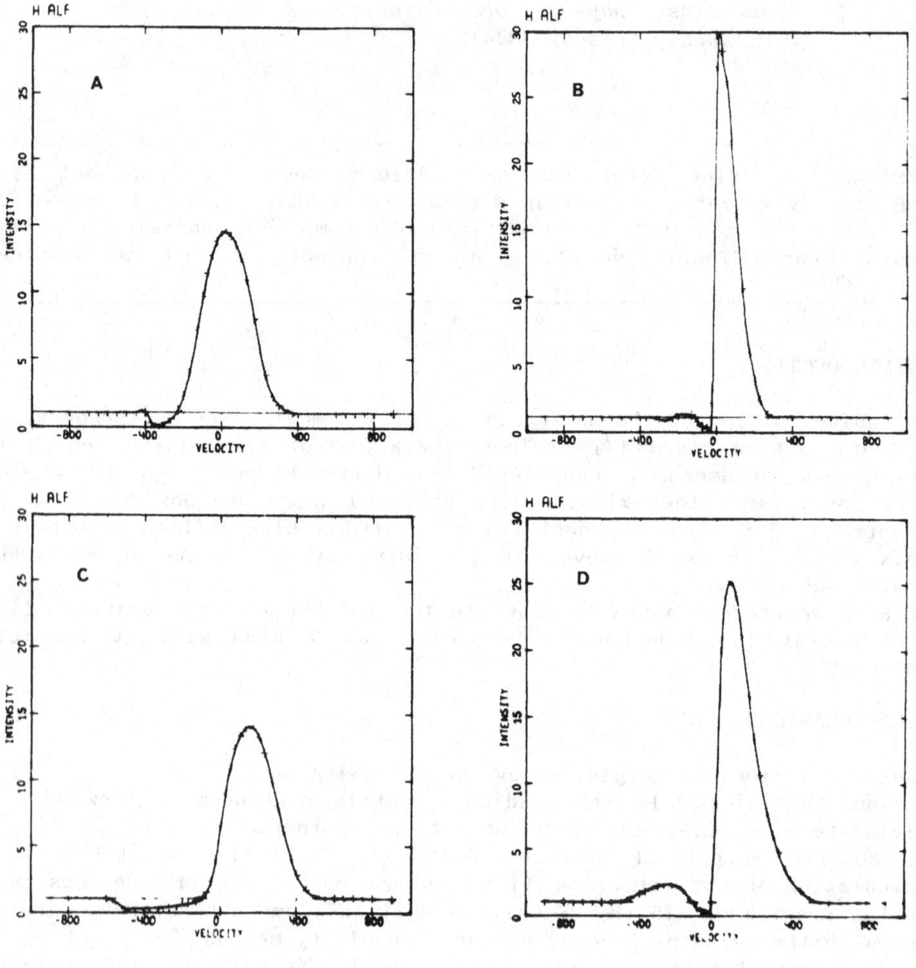

Figure 1. Theoretical profiles of Hα -line for the case of continuous outflow ($\dot{M}=3*10^{-7}$ $M_\odot$/year) with acceleration (a,c) and deceleration (b,d), with (c,d) and without (a,b) turbulent velocity.

Two different types of kinematical models were considered :

1. outflow with acceleration (A- wind):

$$V(r) = V_o + V_1 \ (1-(r/R_*)^{-1/2}) \ ;$$

2. outflow with deceleration (D- wind):

$$V(r)=V_o*(r/R_*)^{-1/2} \ ;$$

were maximum velocity is V_m =400 km/s in both cases.

All calculations of lines profiles were provided with an accurate equation for the intensity of escaping radiation. It allowed to consider two limiting cases of the motion with and without turbulent velocities: $V_{turb}(r)$= A*V(r) were scaling factor was A= 0. and 0.3 .

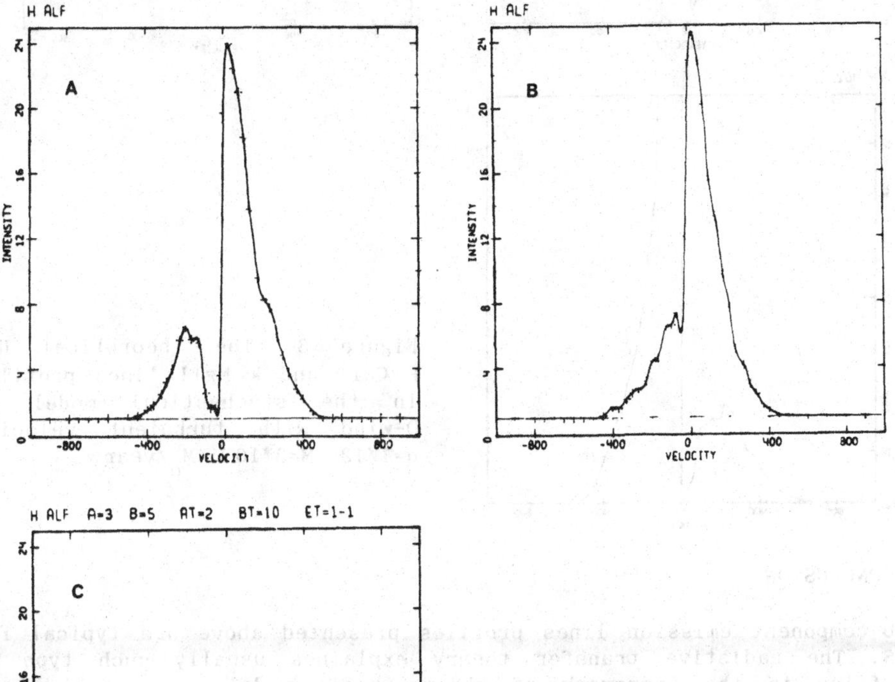

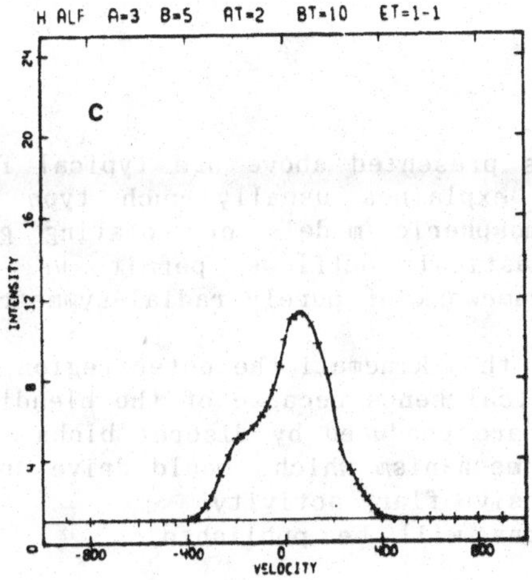

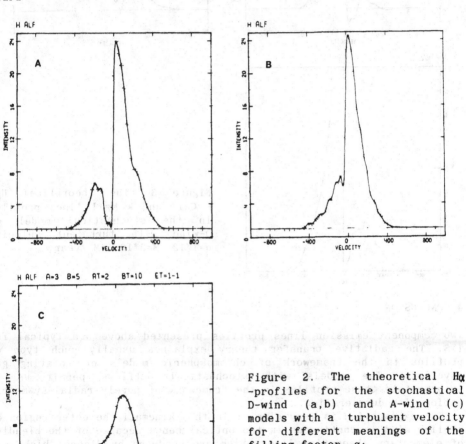

Figure 2. The theoretical Hα -profiles for the stochastical D-wind (a,b) and A-wind (c) models with a turbulent velocity for different meanings of the filling factor q:
a) q=1/12, b) q=1/30, c) q=1/12.
In all cases $\dot{M}$ = $3*10^{-7}$ M_o/year.

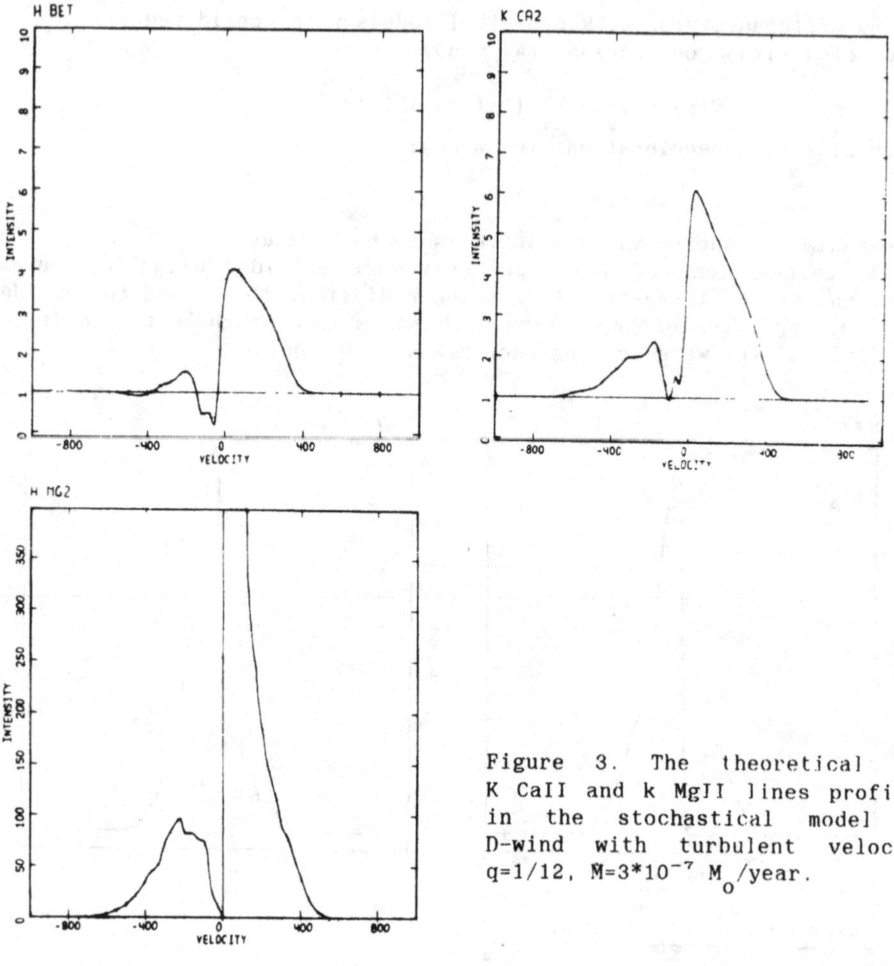

Figure 3. The theoretical Hβ, K CaII and k MgII lines profiles in the stochastical model of D-wind with turbulent velocity q=1/12, $\dot{M}=3*10^{-7}\ M_{o}$/year.

4.CONCLUSION.

Two-component emission lines profiles presented above are typical for TTS. The radiative transfer theory explaines usually such type of profiles in the framework of chromospheric models or rotating gas envelopes. The models with stochastical outflows permit us to interpretate these profiles in the framework of purely radial-symmetric motion with deceleration.

The explanation is very simple: in this kinematicthe outer region of stellar wind is continuous in the optical sence because of the blending of elementary absorption lines which are produced by discret blobs. It is naturally to connect the physical mechanism which could drive such stellar wind from TTS with their intensive flare activity.

More discussions about calculations will be published.

On The Origin Of Dwarf Stars

Charles J. Lada
Steward Observatory, University of Arizona, Tucson, Arizona, USA 85721

The origin of dwarf or low mass stars is one of the most interesting and challenging problems of modern astrophysics. In recent years advances in observational technology particularly at infrared and millimeter wavelengths, have produced an avalanche of revealing new data, unexpected discoveries and new mysteries about the process of star formation. From this new knowledge a complete empirical picture of stellar origins is being synthesized and a more profound and penetrating understanding of the physical process of star formation in our galaxy is beginning to emerge. It is now apparent, for example, that energetic bipolar outflows are a fundamental aspect of the formation of low mass stars and understanding how a star can form by the act of ejecting mass may be the key to unlocking the secrets of stellar genesis.

1. INTRODUCTION

Stars are the basic objects of the universe. Understanding how they form from interstellar clouds of gas and dust is one of the most fundamental unsolved problems of modern astrophysics. Of course, particular interest is evoked in attempts to decipher the origins of dwarf or low mass stars, like the sun, which are the most populous stars in the galaxy. Indeed, most stars which form from interstellar clouds are probably *less* massive than the sun. However, studying the formation and early evolution of stars has turned out to be a formidable challenge for astronomers. This is because stars are born within the dust enshrouded cores of giant molecular clouds. Here newly forming stars (protostars) are rendered completely invisible by the obscuration provided by the visually opaque dust which permeates these clouds. Moreover, the molecular gas which forms stars is extremely cold (i.e., 10–20 Kelvins) and can only be observed at millimeter and submillimeter wavelengths, spectral windows not opened by radio astronomers until the 1970s. As a result of these facts, the star formation process is veiled from direct observation at visual wavelengths and the classical tools of optical astronomy are ineffective probes of regions where stars are being formed. On the other hand, although dust effectively absorbs visible light emitted by buried stars and protostars, this absorbed light

L. V. Mirzoyan et al. (eds.), Flare Stars in Star Clusters, Associations and the Solar Vicinity, 347–361.

heats up the initially very cold dust and is eventually re-radiated at longer (infrared) wavelengths and escapes the cloud. Consequently, it is possible to directly probe star forming regions with observations made at infrared (and millimeter) wavelengths. Indeed, to test even the most basic hypotheses concerning stellar origins *requires* the acquisition of such long wavelength empirical data. For the most part this has only been technologically possible during the last two decades. The direct investigation of the star formation problem is, therefore, a relatively recent development of astronomical science.

During this time considerable progress toward understanding the physical process of stellar formation has been achieved. In particular, during the last 5 years fundamental advances in our knowledge of low mass star formation has taken place. Dwarf or low mass stars (i.e., $M \leq 3M_{\odot}$) provide important advantages for star formation investigations. For such stars the Kelvin-Helmholtz contraction time is considerably greater than the free-fall time of the gaseous cores from which they form. Therefore, these stars emerge from their dusty embryonic wombs before they begin to burn hydrogen in their cores, that is, well before they have reached the main sequence. Thus these stars can be observed even in the optical region of the spectrum while still in their (late) formative stages. Moreover, dwarf stars can form in isolation and are much less destructive of their natal environments than massive stars. On the other hand, dwarf stars are much fainter than high mass stars and can only be practically studied in nearby molecular clouds with the most sensitive instruments. In this paper I will review some of the more interesting new findings concerning low mass star formation that have been obtained as a result of both new observations at infrared and millimeter wavelengths and the development of new theoretical insights concerning this problem.

2. SITES OF LOW MASS STAR FORMATION

2.1 GMCs and OB Associations

In the current epoch of galactic history the vast majority of stars, both low and high mass, form in giant molecular clouds (GMCs). With extents on the order of 100 parsecs and masses often in excess of 10^5 $M_{\odot}$, GMCs are the largest objects in the galaxy and rival globular clusters as the most massive objects in the Milky Way. GMCs are clearly localized objects in the interstellar medium. They have well defined boundaries and are gravitationally bound, that is they are systems with negative total energy. *The statement that stars form in GMCs is equivalent to the statement that stars form in groups or associations.* It has long been suspected that most stars formed in the galaxy began their lives as members of associations (e.g., Roberts 1957). OB and T associations exist because stars form in spatially confined parental clouds. The question of the origin of stellar associations is in reality a question of the origin and evolution of GMCs. OB associations are formed when massive stars which form in a GMC erode, dissipate and disrupt the cloud, leaving behind the stars which formed during the cloud's lifetime. Once the gas has been cleared and removed by the O stars, the remaining stellar association becomes a fossil record of the original GMC (e.g., Duerr, Imhoff and Lada 1982). OB associations and T associations have often been thought of as being physically different types of stellar aggregates. But this is largely due to observational selection effects. Since both high mass and low mass stars predominately form in GMCs, GMCs ultimately produce stellar associations which contain a mixture of both OB and T Tauri stars (e.g., Elmegreen and Lada 1977; Lada 1987).

2.2 The Dynamical Nature of OB Associations

Ambartsumian (1947) was the first to recognize that the space densities of OB associations were well below the critical density for stability against disruption due to galactic tidal forces and that this had the important implication that such associations were considerably younger than the age of the galaxy. This observation provided one of the first fundamental proofs that star formation was occuring in the present epoch of galactic history. Ambartsumian showed that OB associations were gravitationally unbound and systems of *positive* total energy with expansion lifetimes on the order of 10^7 years. Based on early proper motion measurements, Ambartsumian also suggested that the associations had expansion velocities (≈ 10 km s^{-1}) well above that which could be produced by galactic tidal forces alone (Ambartsumian 1955). This lead to a number of hypotheses to explain the origin of these stellar systems of positive total energy. Ambartsumian proposed that associations were formed when massive, super-dense, "proto-stellar" bodies disintegrated producing both expanding groups of stars and their associated gas and dust (Ambartsumian 1955). Opik (1953) suggested that the ejecta from a supernova explosion could sweep up and compress interstellar matter into an expanding shell of gas which could then form an association of stars which would "retain the outward motion of the material of which they were built". Oort (1954) proposed a similar solution using the expansion and compression of an HII region to create expanding clouds of gas from which new stars would form and subsequently "share the outward motions that the HII regions had imposed on these clouds". All these ideas were based on the assumption that an unbound group of stars must have formed from expanding clouds of gas.

During the last 15 years millimeter-wave CO observations have shown that OB associations form from GMCs (e.g., Elmegreen and Lada 1976, 1977; Blitz 1980) which are gravitationally bound systems with negative total energy. Clearly this invalidates the basic assumption on which the solutions proposed by Ambartsumian, Oort and Opik were based. How do GMCs with negative total energy produce unbound OB associations? The key to answering this question is provided by recent observations of star formation activity in molecular clouds. Although GMCs are extremely massive, observations indicate that during their lifetime they convert only a small fraction of their gaseous mass into stars (e.g., Duerr, Imhoff and Lada 1982; Myers *et al.* 1986). In other words the global star forming efficiency of GMCs is low, probably on the order of a few percent or less. Clearly these large clouds must be a *source* of star formation and not the product of stellar creation that Ambartsumian had originally envisioned. At the same time, GMCs are not stable to destruction and dissipation by OB stars which generate HII regions, powerful winds and possibly even supernovae while still embedded in a cloud. Indeed, calculations by Whitworth (1979) showed that O stars could disperse an entire GMC if only 4% of the cloudy material was converted to stars with an IMF typical of field stars. These considerations led Duerr, Imhoff and Lada (1982) to propose that the unbound state of associations is a natural consequence of star formation in a giant molecular cloud with a low conversion efficiency of gas into stars, followed by a rapid destruction and removal of the unprocessed gas from the system. This hypothesis predicts as a consequence that the velocity dispersion of association stars is on the same order as that of molecular gas in GMCs (i.e., 2-4 km s^{-1}). This prediction appears to be confirmed for the λ Ori association studied by Duerr, Imhoff and Lada by a recent measurement of the velocity dispersion of association members of about 2 km s^{-1} made by Mathieu and Latham (1990). In addition it is interesting to recall that the systematic increase in size with age of the subgroups in the Ori OB1 association (Blauuw 1964) is consistent with an expansion velocity of a few km s^{-1} for the stars. These results support the contention that the unbound state of OB associations is a result of the combination of low star formation efficiency and rapid and efficient gas dispersal. These considerations also suggest that the early proper motion measurements of associations overestimated their expansion velocities.

Figure 1 is a sketch which depicts the evolution of star formation in a GMC and the creation of an expanding association as disucssed above and by Lada (1987). First, low mass stars form throughout the cloud converting roughly 1-3% of the gaseous mass into stars. At some point massive OB stars form in the cloud and heat, ionize and disrupt the molecular gas. In a relatively short time ($\approx 10^6$ years), the OB stars disrupt the entire complex and remove the vast majority of the original binding mass of the system. The stars in the cloud, which were originally orbiting in virial equilibrium with the deep potential well of the massive GMC, respond to the rapid removal of the majority of the binding mass by freely expanding into space with their initial virial velocities. This idea that unbound associations could form from bound clouds as a result of gas dispersal was originally suggested in a lecture by Zwicky (1953) and later independently mentioned by McCrea (1955) and von Hoerner (1968) who was the first to suggest (quantitatively) that OB stars could effectively disperse star forming gas and unbind a forming stellar system. However, these proposals were not given much attention because the large expansion velocities suggested by early proper motion measurements were not easily explained in the gas dispersal scenarios (e.g., Oort 1954). With the ability to directly observe molecular clouds and their embedded populations at millimeter and infrared wavelengths, it has now become clear that the origin of expanding associations is a result of the combination of low star formation efficiency and the efficient destruction of of giant molecular clouds by OB stars. The answer to the question of the origin of expanding associations requires understanding why star formation efficiency in molecular clouds is so low (Lada 1987).

2.3 Dense Cores and Embedded Clusters

It has long been suspected that stars form in the dense cores of giant molecular clouds. In the nearest star forming regions: Taurus and Ophiuchus, comparison of millimeter-wave and infrared data has suggested that dense cores both small and large are often associated with extremely young embedded objects and are therefore often sites of recent low mass star formation (e.g., Myers 1987; Wilking and Lada 1983). Observations of these two regions further suggests that the formation of dwarf stars from dense gas occurs in two modes. In Taurus individual, relatively low mass (i.e., 2-10 $M_\odot$), cores are producing individual young stellar objects, while in Ophiuchus a single massive core ($M \approx 500\ M_\odot$) accounts for almost all the star forming activity in the region. The relative strength of magnetic fields in the dense gas may be the factor which determines which mode of star formation is dominant in a given region or cloud core (Shu, Adams and Lizano 1987).

The extent to which either one of these two regions is representative of galac-

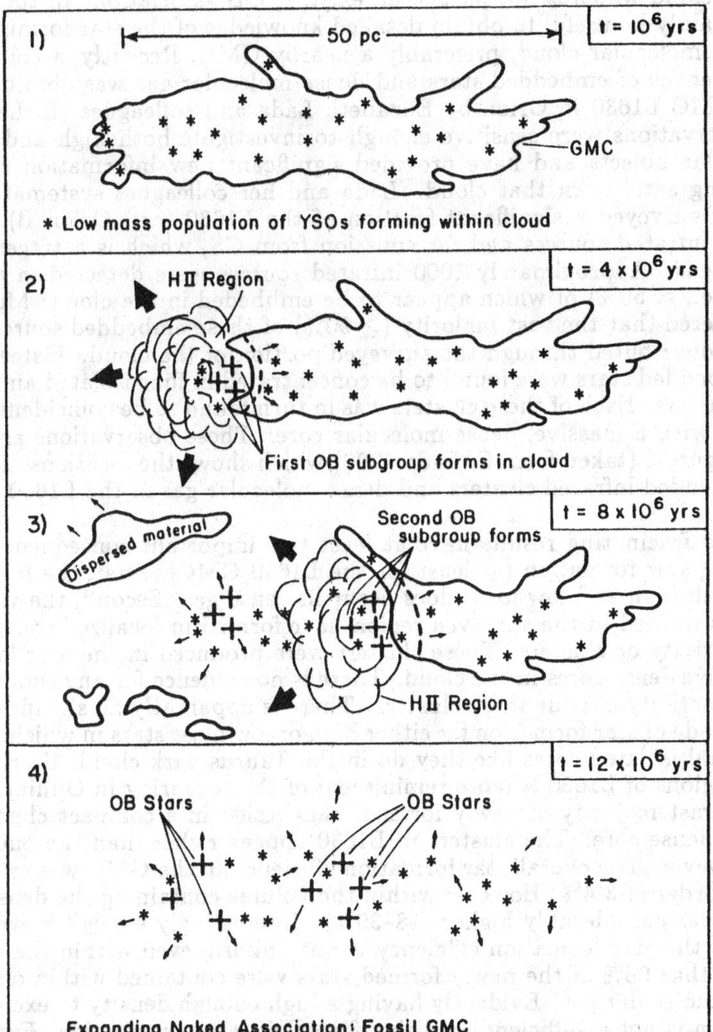

Figure 1. Probable stages in the origin of an expanding OB association from a giant molecular cloud. (Lada 1987).

tic star formation in general is unclear. Neither region is currently part of a GMC. However, the Ophiuchi cloud core is part of the Sco-Cen OB association, and perhaps the last remnant of a once larger GMC which has been mostly dissipated by the OB stars of this association. The Taurus cloud is an intermediate sized

molecular cloud which is not part of an existing OB association. In this regard it would certainly be useful to obtain detailed knowledge of the star forming activity in another molecular cloud, preferably a nearby GMC. Recently, a complete and unbaised census of embedded stars and dense molecular gas was obtained for the massive GMC L1630 in Orion by Elizabeth Lada and colleagues (E. Lada 1990). These observations were sensitive enough to investigate both high and low mass young stellar objects and have provided significant new information concerning star forming activity in that cloud. Lada and her colleagues systematically and completely surveyed a significant fraction of the L1630 (i.e., Orion B) GMC for embedded infrared sources and for emission from CS, which is a tracer of dense molecular gas. Approximately 1000 infrared sources were detected, a significant fraction (i.e., $\approx 50\%$) of which appear to be embbeded in the cloud. Moreover, it was discovered that the vast majority ($\geq 90\%$) of these embedded sources are not uniformly distributed through the surveyed portion of the cloud. Instead, almost all the embedded stars were found to be concentrated in four isolated and spatially distinct clusters. Each of these clusters was in turn found to be coincident or nearly coincident with a massive, dense molecular core. These observations are summarized in Figure 2 (taken from E. Lada 1990) which shows the locations and extents of the embedded infrared clusters and dense molecular gas in the L1630 GMC.

These fascinating results have at least two important consequences for understanding star formation (at least for the L1630 GMC). First, the formation of stars of both high and low mass does occur in dense gas. Second, the vast majority of stars formed in the surveyed region have formed in localized centers of star forming activity or clusters. These clusters were produced in the four largest and most massive dense cores in the cloud. There is no evidence for any significant star formation activity outside these clusters. There is apparently no significant "background" mode of star formation for either high or low mass stars in which stars form in individual isolated cores like they do in the Taurus dark cloud. Star formation in these regions of L1630 is more reminiscent of that occuring in Ophiuchus where again the vast majority of newly formed stars reside in a compact cluster within a massive dense core. The clusters in L1630 appear richer than the one in Ophiuchus, however. The overall star formation efficiency in the GMC was estimated to be on the order of 3-4%. However within the volume containing the dense gas, the efficiency was considerably higher (18–30%). Interestingly enough however, Lada found that the star formation efficiency is not uniform even within the dense gas. She found that 90% of the newly formed stars were contained within only 30% of the dense molecular gas! Evidently having a high enough density to excite $J=2\rightarrow1$ CS emission is not a sufficient condition for efficient star formation. Future study of the dense gas in this GMC, where the star formation efficiency varies so much between the dense cores, could provide important clues for understanding why star formation efficiency is generally so low in GMCs.

The sites of the embedded clusters in L1630 are likely the future sites of OB subgroups which will appear when the cloud is dissipated. When they emerge, these subgroups will contain the vast majority of both high and low mass stars formed in the cloud. Whether any of these embedded clusters will ultimately form bound clusters like the Pleaides depends on the the star formation efficiency achieved in the dense core at the time of destruction and the rate at which the gas is dispersed (Tutukov 1978; Lada, Margulis and Dearborn 1984). However, if this is a typical association, most of the embedded clusters will become unbound subgroups. The

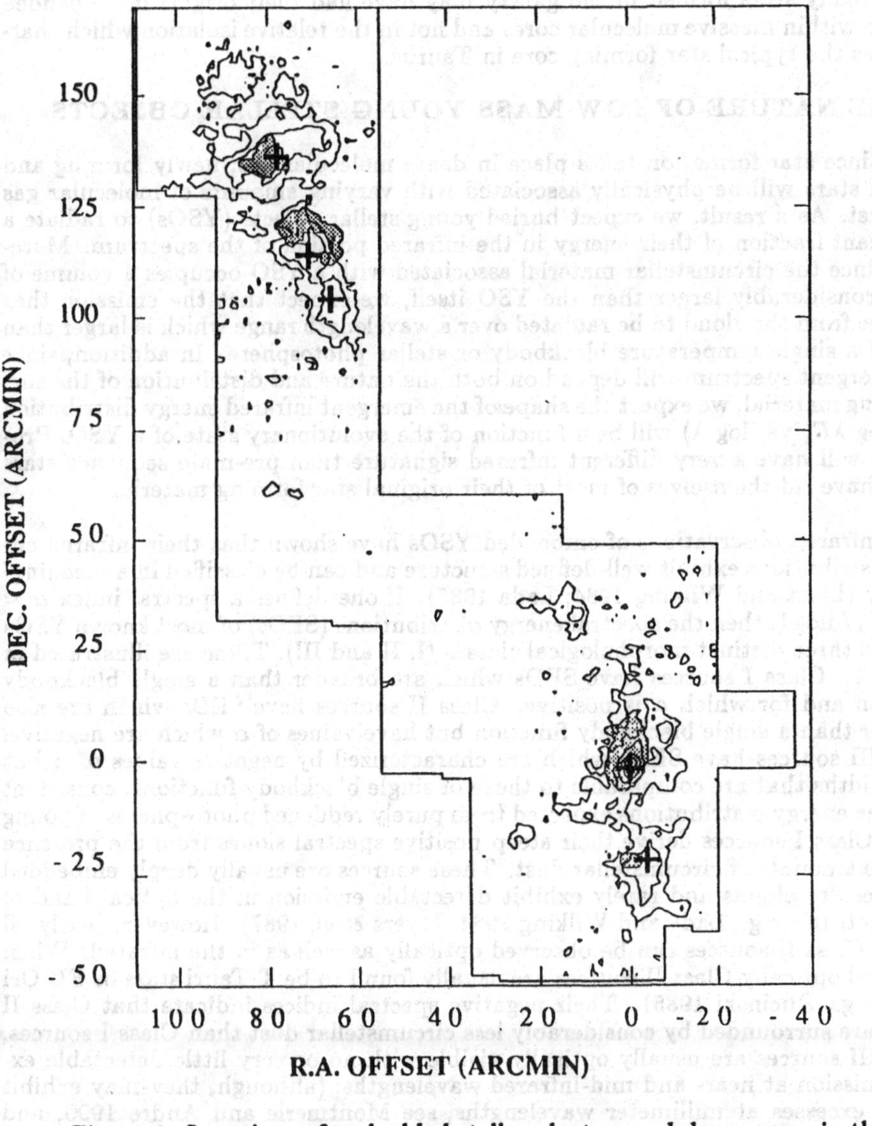

Figure 2. ***Locations of embedded stellar clusters and dense cores in the L1630 GMC (from E. Lada 1990). The contours represent isointensity contours of CS emission from the dense molecular gas in the cloud. The shaded regions are the extents of the embedded clusters in the cloud.***

extent to which L1630 is typical of star formation in the galaxy is yet to be determined. But if studies of nearby OB associations are an indication (e.g., Blauuw

1964), many stars formed in the galaxy may have had their origins in embedded clusters within massive molecular cores and not in the relative isolation which characterizes the typical star forming core in Taurus.

3. THE NATURE OF LOW MASS YOUNG STELLAR OBJECTS

Since star formation takes place in dense molecular gas, newly forming and formed stars will be physically associated with varying amounts of molecular gas and dust. As a result, we expect buried young stellar objects (YSOs) to radiate a significant fraction of their energy in the infrared portion of the spectrum. Moreover, since the circumstellar material associated with a YSO occupies a volume of space considerably larger than the YSO itself, we expect that the emission that emerges from the cloud to be radiated over a wavelength range which is larger than that of a single temperature blackbody or stellar photosphere. In addition, since the emergent spectrum will depend on both the nature and distribution of the surrounding material, we expect the shape of the emergent infrared energy distribution (i.e., log λF_λ vs. log λ) will be a function of the evolutionary state of a YSO. Protostars will have a very different infrared signature than pre-main sequence stars which have rid themselves of most of their original star forming material.

Infrared observations of embedded YSOs have shown that their infrared energy distributions exhibit well-defined structure and can be classified in a meaningful way (Lada and Wilking 1984; Lada 1987). If one defines a spectral index $\alpha = \mathrm{dlog}\lambda F_\lambda/\mathrm{dlog}\lambda$, then the spectral energy distributions (SEDs) of most known YSOs fall into three distinct morphological classes (I, II and III). These are illustrated in Figure 3. Class I sources have SEDs which are broader than a single blackbody function and for which α is positive. Class II sources have SEDs which are also broader than a single blackbody function but have values of α which are negative. Class III sources have SEDs which are characterized by negative values of α but have widths that are comparable to those of single blackbody functions, consistent with the energy distributions expected from purely reddened photospheres of young stars. Class I sources derive their steep positive spectral slopes from the presence of large amounts of circumstellar dust. These sources are usually deeply embedded in molecular clouds and rarely exhibit detectable emission in the optical band of the spectrum (e.g., Lada and Wilking 1984; Myers *et al*, 1987). However, nearly all known Class II sources can be observed optically as well as in the infrared. When classified optically Class II sources are usually found to be T Tauri stars or FU Ori stars (e.g., Rucinski 1985). Their negative spectral indices indicate that Class II YSOs are surrounded by considerably less circumstellar dust than Class I sources. Class III sources are usually optically visible with no or very little detectable excess emission at near- and mid-infrared wavelengths, (although, they may exhibit strong excesses at millimeter wavelengths, see Montmerle and Andre 1990, and this conference) and therefore little or no close-in circumstellar dust. Class III objects include both young main sequence stars and pre-main sequence stars, such as the so-called "post"-T Tauri stars (e.g., Lada and Wilking 1984) and the recently identified "naked"- T Tauri stars (e.g., Walter 1987). It is apparent from existing studies of YSOs that there is a more or less continuous variation in the shapes of SEDs from Class I to Class III (e.g., Myers *et al.*,1987, Wilking Lada and Young 1989).

It is extremely tempting to hypothesize that the empirical sequence of YSO

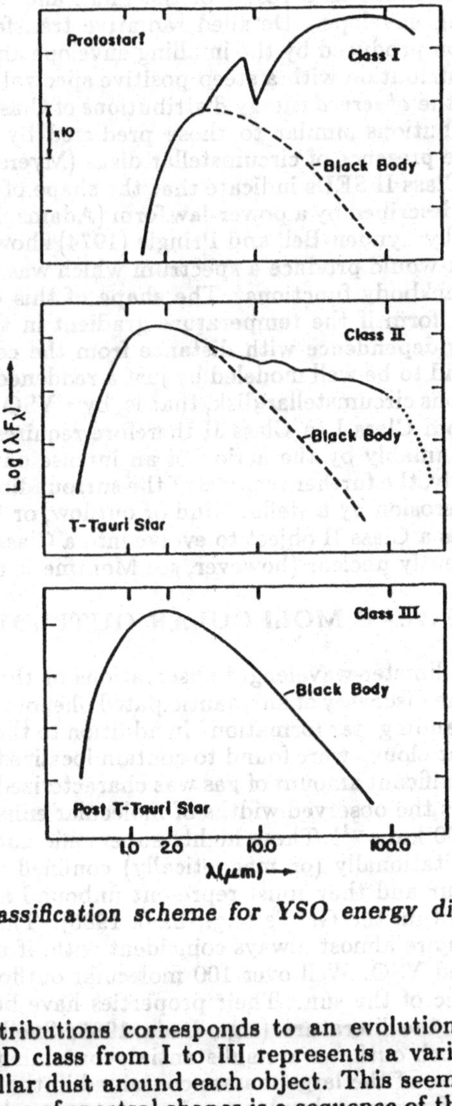

Figure 3. Classification scheme for YSO energy distributions (Lada 1987).

spectral energy distributions corresponds to an evolutionary sequence. Indeed, the variation in SED class from I to III represents a variation in the amount of luminous circumstellar dust around each object. This seems to suggest, then, that the empirical sequence of spectral shapes is a sequence of the gradual dissipation of dust and gas envelopes from around the newly forming or formed stars (Lada 1987). Recently, Adams, Lada and Shu (1987) have been able to theoretically model this empirical sequence as a more or less continuous sequence of early stellar evolution from protostar to young main sequence star using a self-consistent physical theory originally developed by Shu (1977). In this theoretical picture Class I sources are true protostars, objects undergoing accretion and assembling the bulk of the mass they will ultimately contain when they arrive on the main sequence. In particular, it is assumed that low mass protostars form from the nonhomologous, inside-out collapse of a rotating, isothermal cloud core (i.e., Shu 1977; Adams and Shu 1986). At the center of this unstable cloud a dense stellar-like core and disk

develop and become luminous as a result of the infall and accretion of material from the outer infalling envelope. Detailed radiative transfer calculations show that density distribution produced by the infalling envelope and rotation results in an emergent energy distribution with a steep positive spectral index (i.e., Class I). Moreover, modeling of the *observed* energy distributions of Class I sources in Taurus suggests density distributions similar to those predicted by theory for infalling envelopes as well as the presence of circumstellar disks (Myers *et al.* 1987).

Observations of Class II SEDs indicate that the shape of the infrared portion of the spectrum is well described by a power-law form (Adams, Lada and Shu 1988). Early theoretical work by Lynden-Bell and Pringle (1974) showed that an optically thick circumstellar disk would produce a spectrum which was the superposition of a series of different blackbody functions. The shape of this composite spectrum would be power-law in form if the temperature gradient in the disk was characterized by a power-law dependence with distance from the central star. Class II SEDs are therefore found to be well modeled by just a reddened stellar photosphere surrounded by a luminous circumstellar disk, that is, by a YSO without its infalling envelope. To evolve from Class I to Class II therefore requires the removal of the infalling envelope, presumably by the action of an intense outflow as will be discussed later. Presumably, the further removal of the surrounding disk, via accretion onto the star itself, by erosion by a stellar wind or outflow, or by incoporation into planetary bodies, causes a Class II object to evolve into a Class III source. Exactly how this occurs is presently unclear (however, see Montmerle and Andre 1990).

4. ENERGETIC BIPOLAR MOLECULAR OUTFLOWS

Ten years ago millimeter-wavelength observations of the molecular gas surrounding YSOs led to the discovery of an unanticipated phenomenon of fundamental importance for understanding star formation. In addition to their global supersonic velocity fields, molecular clouds were found to contain localized regions (0.1–3 parsecs in size) where a significant amount of gas was characterized by hypersonic bulk motion. In these regions the observed widths of molecular emission lines are found to range between 10–100 km s^{-1}! These highly supersonic and super-Alfvenic velocities cannot be gravitationally (or magnetically) confined within the localized regions where they occur and they must represent unbound and expanding flows of cold molecular gas within the GMCs (e.g., Lada 1985). The regions containing the hypersonic outflows are almost always coincident with, if not centered on, the position of an embedded YSO. Well over 100 molecular outflows are now known, most within a kiloparsec of the sun. Their properties have been extensively and thoroughly reviewed in the literature (e.g., Lada 1985; Snell 1987; Fukui 1990) Briefly, the masses of such outflows are substantial, containing anywhere between 0.1 and 100 $M_\odot$. Because of the large masses contained in the molecular outflows, it is likely that the outflowing molecular gas is swept-up ambient cloud material rather than original ejecta from the driving source. More significantly, the corresponding kinetic energies of the flows are enormous, ranging between 10^{43} and 10^{47} ergs! The dynamical timescales of the flows are estimated to be between 10^3 and 10^5 years and their local formation rate is estimated to be roughly comparable to the formation rate for stars of a solar mass or greater. Taken together, these facts suggest that molecular outflows play a fundamentally important role in the star formation process.

Perhaps the most intriguing property of the molecular outflows is their tendency to appear spatially bipolar. That is, they often consist of two spatially separate lobes of emission, with one lobe containing predominantly blueshifted gas and the other predominantly redshifted gas. Furthermore, the two separating lobes are almost always more or less symmetrically situated about an embedded infrared source or young stellar object. About 75% of the known outflows are bipolar; the rest are either single-lobed, (i.e., one lobe of either predominately red or blue-shifted emisison), isotropic (i.e., one lobe but with both red and blue-shifted high velocity emission spatially coincident) or of complex morphology.

5. ENERGETIC OUTFLOWS AND STAR FORMATION

Bipolar molecular outflows are indivdually energetic enough to disrupt cloud cores and collectively powerful enough to have a significant impact on the dynamics and structure of a an entire GMC (e.g., Margulis Lada and Snell 1988). In fact the molecular outflows generated by a population of embedded YSOs may be able to generate the turbulent pressure that keeps GMCs from global collapse, thereby solving one of the outstanding problems of cloud dynamics. In any event, it is clear that molecular outflow is the likely agent that removes circumstellar material and drives the evolution of an embedded young stellar object from the Class I to the Class II stage. In this regard it is interesting to determine by direct observation the nature of the embedded sources which drive cold molecular outflows. A growing body of observational data now clearly shows that molecular outflows are most frequently associated with Class I type sources and only rarely with Class II or III objects (e.g., Lada 1985, 1988; Berrill *et al.* 1989; Snell *et al.* 1988; Margulis, Lada and Young 1989). In fact survey observations of both embedded source populations within individual clouds (Margulis, Lada and Young 1989) and among all molecular clouds (Berrill *et al.* 1989; Snell *et al.* 1988) indicate that at least half of all studied Class I objects are sources of molecular outflow. On the other hand, less than 10% of Class II and III objects are associated with molecular outflow, although many of these may still drive stellar winds (Lada 1988). This suggests that outflow activity is ignited during the Class I or protostellar phase and continues into the Class II phase where it subsequently dies out.

Although molecular outflows appear to provide the key to understanding how a Class I source removes surrounding material and in doing so evolves into a Class II source, the high frequency of association between Class I infrared sources and molecular outflows poses a paradox. The statistics suggest that a significant fraction of the lifetime of a Class I object is spent in the outflow phase. Yet, if Class I sources are true protostars, their evolution should be characterized by the *infall* of surrounding material. How can a protostar for most of its existence be simultaneously a source of infall and outflow? How can a star form by losing mass? The answer to this question is the key to understanding the basic physics of the star formation process.

The solution to this paradoxical problem may contain two crucial ingredients: angular momentum and magnetic fields. The fact that disks are found around most YSOs implicates an important role for angular momentum. The formation of a disk around a young stellar object is the natural consequence of the presence of angular momentum (even in small amounts) and its conservation in dynamically evolving cloud cores. For a rotating protostar most of the mass that ends up on the star must be accreted from the surrounding disk. In order for material to flow through the disk and onto the protostar, the material must lose both energy and angular momentum. If the mass of the disk is not much larger than that of the central object, the material in the disk should rotate differentially in Keplerian fashion. Gas falling through such a disk will reach the surface of the central star with an orbital velocity and specific angular momentum which is relatively high compared to that in the star (Shu *et al.* 1989). If this material is added to the star it will spin up the star. The star will quickly reach break up equatorial velocities at which point material can no longer be added to it. A centifugal barrier prevents the further growth of the protostar. Thus the process of star formation can only proceed if the incoming gas somehow can lose additional angular momentum in the process of accreting onto the star or if the star can somehow spin down while accretion is taking place.

Angular momentum can be carried away from a star by a stellar wind. *Consequently, a protostar may be able to gain mass only if it simultaneously loses mass.* To allow star formation to continue the rate of mass loss from the wind should be a fraction of the mass accretion rate i.e.,

$$\dot{M}_{wind} = f\dot{M}_{infall}$$

where the fraction f is determined by the physics of the wind generating mechanism. The ideal protostellar wind is one that carries away little mass but lots of angular momentum. A number of recent investigations have shown that centrifugally-driven hydromagnetic winds are potentially capable of doing the job (e.g., Pudritz 1988). Such winds could be driven from either circumstellar disks (e.g., Uchida and Shibata 1984) or from the surfaces of central protostars (e.g., Shu *et al.* 1988). It may be that star formation in a rotating, magnetic cloud cores results in the formation of protostar-disk systems which can generate powerful outflows and in doing so resolve the paradox of the protostar which gains mass by losing mass.

In conclusion, it is becoming increasingly apparent that the generation of an intense stellar wind is of fundamental significance for any scenario or theory of star formation. The wind is both necessary for star formation to proceed (by enabling accretion of material through a disk) and for providing a natural mechanism for the ultimate reversal of infall from the surrounding infalling envelope. In addition, the stellar wind and the bipolar molecular flow it generates limit the mass available to be accreted onto the protostar by clearing away the surrounding gas and dust. The wind is thus the agent that drives the evolution of a protostar from a Class I to a Class II object and determines the final mass of the forming star.

5. CONCLUDING REMARKS

Stellar formation in our galaxy is indeed a rich and wonderful physical process to investigate and behold. During the last two decades advances in observational technology have lead to remarkable progress in our quest to decipher the mysteries of stellar origins. The questions we ask today are in many ways totally different form those asked by investigators even 20 years ago. Yet, in delivering a lecture on the origin of stars here at the Byurakan Observatory one can hardly escape contemplating the legacy of Academician Ambartsumian to this field of astronomical endeavor. His pioneering work on OB associations began the modern study of star formation in our galaxy. Perhaps most interesting and impressive however, was Ambartsumian's intuition about the importance of the role of expansive motions in the star formation process. Although, many of his ideas about the origin of stars and clouds appear with the light of modern evidence to be wrong (and this may not be too suprising given the paucity of relevant observational information available in the 50's and 60's), his basic belief that expansion, explosion and outflow were fundamental to the phenomenon of stellar origins seems to be borne out by the the most recent knowledge provided by modern observation and theory. As Newton first thought, gravity is, after all, at the heart of the process. However, because nature has also provided magnetic fields, angular momentum and perhaps other ingredients we do not yet fully appreciate, the story of the origin of stars has turned out to be more bizzare, mysterious and interesting than anyone (except perhaps Academician Ambartsumian) could have imagined 50 years ago.

6. ACKNOWLEDGEMENTS

I am grateful to the Organizing Committee of IAU Symposium No. 137 for inviting me to speak at and participate in this conference and for providing the basic financial support which made my participation possible. I also thank the University of Arizona Committee For Foreign Travel for assistance with travel support. I thank Frank Shu and Elizabeth Lada for enlightning discussions. Some of the research described here was supported, in part, by NSF Grant AST 8815753.

REFERENCES

Adams, F.C. and Shu, F.H. 1986,*Ap.J.*,**296**,655.
Adams, F.C., Lada, C.J., and Shu, F.H. 1987,*Ap.J.*,**213**,788.
Adams, F.C., Lada, C.J., and Shu, F.H. 1988,*Ap.J.*,**326**,865.
Ambartsumian, V.A. 1947, *Stellar Evolution and Astrophysics*, (Armenian Acad. of Science)
Ambartsumian, V.A. 1955, *Observatory*, **75**,72.
Berrill, F. *et al.* 1989,*M.N.R.A.S.*,**237**,1.
Blaauw, A. 1964,*Ann.Rev.Astr.Ap.* ,**2**,213.
Blitz, L. 1979, *in Giant Molecular Clouds in the Galaxy*, eds. P.M. Solomon and M.G. Edwards, (Oxford:Pergamon),p. 211.
Duerr, R., Imhoff, C., and Lada, C.J. 1982,*Ap.J.*,**261**,135.
Elmegreen, B.G, and Lada, C.J. 1976,*A.J.*,**81**,1089.
Elmegreen, B.G, and Lada, C.J. 1977,*Ap.J.*,**214**,725.
Fukui, Y. 1990,*in Low Mass Star Formation and Pre-Main Sequence Objects*, ed. B Reipurth, (Garching:ESO), p. 95.
Lada, E.A. 1990, PhD Dissertation, Univeristy of Texas.
Lada, C.J. 1987,*in IAU Symposium No. 115: Star Forming Regions*, eds. M. Piembert and J. Jugaku (Dordrecht: Reidel), p.1.
Lada, C.J. 1988, *in Galactic and Extragalactic Star Formation*, eds. R. Pudritz and M. Fich, (Dordrecht: Reidel), p. 1.
Lada, C.J., Margulis, M. and Dearborn, D. 1984,*Ap.J.*,**285**,141.
Lada, C.J. and Wilking, B.A. 1984,*Ap.J.*,**287**, 610.
Lynden-Bell, D., and Pringle, J.E. 1974, *M.N.R.A.S.*, **168**, 603.
Margulis, M., Lada, C.J. and Snell R. 1988,*Ap.J.*,**333**, 316.
Margulis, M., Lada, C.J. and Young, E. 1989,*Ap.J.*,**345**,906.
Mathieu, R. and Latham D., 1990, unpublished observations.
Montmerle, T. and Andre P. 1990, *in Low Mass Star Formation and Pre-Main Sequence Objects*, ed. B. Reipurth, (Garching:ESO), p.407.
McCrea, W.H. 1955, *Observatory*, **75**,206.

Myers, P.C. *et al.* 1986,*Ap.J.*,**301**, 398.
Myers, P.C. *et al.* 1987,*Ap.J.*,**319**, 340.
Myers, P.C. 1987,*in IAU Symposium No. 115: Star Forming Regions*, eds. M. Piembert and J. Jugaku (Dordrecht: Reidel), p. 33.
Oort, J.H. 1954, *B.A.N*, **12**,177.
Opik, E.J. 1953, *Irish Astron. J.*, **2**, 219.
Pudritz, R. 1988,*in Galactic and Extragalactic Star Formation*, eds. R. Pudritz and M. Fich, (Dordrecht: Reidel),p.135.
Roberts, M.S. 1957,*Publ.Astr.Soc.Pacific*,**69**,59.
Rucinski, S.M. 1985,*A.J.*,**90**, 2321.
Shu, F.H. 1977,*Ap.J.*,**214**, 488.
Shu, F.H., Adams, F.C., and Lizano, S. 1987, *Ann.Rev.Astr.Ap.* ,**25**, 23.
Shu, F.H., Lizano, S., Ruden, S.P. and Najita, J. 1988,*Ap.J.(Letters)*,**328**, 19.
Snell R. 1987,*in IAU Symposium No. 115: Star Forming Regions*, eds. M. Piembert and J. Jugaku (Dordrecht: Reidel), p.213.
Snell R., Huang, Y-L, Dickman, R.L. and Claussen, M. 1988,*Ap.J.*,**325**,853.
Tutukov, A.V. 1978,*Astr.Ap.*,**70**,57.
Uchida, Y. and Shibata, K. 1984,*P.S.A.J.*, **36**, 105.
von Hoerner, S. 1968, *in Interstellar Ionized Hydrogen.* ed. V. Terzian, (New York: Benjamin), p. 101.
Walter, F.M. 1987,*Publ.Astr.Soc.Pacific*,**99**, 31.
Whitworth, A.P. 1979, *M.N.R.A.S.*,**186**, 59.
Wilking, B.A., and Lada, C.J. 1983, *Ap.J.*,**274**,698.
Wilking, B.A., Lada, C.J. and Young E.T. 1989, *Ap.J.*,**340**,823.
Zwicky. F. 1953, *Publ.Astr.Soc.Pacific*,**65**,205.

BLAAUW: Do you consider the formation of clusters like the Pleiades, h and chi Persei etc. to fit in with the general scenario you described, i.e. extremes of a more or less continuous spectrum of formation efficiency, or should we invoke a different mechanism?

LADA: Bound clusters must form in cores which have high efficiencies and which are more gently disrupted than OB starforming cores, that is, bound clusters cannot contain O stars in them when they form. Clearly, since only 10% of all stars were formed in bound clusters,the conditions that give rise to bound cluster formation must be in some sense "special". So far we know of only one region where the star formation efficiency (SFE) is large. In Ophiuchus the SFE is about 20%. The general conditions there are much different than those in Taurus where the SFE is about 2%, so indeed different mechanisms are operating. However, E. Lada's observations of Ori B suggests that most stars are formed in cores similar to those in Ophiuchus, unlike the conditions in Taurus. But only a few of these cores will ever produce bound groups. The formation of massive stars in most cores may explain why they do not end up with bound groups. Tidal disruption by the parental GMC may also disrupt young clusters at an early stage.

BLAAUW: Might not the high efficiency just referred to be counteracted by the violent outflow phenomena you described at the end?

LADA: Evidently in cores which produce bound clusters, factors which give rise to disruption are suppressed. For example, O stars cannot have formed in bound clusters: they are too disruptive.

BLAAUW: Might the outflow phenomena give rise to secondary star formation?

LADA: Yes, it is certainly possible although we do not yet have a clear example of this happening.

GIAMPAPA: What is the mass of the disk compared to the mass of the star? How does that fraction vary from high mass to low mass stars? (Do massive stars have massive disks and are low mass stars characterized by very low mass disks?)

LADA: For T Tauri stars disk masses are on the order of 0.1 Mo or less (i.e. they are generally less massive than the star itself). How that fraction varies between stars of different masses is not yet clear.

Magnetic activity and evolution of low-mass young stars

Thierry Montmerle
Service d'Astrophysique, Centre d'Etudes Nucléaires de Saclay
91191 Gif-sur-Yvette, France

1. X-rays from star-forming regions

1.1. *Main observational results*

X-rays turned out rather unexpectedly to be of fundamental importance in studying, and especially in discovering, pre-main sequence stars. The bulk of the data we now have comes from observations using the *Einstein* satellite, operating in the imaging mode between ~0.4 to ~4 keV (for reviews, see, e.g., Feigelson 1984, Feigelson, Giampapa, and Vrba 1990); because many of the detected sources turned out to suffer a relatively large absorption (A_V up to several magnitudes, see below), *EXOSAT*, sensitive to softer X rays, unfortunately proved to be of little use.

Most of the observed regions were nearby dark clouds, well known to undergo active star formation and containing many T Tauri stars (low-mass pre-main sequence stars): ρ Ophiuchi (Montmerle et al.1983), Taurus-Auriga (Feigelson et al. 1987, Walter et al. 1988), Chameleon (Feigelson and Kriss 1988), all at distances ~ 160 pc, and Orion (Ku, Righini-Cohen, and Simon 1982, Caillault and Zoonematkermani 1987) at ~ 450 pc. Note that ρ Oph and Orion in addition contain stars of earlier spectral types, the earliest being B3 and O7, respectively, themselves X-ray emitters. Note also the late publication date of some papers, which reflects the very contribution of the X-rays: many of the stellar X-ray sources discovered were previously unknown, and it took often several years to establish (spectroscopically) their pre-main sequence nature.

All these observations revealed dozens of new sources, and in many cases, the X-rays revealed more PMS stars (by a factor $f > 1$) than were discovered by the usual criteria based on the presence of strong emission lines, notably Hα. Although a definite figure cannot be given for lack of completeness in the various surveys, some authors (e.g., Walter et al. 1988) argue that $f \lesssim 10$.

1.2. *Emission mechanism*

When a good X-ray spectrum of a PMS star is available (which is not always the case at the level of sensitivity of *Einstein* because in general too few counts are recorded), it is consistent with bremsstrahlung emission from a hot (kT ~ 1 keV) plasma, significantly absorbed along the line of sight ($N_H \lesssim$ a few 10^{21} cm^{-2}, or $A_V \lesssim 10$). The typical luminosities are in the range $\langle L_X \rangle \sim 10^{31-32}$ erg.s^{-1}.

But more significant results can be gained from the study of time variability. Much evidence has accumulated in favor of the existence of large X-ray flares (e.g., Feigelson and DeCampli 1981, Walter and Kuhi 1984, Montmerle et al. 1983, 1984), which in fact dominate the X-ray emission. Within the still modest precision of the data, these flares are solar-like, but enhanced by factors up to 10^6 with respect to the Sun, also enhanced with respect to flares on dMe "flare stars" (see discussion in Montmerle et al. 1983), and quite similar to flares observed on RS CVn close binary systems (e.g., Mutel et al. 1987). In addition to the

L. V. Mirzoyan et al. (eds.), Flare Stars in Star Clusters, Associations and the Solar Vicinity, 363–370.

solar analogy, many factors, including the presence of large spots and a correlation with rotation based on them (Montmerle 1987, Bouvier and Bertout 1989), argue in favor of a *magnetic* origin for this X-ray activity. However, strong flaring activity takes place ~ 5 % of the time on average (Gahm 1989).

1.3. *Implications on the stellar environment*

Interpreting the X-rays from PMS stars in terms of a magnetically confined plasma, the emission measure yields the typical size H_X of the magnetic loops: one finds $H_X \lesssim$ 2-3 R_*. In turn, one deduces a typical value for the surface magnetic field: $B_{*,X} \sim$ a few 100 G. Whereas the size of the loops is large when compared to the Sun, $B_{*,X}$ is on the contrary quite representative of the solar active regions.

The width of the Hα line associated with chromospheric activity is small, a few Å at most, and this is indeed the case for most of the X-ray discovered PMS stars. But in this context, it is surprising that the width of the Hα line turns out to be much larger (up to a few 100 Å) for all the previously known PMS stars like T Tauri stars, for a comparable X-ray flux (e.g., Bouvier 1987, shown in Montmerle 1987). Such an Hα "excess" goes along with other excesses (over a normal, late-type photosphere), in the UV and the near-IR, observed in the continuum spectra of these stars. These excesses are clearly "non-solar", and point to the existence, in the vicinity of the star, of circumstellar material, made up of hot (~ 10^4 K) ionized gas, as well as warm ($\lesssim$ 1000 K) dust.

Current models (Kenyon and Hartmann 1987, Bertout, Bouvier, and Basri 1988, Bertout 1989) attribute these excesses to the presence of an *accretion disk*, of size $\lesssim$ 1000 R_*. This disk is likely in keplerian rotation; in the immediate vicinity of the star, the corresponding orbital velocity is typically ~ 250 km.s^{-1}, whereas the stellar equatorial velocity is ~ 20 km.s^{-1}. This introduces a new, important component in the circumstellar environment, in the form of a hot *boundary layer*. It is this layer which is thought, in these models, to be responsible for the strong emission lines, including Hα. The corresponding stars are now called "classical" T Tauri stars ("CTTS").

By contrast, the X-ray discovered PMS stars, which do not show these excesses, were named "naked" T Tauri stars (Walter 1987), but the term "weak-line" T Tauri star ("WTTS") is now to be preferred, since it does not imply that the absence of IR and/or UV excess means an absence of circumstellar material (see §3.1).

2. Radio observations in the cm range

2.1. *Rationale*

Given the links between X-rays and magnetic activity, it is natural to look for non-thermal radio emission from the flare-associated electrons in the magnetic loops. This emission is observed on the Sun, and the above evidence for solar-type activity reinforces the potential interest of radio observations. But another type of radio emission is also expected: thermal free-free emission associated with the hot circumstellar environment.

Taking advantage of the increased sensitivity and the imaging capability in the cm (GHz) range offered by the Very Large Array in New Mexico, a number of workers have done radio observations following two approaches. In the first approach, known objects, selected according to a variety of criteria, are the targets of *pointed* observations: CTTS (e.g. Bieging, Cohen, and Schwartz 1984), sources of molecular outflows (related PMS objects, see Snell and Bally 1986; Lada 1985). In the second approach, a whole region is subject to an unbiased *survey*, without any prior selection of targets.

The first approach gives evidence that a (small) number of CTTS or molecular flow sources have a detectable radio flux, of thermal origin, and confirms the existence of ionized stellar winds, already suspected to be present on the basis of some Hα and NaD emission line profiles (e.g., Mundt 1984). The observed mass-loss rates are typically $\dot{M} \sim 10^{-8}\ M_{\odot}\ yr^{-1}$, velocities $v_W \sim$ 200-300 km. s^{-1}, and the size of the emitting region is $R \sim 1000\ R_*$ (e.g., Panagia and Felli 1975).

The second approach makes no assumption about the nature of the possible radio emitters, and has revealed the existence of hitherto unknown objects, in ρ Oph (André, Montmerle, and Feigelson 1987), CrA (Brown 1987), and Orion (Trapezium region, Garay, Moran, and Reid 1987, Churchwell et al. 1987). Several stellar radio sources had been previously found in X-rays, but most were discovered deeply embedded in the clouds, so that even the X-rays, if present, would be absorbed. Here again, the study of time variability proved crucial in understanding the nature of the radio emission: strong variability was noticed as early as 1985, with the discovery of the first radio flare in a PMS star (Feigelson and Montmerle 1985), and was later established to be rather frequent among PMS radio emitters. The typical timescale for variability is unknown, but variability of factors of at least 2 have been found in a few hours, and up to 10 between observations separated by a few months (e.g., Cohen and Bieging 1986). Such timescales can be explained only in terms of a nonthermal emission mechanism, because the interpretation in terms of an ionized wind implies timescales ≳ years. Note that the commonly used criteria based on the spectral index α (where $S_\nu \propto \nu^{\alpha}$), namely that if $\alpha > 0$ the emission is thermal and if $\alpha < 0$ the emission is nonthermal, cannot reliably indicate the nature of the emission mechanism, but simply indicates the opacity (resp. optically thick, and optically thin). (For details, see André 1987.)

2.2. *Non-thermal radio emission*

As mentioned above, there are indications that nonthermal radio emitters are not unusual among the whole population of PMS radio emitters. But the only systematic search to date has been undertaken in the ρ Oph cloud, over a ~ 2° x 2° area, following the *Einstein* results (André, Montmerle, and Feigelson 1987, Stine et al. 1988). Forgetting about spectra, the nonthermal nature of a source can be established in three ways:

- large-amplitude variability over short timescales (≲ years, see above);
- polarization;
- unresolved size (i.e., < 0.1" at 15 GHz for the VLA in A configuration at the distance of the nearest dark clouds).

As mentioned above, variability has been established in many cases, and, in ρ Oph, appears to be widespread among the 9 sources found above 1-2 mJy (Stine et al. 1988). Also, none of these sources is resolved, as shown by a visibility analysis (see André 1987). More remarkably, in one instance, and for the first time, polarization (at a level of ~ 7%, and seen repeatedly) has been found in S1, a very young (≳ 2000 yrs) B3 star in ρ Oph..

In all these examples, the emission mechanism is likely to be gyrosynchroton radiation from ~ MeV electrons spiraling in fields ~ 1 G; at least, quantitative models exist for a few well-documented cases (DoAr21, see André 1987; S1, André et al. 1988), using non-homogeneous emission models featuring dipolar magnetic loops, and developed for the

Sun (e.g., Klein and Chiuderi-Drago 1987). The magnetic field at the stellar surface is in general $B_{*,r}$ ~ a few 100 G, consistent with the values $B_{*,X}$ deduced from the X-rays; in the case of S1, however, a higher value is required (≳ 10 kG), about the same as prevails in the well-known magnetic B stars.

All these stars share a remarkable feature: interpreting the radio emission in terms of the gyrosynchrotron mechanism, implies that they must possess *extended magnetic structures,* the size of which H_r may extend up to ~10 stellar radii (see Fig. 4 of Montmerle and André 1988), which is significantly larger than the size H_X deduced from the X-rays. They therefore make up a *new population* of PMS stars, seen in several dark clouds but best known in ρ Oph, and comprising to date about 15 members.

3. Implications on evolution

3.1. *Young stars and circumstellar material*

Contrary to the X-rays, which are spread between CTTS and WTTS (even though X-ray selected WTTS largely outnumber CTTS, as discussed in § 1.1), there is in the radio range a clear distinction between emitters: the thermal mechanism concerns only CTTS and similar objects, and the nonthermal mechanism concerns only WTTS and related objects. There is also another strong difference between the X-rays and the radio: whereas the X-rays come from stars spread over a large area in the vicinity of a dark cloud, the *non-thermal* radio emitters (equivalently, those which have large magnetic structures) are found to be embedded in the clouds. The case of ρ Oph is particularly striking: the 9 sources are strongly concentrated near the cloud core, whereas a large area (~ 2° x 2°) has been searched with a good sensitivity (~ 2 mJy).

An immediate conclusion is that these objects must be very young, although, strictly speaking, no age can be attributed to them because most are not seen in the optical and hence cannot be placed on an HR diagram. But they are visible in the near-IR (e.g., Wilking, Lada, and Young 1989), and show no excess in this range. According to the IR classification (Lada 1988), they are "Class III" objects. So we are faced with a problem: how is it that very young stars, still embedded in cloud cores, apparently do not show evidence for circumstellar material?

A related problem exists when considering optically visible CTTS and WTTS as a whole. Indeed, when put on an HR diagram, these two classes of PMS stars appear mixed (e.g., Walter et al 1988, Edwards et al. 1987), whereas, according to "standard" evolutionary models (e.g., Adams, Lada, and Shu 1987), CTTS, being surrounded by accretion disks, should be younger, and WTTS, deprived of such disks, should represent a more advanced evolutionary stage.

3.2. *Cold circumstellar material and magnetic fields*

The probable answer to this problem lies in the fact that the "standard" scenario of early stellar evolution (Lada 1988, Adams et al. 1987) is based on IR data, from the near-IR (≳ 2 μm), which traces warm material (≳ 1500 K), to the far-IR (≲ 100 μm, IRAS data), which traces colder material (few 100 K). In current disk models (see above, § 1.3), the temperature decreases outwards with radius according to a power law (index –3/4 in the case of a keplerian spatially thin viscous disk): the warm regions seen in the IR are comparatively close to the star (≲ 1 AU), and the boundary layer exists only if there is a physical contact between the disk and the star. Outer (≳ 10 AU), cold (<< 100 K) material is visible only in the mm range, and recent work in this band has given new insight into the presence of circumstellar matter around young stars, and into the possible role of magnetic fields (Montmerle and André 1989).

Using the 30-m dish of the French-German IRAM telescope located on Pico Veleta near Granada (Spain), equipped with the sensitive MPIfR 1.3 mm bolometer, Beckwith et al. (1990) and André et al. (1990) have shown that about half of PMS stars, irrespective of their optical or IR classification, are strong emitters in the mm range, hence are surrounded by cold dust. More precisely, this is true for CTTS and WTTS (Beckwith et al. 1990), as well as for embedded IR sources in ρ Oph, having strong IR excess to no excess (André et al. 1990). Combining the IR data with the mm data shows that a variety of circumstellar material (presumably disks) may exist around similar stars: mm emission without IR excess indicates the presence of "hollow", cold disks, whereas IR excess without mm emission shows the existence of "compact", warm disks; "normal", i.e., extended disks, are also present (see Fig. 1 of André et al. 1990).

Put in an evolutionary perpective, it therefore appears qualitatively that stars of (roughly) the same mass and age may be surrounded by mophologically different disks. The reason for this difference can be linked with the question of *disk stability* on timescales 10^5 - 10^7 yrs, and magnetic fields may there play a crucial role (Montmerle and André 1989). For instance, Tagger et al. (1990) have recently shown that spatially thin accretion disks threaded by vertical magnetic field lines (as is the case for dipolar fields) are subject to efficient dynamical instabilities. Some of these instabilities may cause the inner parts of the disk, out to the corotation radius, to fall quickly on the star, resulting in the formation of an inner "hole". Much work remains to be done, however, before this can be quantitatively proven.

4. Concluding remarks

❑ The environment of young stellar objects is complex, and involves a variety of interactions with them, on very different length scales: boundary layers ($r \gtrsim R_*$), magnetic activity and magnetic fields ($r \lesssim 1$–$10\ R_*$), disks ($r \lesssim 1000\ R_*$), ionized winds ($r \gtrsim 1000\ R_*$) and molecular outflows ($r \lesssim 10^6\ R_*$).

❑ X-rays and cm radio observations give crucial information on magnetic fields and on the existence of ionized circumstellar material; the UV range appears to be connected mainly with the existence of boundary layers.

❑ IR and mm radio observations give crucial information on the nature and extent of the outer, colder material; this material, presumably in the form of circumstellar disks, may be unstable on a wide range of timescales, from ~ 10^5 to ~ 10^7 years, plausibly owing to the presence of magnetic fields.

❑ One probably has to put together the information contained in all these energy domains, and the existence of magnetic fields, to better understand the mechanism(s) of star formation.

References

Adams F.C., Lada C.J., Shu F.H. 1987, *Ap.J.* **312**, 788.

André P., Montmerle T., Feigelson E.D. 1987, *Astr.J.* **93**, 1182.

André P., Montmerle T., Steppe H., Feigelson 1990, *Astr.Ap.*, submitted.

André P., Montmerle T., Stine P.C., Feigelson E.D., Klein K.L. 1988, *Ap.J.* **335**, 940.

Beckwith S.V.W., Sargent A.I., Chini R.S., Güsten R. 1990, *Astr.J.* in press.
Bertout C. 1989, *Ann.Rev.Astr.Ap.*, **27**, 351.
Bertout C., Basri G., Bouvier J. 1988, *Ap.J.***330**, 350.
Bieging J.H., Cohen M., Schwartz P.R. 1984, *Ap.J.***282**, 699.
Bouvier J. 1987, Ph.D. Thesis, University Paris 7.
Bouvier J., Bertout C. 1989, *Astr.Ap.***211**, 99.
Brown A. 1987, *Ap.J.(Letters)***322**, L31.
Caillault J.P., Zoonematkermani S. 1987, in *Circumstellar Matter,* ed. I. Appenzeller and C. Jordan (Dordrecht: Reidel), p. 119.
Churchwell E., Felli M., Wood D.O.S., Massi M. 1987, *Ap.J.***321**, 516.
Cohen M., Bieging J.H. 1986, *Astr.J.***92**, 1396.
Edwards S., Cabrit S., Strom S.E., Heyer I., Strom K.M., Anderson E. 1987, *Ap.J.***321**, 473.
Feigelson E.D., DeCampli W.M. 1981, *Ap.J.(Letters)***243**, L89.
Feigelson E.D., Giampapa M.S., Vrba F.J. 1990, in *The Sun in Time,* eds. C.P. Sonett and M.S. Giampapa (Tucson: University of Arizona Press), in press.
Feigelson E.D., Kriss G.A. 1981, *Ap.J.(Letters)***248**, L35.
Feigelson E.D., Jackson J.M., Mathieu R.D., Myers P.C., Walter F.M. 1988, *Astr.J.***94**, 1251.
Feigelson E.D., Montmerle T. 1985, *Ap.J.(Letters)***289**, L19.
Gahm G.F. 1990, This volume.
Garay G., Moran J.M., Reid M.J. 1987, *Ap.J.***314**, 535.
Kenyon S.J., Hartmann L. 1987, *Ap.J.***323**, 714.
Klein K.L., Chiuderi-Drago F. 1987, *Astr.Ap.***175**, 179.
Ku W.H.-M., Righini-Cohen G., Simon M. 1982, *Science***215**, 61.
Lada C.J. 1985, *Ann.Rev.Astr.Ap.* **23**, 267.
Lada C.J. 1988, in *Formation and Evolution of Low Mass Stars,* ed. A.K. Dupree and M.T.V.T. Lago, NATO ASI Series (Dordrecht: Kluwer Academic Publishers), p.93.
Montmerle T. 1987, in *Solar and Stellar Physics,* Proc. 5th Eur. Solar Meeting, ed. E.H. Schröter and M. Schüssler (Berlin: Springer), *Lect. Notes in Phys.***292**, 117.
Montmerle T., André P. 1988, in *Formation and Evolution of Low Mass Stars,* ed. A.K. Dupree and M.T.V.T. Lago, NATO ASI Series (Dordrecht: Kluwer Academic Publishers), p. 225.
Montmerle T., André P. 1989, in Proc. ESO Workshop *Low Mass Star Formation and Pre-Main Sequence Objects,* ed. B. Reipurth (Garching: ESO), p. 407.
Montmerle T., Koch-Miramond L., Falgarone E., Grindlay J.E. 1983, *Ap.J.***269**, 182.
Montmerle T., Koch-Miramond L., Falgarone E., Grindlay J.E. 1984, in Proc. *Very Hot Astrophysical Plasmas,* ed. L. Koch-Miramond and T. Montmerle, *Phys. Scripta,* **T7**, 59.
Mundt R. 1984, *Ap.J.* **280**, 749.
Mutel R.L., Morris M.H., Doiron D.J., Lestrade J.-F. 1987, *Astr.J.***93**, 1220.
Panagia N., Felli M. 1975, *Astr.Ap.* **39**, 1.
Snell R.L., Bally J.B. 1986, *Ap.J.* **303**, 683.
Stine P.C., Feigelson E.D., André P., Montmerle T. 1988, *Astr.J.* **96**, 1394.
Tagger M., Henriksen R.N., Sygnet J.F., Pellat R. 1989, *Ap.J.(Letters),* in press.
Walter F.M. 1987, *Pub.A.S.P.***99**, 31.
Walter F.M., Kuhi L.V. 1984, *Ap.J.***284**, 194.
Walter F.M., Brown A., Mathieu R.D., Myers P.C., Vrba F.J. 1988, *Astr.J.***96**, 297.
Wilking B.A., Lada C.J., Young E.T. 1989, *Ap.J.* **340**, 823.

LANG: Could the VLBI be resolving the H II region, and why would you expect thermal emission at radio wavelengths?

MONTMERLE: The H II region is much bigger (about 3"), and some thermal radio emission due to stellar winds have been detected from some T Tauri stars.

BENZ: The interaction of an accretion disk and a magnetic field would be a possible way to accelerate the radio emitting electrons (example: Jupiter's moon Io). Why don't you have an accretion disk in your model?

MONTMERLE: In the case of the Rho Oph radio sources, we have no observational indication of the presence of warm circumstellar material (i.e. near-IR excess). So we have proposed (see Andre et al., 1988, Ap. J, 335, 940) that these electrons could come from a weak ionized wind, combined with the star's rotation (see also the case of Jupiter's magnetosphere, rotating in the solar wind). This is very likely in the case of S1, which is a B star, more speculative in the case of DoAr 21, which is of a later type. But any weak wind of solar type (i.e. larger than 10**(-14) solarmasses per year) would presumably be sufficient.

PALLAVICINI: Knowing only the temperature and emission measure of the X-ray source, how can you infer the size of the X-ray emitting region?

MONTMERLE: In the particular case of the Rho Oph X-ray source ROX-20, this is possible because we know, in addition, the decay time of the flare (cooling by radiation); see Montmerle et al. 1983, Ap.J. 269, 182.

PALLAVICINI: Could you explain the quiescent radio emission by thermal gyrosynchrotron by the same electrons responsible for the X-ray corona?

MONTMERLE: This idea has recently been proposed by Linsky (see Drake et al., Ap.J. suppl., in press), on the basis of a correlation between the 6 cm radio flux and the X-ray flux for RS CVn stars. However, recent work by Chiuderi-Drago and Klein (11th IAU European Regional Astronomy Meeting, 1989) has shown that, unless very high surface magnetic fields (several kG at least) are present, it is not possible to account simultaneously for other radio data (for instance at 20 cm and 2 cm in the case of HR 1099 and UX Ari) when available. Such high magnetic fields are too large compared with the values of less than 1 kG inferred from other works on these stars. So we stick to a nonthermal electron distribution.

GAHM: Did I understand correctly that you make a distinction between the nonthermal extended radio stars and the naked T Tauri stars?

MONTMERLE: Yes and no. In our Rho Oph sample we found radio emission only from stars embedded in the cloud core, which, presumably because of the heavy extinction, have little or no X-ray emission. Conversely, down to a sensitivity of 2 mJy (see Stine et al. 1988, Astr.J. 96, 297) we have not detected any of the ROX sources which are naked (or preferably "weak-line") T Tauri stars.

On the other hand, to this sensitivity in radio, only large emission measures (i.e. large sizes) are detectable. So it is likely that the present distinction only means that naked TTS have in general smaller magnetic loop sizes. Since we have noted (see Montmerle and Andre in "Formation and Evolution of Low-Mass Stars", NATO-ASI, Kluwer 1988, p. 225) that nonthermal radio emitters have high rotation velocities (more than 80 km/s, although measured on an admittedly small sample) it is quite possible that the bulk of the naked TTS have smaller loop sizes because they are slow rotators. However, a deeper survey of radio emission from NTTS is still missing.

TSVETKOV: Is the object DoAr 21 a T Tauri star?

MONTMERLE: Strictly speaking, no, since it appears to be more massive (about 2 solar masses). But it seems to be sufficiently high above the main sequence that it is still on a convective track, hence its basic properties should not differ drastically from those of standard T Tauri stars (about 1 solar mass or less).

PROSPECTS FOR STUDIES OF UV CET-TYPE FLARE STARS

M. RODONO'
Astronomy Institute of Catania University,
and Catania Astrophysical Observatory
Viale A. Doria 6
I-95125 Catania
Italy

ABSTRACT. The present review addresses selected questions on UV Cet-type red-dwarf stars primarily concerning the physics involved in the various aspects and phases of the flare phenomenon, rather than the average activity behaviour of the flare stars. In fact, while flare activity level and general trend are reasonably well established, a fully consistent physical picture of both solar and stellar flare events is still missing. Some recent results are presented with the aim of showing which observations are needed and the relevant role that coordinated multiband studies can play.

1. INTRODUCTION

The UV Cet-type stars in the solar neighbourhood play a fundamental role in the study of flare stars located in different clusters and associations, a role comparable to that of studying solar flares for the purpose of understanding the physics of the flare phenomenon in stars. The latter is a heuristic approach because it is still to be demonstrated that flares on UV Cet-type stars are solar analogues and their flaring behaviour could be considered only a variant of the flaring behaviour of cluster stars, as modified by age or other parameters' effects. On the other hand, it is quite obvious that "interdisciplinary solar - UV Cet - cluster star" flare studies could be mutually profitable, even if, eventually, we shall only be able to prove that we are dealing with completely different phenomena.

The flare activity of UV Cet-type stars has been the subject matter of several internationally coordinated

L. V. Mirzoyan et al. (eds.), Flare Stars in Star Clusters, Associations and the Solar Vicinity, 371–391.

studies and dedicated observations in the past twenty years. Their activity behaviour, as a group, may be considered well established thanks to the statistical investigations by the Crimean astronomers (Gershberg and Shakhovskaya 1983, Shakhovskaya 1989, and references therein), who have carried out fundamental works in this field by analysing more than 2000 optical flares in a few dozen of stars observed during the course of several thousand hours of photoelectric monitoring obtained at several observatories.

The following general results appear to be well established.

a) UV Cet-type flare stars in the solar vicinity form a rather mixed age group, which includes young and old disk, and, possibly, even halo objects.

b) The distribution of the total energy emitted in the B-band during flares E(B), versus mean occurrence rate of flares with energies exceeding E(B), shows a linear correlation in a log-log scale, i.e., the so called "flare energy spectra" can be represented by a power-law with spectral index $\beta \approx 1.0 \pm 0.5$ given by the slope of the linear dependence in the log-log representation. The β index shows a moderate tendency to increase toward fainter stars. A similar trend, spanning over several decades, is shown by Orion and Pleiades flare stars and the Sun, at the high and low energy ends of the distribution, respectively. Therefore, stars of quite different ages, masses and physical conditions, behave very similarly. This strongly suggests that the flare triggering agent, and especially the involved mechanisms, though obviously dependent on the physical characteristics of the flaring stars, cannot be substantially different, otherwise quite different β indices should result.

c) The maximum values of the time averaged luminosity due to flares (L_f) versus M_V show a systematic decrease toward fainter flare stars, while the maximum values of the normalized ratio L_f/L_{bol} show an upper limit of about 10^{-3}, which is independent of M_V. However, both L_f and L_f/L_{bol} values show a large spread up to three orders of magnitude indicating that stars of almost equal mass and luminosity display quite different activity levels.

d) There is clear evidence that the activity phenomena occurring on a given star at different atmospheric levels are closely connected. Chromospheric Balmer and transition region lines, such as N V, C IV, and Si IV, enhance simultaneously during stellar flares (Agrawal et al. 1986).

A well defined proportionality exists between integrated Hγ and coronal soft X-ray flux for stellar and solar flares (Butler et al. 1988; Haisch 1989) implying a close link between flare emission from relatively cooler and denser chromospheric regions ($T \approx 10^4$ K) and emission from hot ($T \approx 10^7$ K) and thin coronal plasma. Moreover, Butler et al. (1986) found a close time correlation between coronal X-ray and chromospheric Hγ line flux enhancements during UV Cet flares observed simultaneously with the EXOSAT satellite and the ESO 3.6 m telescope, respectively. This correlation suggests a close connection between flares and coronal heating. In this context it is important to point out that the mean flare luminosity (L_f) directly correlates with activity indicators at quiescent level, such as the luminosity of Balmer lines (Shakovskaya 1989) and soft X-rays (Doyle and Butler 1985). This means that the atmospheres of flare stars, also in their "quiescent" state, possess the embryonic physical signatures of activity.

Despite the observed stellar flares are often 1-3 orders of magnitude more energetic than solar flares, their characteristic behaviour and time evolution are basically similar. Solar flare data generally follow the empirical correlations between flare parameters found for stars, with the Sun occupying the low energy part of the correlation. For this reason, stellar flares are believed to be scaled up versions of solar flares (see Mullan 1989), though several questions concerning the physical interpretation of the flare phenomenon still remain to be answered.

In order to address basic questions on the so-called solar-stellar connection, the most recent studies of UV Cet-type stars, more than being concerned with the collective properties of flare stars, as a group, have tuned in to the physics involved in the various aspects and phases of the flare phenomenon (see Haisch and Rodonò 1989a, 1989b). This trend has naturally followed the increasingly converging paths of solar and stellar studies especially because of the improvements of time and spectral resolutions of stellar observations and the ever increasing use of new spectral windows, from X-ray, to near-infrared and microwave bands, which have finally become available to stellar flare studies. On the other hand, solar studies have made important progress in the understanding of the flare physics, due to the feasibility of high spatial, spectral and time resolved observations, which are not achievable in the stellar case. For the above mentioned reasons, the Sun has necessarily taken the role of the "rosetta stone" of stellar activity and particularly of flare studies (Parker 1989).

Several thorough reviews on flare stars, flare observations and theory have been given at recent meetings

(Byrne and Rodonò 1983, Gondalhekar 1986, Mirzoyan 1986, Haisch and Rodonò 1989a, 1989b). Therefore, the present paper will not duplicate nor attempt to summarize existing reviews on UV Cet-type flare stars, but will concentrate on selected specific aspects raised by recent observations of stellar flares on cool red-dwarfs and subgiants.

More specifically, the following topics, which are relevant in the interpretation of stellar flares, will be discussed:

a) the inhonogeneous structure of the magnetically controlled atmospheric plasma, where stellar flares occur;

b) the role of wide-band high-speed optical and IR photometry in evaluating the flare energy budget and testing the proposed flare mechanisms;

c) the importance of time resolved flare spectra in investigating the physical conditions of the flaring plasmas;

d) the essential requirement of conducting simultaneous multiwavelength observations to study these short, fast-evolving and non-repeating unique events and the research perspectives.

2. THE INHOMOGENEOUS STRUCTURE OF ACTIVE STAR ATMOSPHERES AND THE FLARE PHENOMENON

One of the most recent and significant progress in the study of stellar atmospheres has been the conclusive evidence that plane-parallel and homogeneous models are only first order approximations especially in the case of active stars. In fact, overwhelming evidence has been collected on the ubiquitus nature of magnetic fields, which play a fundamental role not only in the energy storage, energy release and activity mechanisms, but also in structuring the atmospheric plasma according to its complex and inhomogeneous structure, from photospheric up to coronal levels. As well known for the Sun, the emergence of magnetic flux tubes into the photosphere presides over the formation of sunspots, which are the footpoint of huge magnetic loops anchored into the photosphere and extending up to coronal levels. Flares generally originate in complex magnetic regions, as a result of magnetic loop interactions and energy dissipation, following magnetic field reconnections (see Forbes 1988). This qualitative picture is still far from being understood in detail, despite the huge amount of available solar flare data. However, even a simple, i.e., one-loop flare model can account for most of the observed solar flare phenomenology (see Emslie 1989, Dennis and Schwartz 1989 and references therein), especially as far as the initial and impulsive (= short duration) flare phase, which clearly demonstrates that magnetic loops are the fundamental constituents of flares.

Coronal mass ejection (CME) events generally occur at or before flare beginning, when the amount of energy deposition is so large that cannot be dissipated fast enough by the radiating plasma. CMEs may involve kinetic energies larger that subsequently radiated from the flare site. The formation of a very hot plasma with $T \approx 10^7$–10^8 K, from where CMEs often originate, marks the beginning of a flare at the top of a coronal loop (Figure 1) with the emission of microwave, mainly due to gyrosynchrotron emission, hard X-ray and γ-ray radiation, due to electron-ion bremsstrahlung. The acceleration of electrons and protons along field lines toward the chromosphere leads to plasma evaporation in this relatively denser region (thick target model), as demonstrated by the enhancements of mainly soft X-ray emission, XUV lines, Hα and other Balmer emission lines, and UV continuum. When the accelerated particles reach the most dense chromospheric or even photospheric regions, nuclear γ-ray lines and neutrons can be produced. If the accelerated particles are not sufficient energetic and are stopped well before reaching the chromosphere, soft X-rays are emitted from intermediate coronal levels and the energy transport toward the chromosphere is ensured by the thermal conduction. Due to the highly anisotropic bremsstrahlung cross section, γ-ray and neutron emissions are highly beamed, so that they are observed only from limb flares.

Several questions, however, remain to be answered, especially as far as it concerns the rapid rate of energy release (up to 10^{30} erg s^{-1}) and the acceleration of particles to relativistic energies in a few seconds by what should be a very efficient mechanism. However, the energy conversion, transport and dissipation are ill defined mechanisms, that, in addition to other simplifying assumptions, disregard possibly important effects of the field geometry and topology and of secondary interactions between plasma hydrodynamics and radiative transfer along the loop legs, up to its footpoints. Moreover, the idealized single-loop flare model, which is basically one dimensional, does not take into account the actual configuration of magnetic field lines that is likely to be much more complex, as shown in Figure 2 (from Parker 1989).

Therefore, though important progress has been made in the study of the flare phenomenon, we may say that some fundamental questions still remain to be answered. In a very synthetic way, we may say that we have learned a lot on "what" is a flare, but little progress has been made on "how and why" flares occur. In order to overcome this potentially stalling situation, solar studies are presently aimed at disclosing the finest spatial and energetic

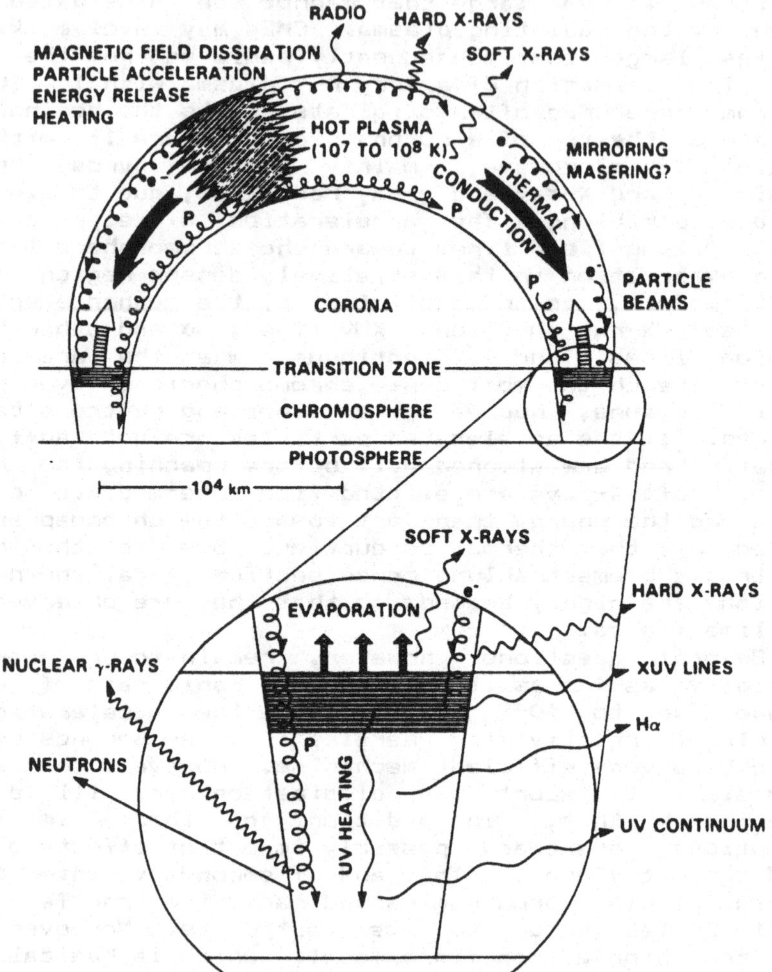

Figure 1. Physical processes and location of the emission sources in the solar atmosphere for a simple one-loop flare model (from Dennis and Schwartz 1989).

details of the flare triggering and emission mechanism, while stellar studies are becoming more concerned with the complementary aspect concerning the development of flares in a large variety of physical ambients, as only other stars, other than the Sun, can offer. For this reason, as anticipated in the Introduction, the present paper will consider only few recent aspects of stellar flare research concerning the flare physics rather than the collective properties of flare stars.

Several shortcomings affect stellar in comparison to solar flare observations, as for example the low level of the available flux, the lack of high spatial and spectral resolution, and our inability to detect γ-rays and neutrons with the presently available instrumentation. However, some specific aspects of stellar flare studies, which are considered in the following paragraphs, are complementary to solar flare studies, in that they allow us to concentrate on the general behaviour of the flare phenomenon versus global stellar parameters and on its cosmic significance.

3. HIGH-SPEED OPTICAL AND IR PHOTOMETRY AND THE FLARE ENERGY BUDGET

Wide-band studies of flares are carried out more efficiently in the stellar than in the solar case because of the much improved contrast between the flare radiation and the quiescent star background. For typical flares on M dwarf stars, the relative flux increase in the U-band ranges from 1 to 100, while for the Sun is much less than unity. This has allowed us to study in detail the energy and time characteristics of stellar flares and their statistical properties, already presented in the Introduction.

Moffett (1972) first pointed out the importance of high time resolution observations of stellar flares. By using time resolutions of 50 ms, he detected short-lived flare structures lasting no more than 2-3 seconds and showing light increase rates as fast as 0.3 magnitudes per second. Subsequent observations with time resolution of 10 ms by Rodonò et al. (1979) with a double-beam photometer, which allowed them to observe simultaneously the variable and a reference star, did give definite evidence on the occurrence of short-lived pre-flare dip and long-term variability of the "quiescent" level. Short-duration spiky flares have been reported by several observers with contraddictory results. (e.g., Zalinian and Tovmassian 1987, and other papers presented at this Symposium). These high-time resolution observations address important questions on the physics of flares, e.g. whether purely

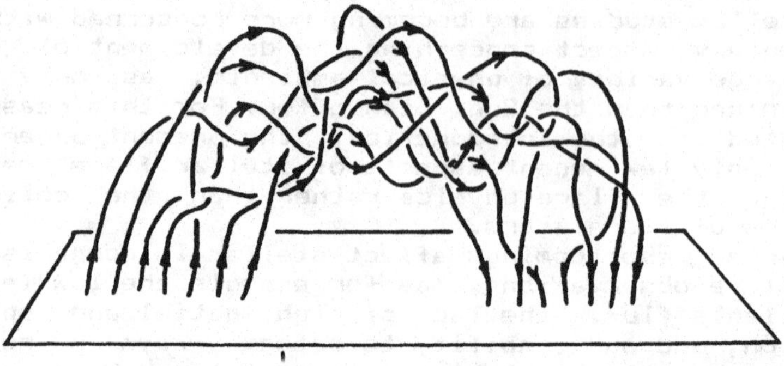

Figure 2. A possible configuration of the lines of force for a bipolar magnetic region (from Parker 1989).

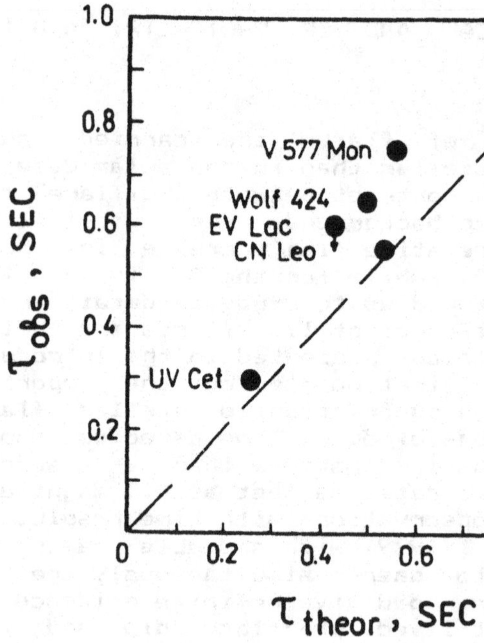

Figure 4. Comparison of observed and theorethical lifetimes of the steepest flare rising phase, the latter from the gas-dynamical model of Katsova et al. (1981).

thermal mechanisms, which successfully predict the overall spectroscopic features of flares, can also account for the observed flaring on time scales of the order of 1 s. The question of short-lived flares has been recently studied by USSR astronomers by using the so called MANIA (Multichannel Analyzer Nanosecond Intensity Amplification) system fed by the 6-m telescope at the Special Astrophysical Observatory (Beskin et al. 1989a). The MANIA system is able to record the time of arrival of individual photons and to achieve time resolutions as short as 3×10^{-7} s. From the observations of about one hundred flares on eight UV Cet-type dwarfs, they found no evidence of fine structures shorter than about 0.3 s. Moreover, considering only the steepest part of the rising flare phases, Beskin et al. (1989b) found that the flare rise times cluster around 2-3 s with minimum and maximum values at 0.3 s and 10 s, respectively. The shortest observed flare had a total duration of 1.75 s (Figure 3a). According to the gas-dynamic model of stellar flares (Katsova et al. 1981, Katsova and Livshits 1986) the flare rising time (τ) is directly related to the time required by a shock wave front to propagate towards the photosphere along a magnetic flux tube, up to a distance of about one scale height. By comparying the theoretical rise times ($\tau_{theor} = V_s$ / M g, where V_s is the sound speed in the chromosphere, M the Mach number and g the gravity acceleration) with the observed values (τ_{obs}), Beskin et al. (1989b) conclude that the thermal gas-dynamic model is adequate to interpret the impulsive start of the optical flare behaviour and that the decay time scales are consistent with the recombination times of optically thin plasma or the cooling times of optically thick plasma heated by the energy dissipation of shock wavesduring the initial phase of flares (Beskin et al. 1989c). The revival of high-time resolution photometry of stellar flares by the USSR astronomers is particularly important also because such data cannot be obtained for solar flares due to the high solar background level.

The existence of short-duration quasi-periodic light oscillation during flares is an additional question that high time resolution photometry can address. After Rodonò's (1974) observations of quasi-periodic 12-14 s light oscillation during the course of a flare on the Hyades star Hertzspung II 2411 and Mullan's (1976a, b) interpretation in terms of electron cyclotron waves ("whistlers") travelling from one magnetic pole to the other, no additional optical evidence has been collected. Oscillatory behaviour also at radio-wavelengths has been reported (see Gibson 1983). The onset of oscillations during stellar flares is a virtually unexplored field, although it is of great interest for the purpose of understanding the flare mechanisms and energy dissipation modes. Some interesting

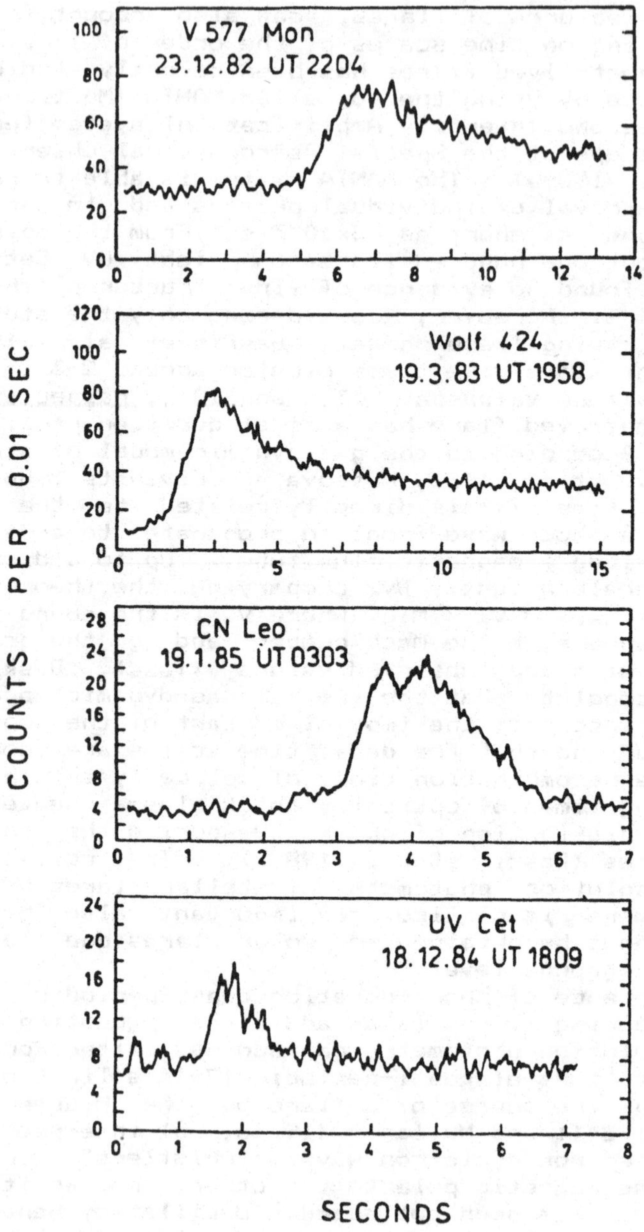

Figure 3. High-speed photometry of stellar flares obtained with the MANIA photometer fed by the 6-m telescope at the Special Astrophysica Observatory, USSR (from Beskin et al. 1989b).

and very promising results have been recently obtained by Andrews (1989, and references therein).

Another important aspect of wide-band flare studies is the energy involved in the various spectral regions. As shown by Gershberg and Shakhovskaya (1983) statistics, the total optical energy output during the course of a stellar flares on UV Cet type stars is in the range from 10^{27} to 10^{34} erg. Taking into account the energy emitted in other wavebands, the flare total energy budget very seldom increases by at most one order of magnitude. From simultaneous optical (U) and infrared (K) photometry, Rodonò and Cutispoto (1988) detected for the first time "negative" infrared events at wavelengths longer than 1 μm in coincidence with optical flares (Figure 5), as predicted by Gurzadian (1980) non-thermal and Grinin (1976) thermal models. At present it is not possible to draw any conclusion about the validity of these flare models because only a few events have been observed. One important question raised by these IR observations concerns the flare energy budget. The missing energy due to the K-band "negative" flare is about one order of magnitude larger than the energy released in the U-band, i.e., the energy missing in the K-band alone can account for the energy released in all other spectral regions. Additional observations are needed to reach a definite conclusion on whether this result is directly linked to the shift of photospheric photons towards shorter wavelengths by their inverse Compton interaction with fast electrons, as first predicted by Gurzadian (1980, and references therein), or can be accounted for by the increase of H^- opacity during the very first phase of flares (Grinin 1976). The amplitudes of the observed "negative" flare are consistent with those predicted by both models, but the infrared flare light curve does not appear to be the mirror image of the optical light curve as far as time evolution is concerned. However, since the observed flares are rather complex, they are not adequate for a stringent test of both Gurzadian's and Grinin's models, as the simultaneous optical and IR observations of intense single-peaked flares would be.

4. TIME-RESOLVED FLARE SPECTRA

The relative importance of the basic emission mechanisms that drive emission line fluxes in the optical region during the course of stellar flares are poorly known, mainly because sufficiently time-resolved spectra of flares are difficult to obtain because of the concurrent negative effects due to the faintness of stars and the rapid development of flares. Large aperture telescopes and adequate

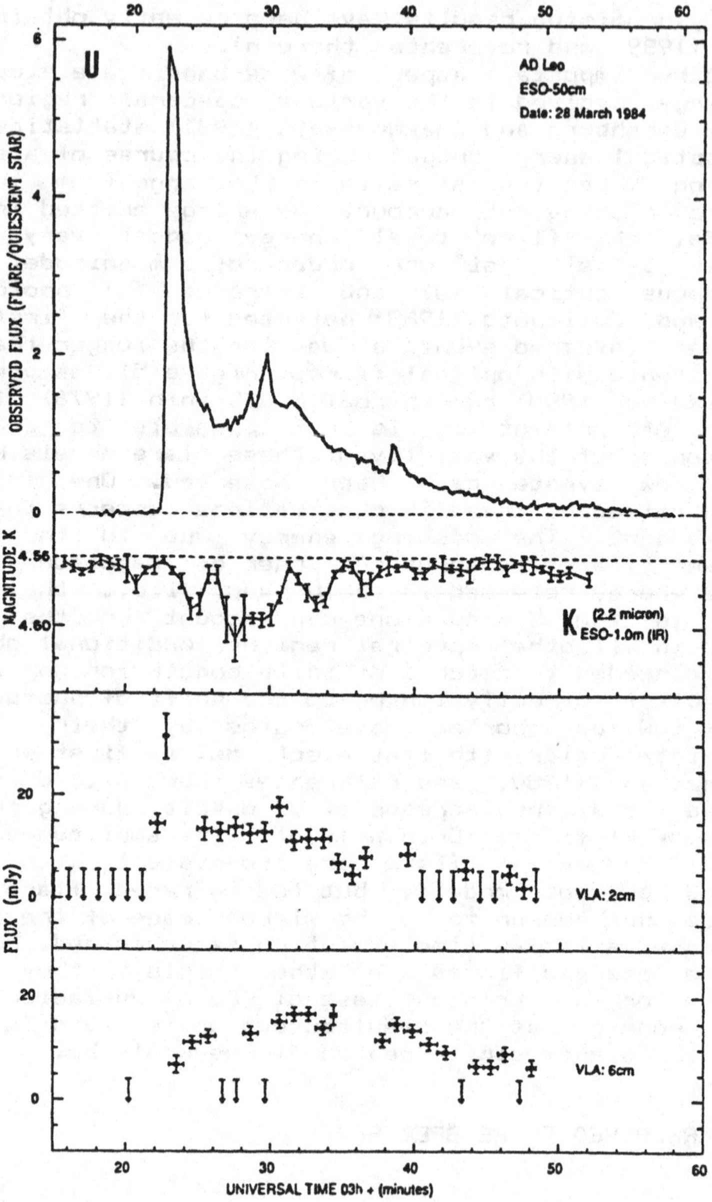

Figure 5. Simultaneous optical (U-band), infrared (K-band), and microwave (2- and 6-cm) observations of a complex flare on AD Leonis (form Rodonò et al. 1989a).

fast detectors are required to achieve that goal.

The best time-resolved and accurate spectra of stellar flares, which are presently available, were obtained by Rodonò et al. (1989a) using the Image Dissector Scanner (IDS) fed by the ESO 3.6-m telescope. These observations show a distinctively different time evolution of Ca II K, He II 4026 A and Hydrogen Balmer lines, the former line showing a more gradual enhancement and decay than He II and Balmer lines (Figure 6). Houdebine et al. (1989a), have developed line models based on the main atomic processes taking place in a typical flare plasma, assumed to be stationary and optically thin. Their calculations provided convincing evidence that the line fluxes are largely influenced by electron temperature variations. They also showed that during flare decay two main phases can be identified: the first phase is dominated by radiative processes, such as photoionization, and the later phase by collisions. The resulting electron densities for the observed flares range from 3×10^{11} to 10^{12} cm^{-3}, much higher than in previous studies (Gurzadian 1977). Shallow opacities and NLTE departure effects on the higher Balmer lines were also pointed out by Houdebine et al. (1989a). By analysing the same time-resolved spectra of flares obtained at ESO, Houdebine et al. (1989b) showed evidence for a high velocity mass ejection event which occurred at the onset of a particularly violent flare on the M dwarf AD Leo. The plasma was ejected at projected line-of-sight speeds of up to 5800 Km s^{-1} (Figure 7). That event appeared to be similar to solar coronal mass ejection (CME) events, but involving kinetic energy (5×10^{34} erg), mass ejection (7.7×10^{14} kg) and speed values ($\approx$ 6000 km s^{-1}) that are 500, 40 and 5 times larger than typical CME solar events, respectively. The estimated mass loss for various sets of electron temperatures and plasma opacities, combined with typical flare occurence rates, indicates that flare activity may have important consequences on the evolution of active stars and on the composition and physical state of the circumstellar and interstellar medium.

As recently reviewed by Byrne (1989), line broadenings of the order of 10-100 A in the optical band and line shifts of the order of 10-100 Km s^{-1} UV bands have been reported by several authors. Although only high time and spectral resolution data can provide definite evidence of mass motions within and from flaring regions on stars, similar to the well ascertained mass motions in solar flare regions, the available stellar data, which are usually derived from low and medium resolution spectra, are very promising and encourage to pursue further the collection of high quality flare spectra with time resolutions of few seconds and spectral resolutions of the order of 10^4, by using large aperture telescopes and modern bidimensional

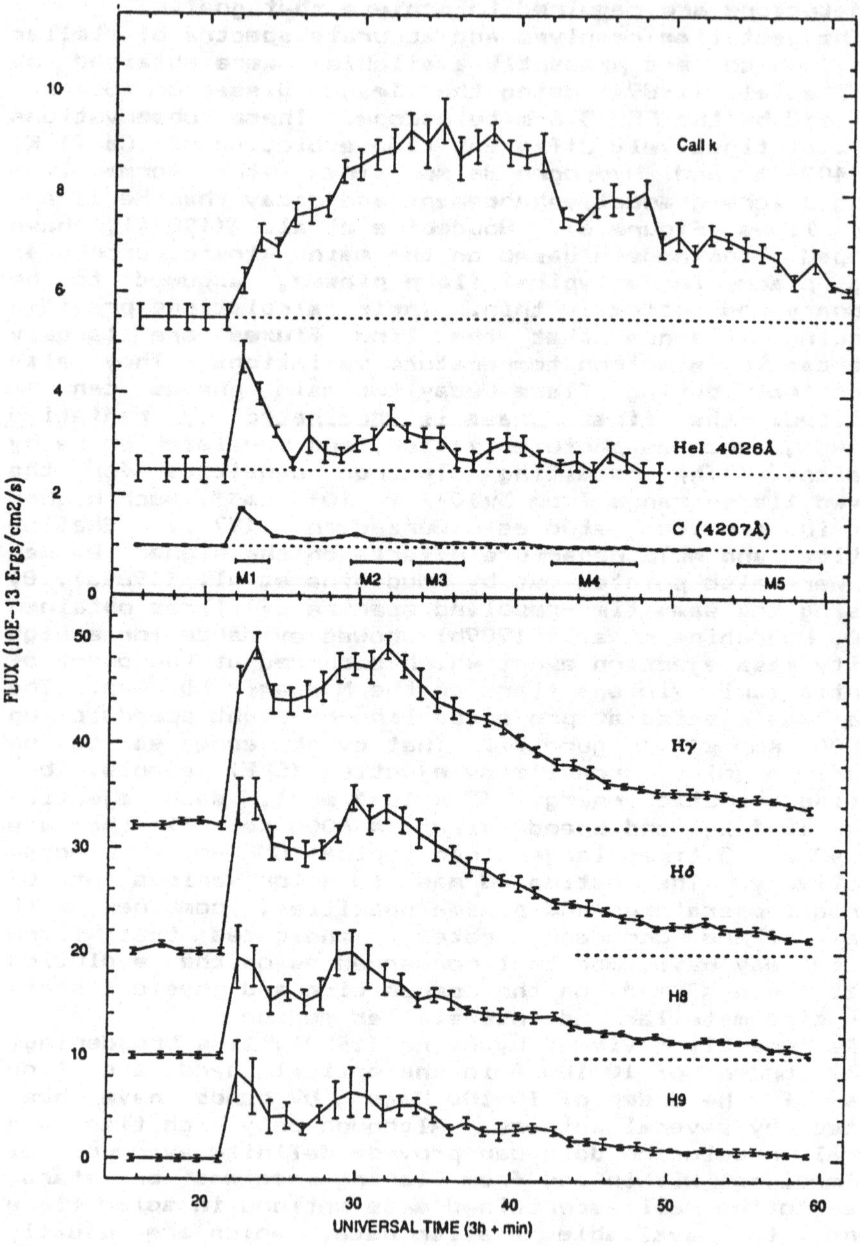

Figure 6. Time development of emission line fluxes during the course of the AD Leo flare, whose light curve is shown in Figure 5 (from Rodonò et al. 1989b).

detectors, such as Charge-Coupled Devices (CCD) or Micro-Channel Plates (MCP).

Some evidence of oscillatory behaviour of Balmer line Doppler shift during the decay phase of a flare on AD Leo was reported for the first time by Houdebine (1989). The question of flux or velocity oscillations, as already quoted in paragraph 3, would deserve to be addressed more systematically with high time resolution photometry and spectroscopy because of its potential importance as a diagnostic tool of the physical conditions of the flaring plasma and flare mechanisms.

In a paragraph devoted to time-resolved flare spectra we should not even mention UV data, the majority of which has been obtained with IUE. As known, this otherwise very productive satellite, does not have the capability of obtaining high resolution UV spectra of flare stars with the required time resolution. In fact, the best time resolved flare spectra have been obtained with the low-dispersion camera and with time-resolutions of the order of minutes. Therefore, only time-averaged flare parameters could be obtained (see Byrne 1989). Instead, long duration flares on bright RS CVn stars have been studied in some details. By using the so called Doppler Imaging Technique, Linsky et al. (1989) were able to extract pure flare spectra and derive line profiles showing clear evidence of mass motion. Nevertheless, IUE has allowed us to make important progress in the study of flares on dMe stars. All choromospheric and transition region lines are strongly enhanced at the time of flares and the degree of enhancement increases with the temperature of line formation, as observed during solar flares. This is primarily due to steepening of the temperature gradient in the upper chromosphere and transition region, which during the course of flares is pushed down towards higher density atmospheric levels. From time integrated flare data, electron densities of the order of 10^{11}-10^{12} cm^{-3} and integrated energies of a few 10^{32} erg in the transition region line, with emission measures in the C IV line of the order of 1-10x10^{47} cm^{-3}, have been derived.

It is evident that the UV part of the spectrum contains very important chromospheric and transition region line diagnostics. Therefore, it is fundamental that a new generation of large aperture UV telescopes, such as the proposed SPECTRUM UV TELESCOPE of 170-cm aperture, becomes available in the nearest future to allow us to obtain high quality data on the UV spectral behaviour of stellar flares with time resolutions of the order of 10 seconds and spectral resolutions of at least a few thousands.

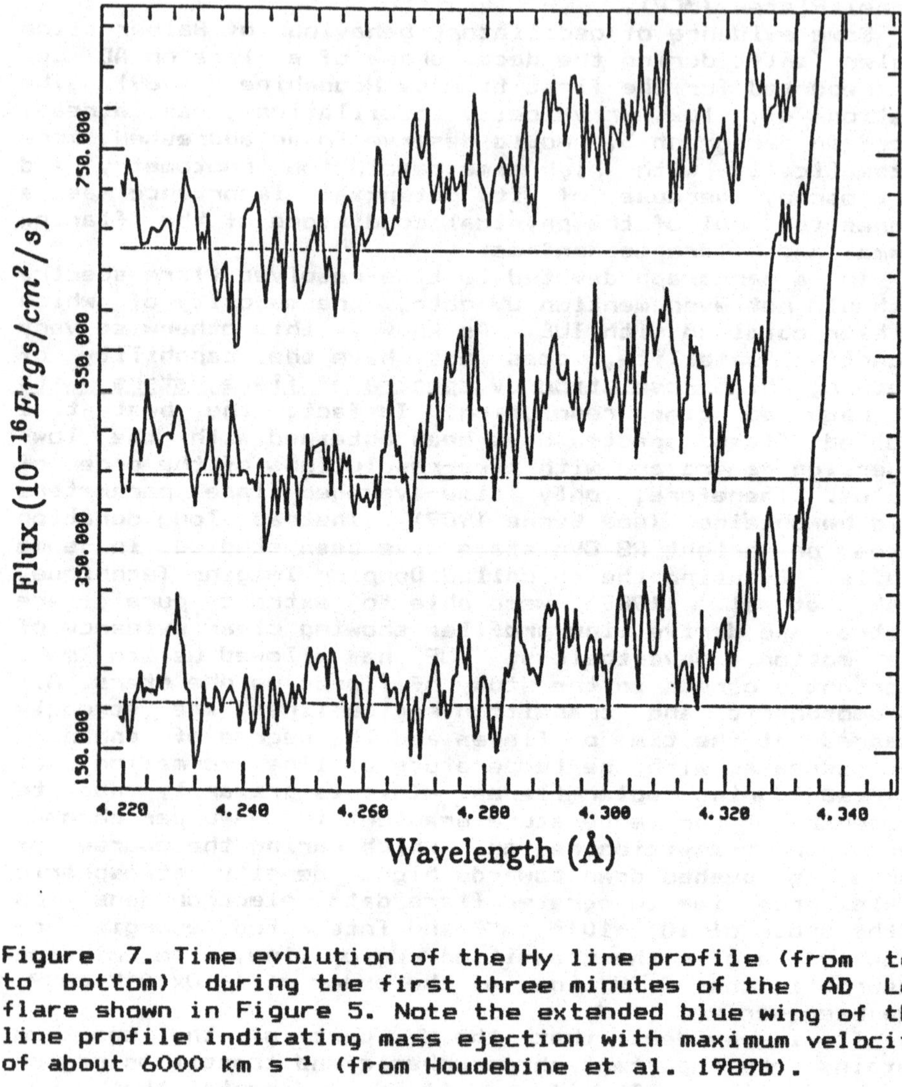

Figure 7. Time evolution of the Hγ line profile (from top to bottom) during the first three minutes of the AD Leo flare shown in Figure 5. Note the extended blue wing of the line profile indicating mass ejection with maximum velocity of about 6000 km s^{-1} (from Houdebine et al. 1989b).

5. SIMULTANEOUS MULTI-WAVELENGTH OBSERVATIONS, AND RESEARCH PROSPECTS

Since flares are non-repeating, short-duration and fast-evolving events, which affect almost any level of stellar atmospheres, simultaneous multi-wavelength data are required in order to derive significant information on the physical state of the flaring plasma and on the flare triggering and energy release mechanisms. As emphasized by Byrne (1989), a complete set of multi-wavelength observations including all potentially available spectral regions - from X-rays to microwaves - does not exist even for one single flare, despite so many years of flare studies (see Rodonò 1986). Some efforts have been hampered by instrument failures, wheather conditions, and, unfortunately, sometimes also by unresponsive Allocation Committees of telescope time. Often, in addition to optical data, only one additional waveband has been covered, such as the soft X-ray, UV or microwave band. The most complete data set, which is available to date, include optical photometry and spectroscopy, IR photometry, microwaves, and, in a few instances, also soft X-ray coverage (Rodonò et al. 1989a, Butler et al. 1986). Some relevant results obtained by using these data have been already presented in the previous paragraph. What I should like to stress at this point is that if we want to progress further in the understanding of both solar and stellar flares we must become able to organise further internationally coordinated observations, to be obtained simultaneously in several spectral regions, including many chromospheric, transition region and coronal diagnostics. This implies to observe stellar flares simultaneously with ground-based and space observatories in the optical (U-B-V), IR (J-K-L bands), microwave (2-6-20 cm), low and medium energy X-ray bands, with high time and spectral resolutions adequate to follow the fast time- and spectral-evolution of flare events. In the future, also hard energy spectral regions, such as hard X-rays and γ-rays, should be explored.

One of the neglected aspects of flare studies is their occurrence in binary systems. Actually, the magnetic field configuration in binary systems may extend over the entire system with magnetic loops connecting the two stars, as proposed by Uchida and Sakurai (1983) and Uchida (1986) and suggested by the IUE observations of an intense flare on the RS CVn-type system UX Ari (Simon et al. 1980). Almost simultaneous flaring of both components of binary systems in the optical (Rodonò 1978) and radio wavelength bands (see Gibson 1983) is another neglected topic that would deserve dedicated studies. In order to make some progress in the study of binary flare stars, we should also aim at high spatial resolutions by using interferometric techniques, both at radio and optical wavelengths.

Finally, long-term flare activity cycles, similar to and likely associated with spot cycles as in the Sun, have not been studied in a systematic way, despite their obvious importance.

The main problems with long-term systematic studies of stellar activity resides over the difficulty of obtaining sufficient telescope time and the heavy work required by dedicated and systematic observations, which sometimes are not sufficiently rewarding. Automatic Photometric Telescopes (APT), being able to perform robotic observations (Genet et al. 1987), will give new impetus to stellar activity studies. An international APT network and coordinated collaborative programs, such as SYNOP promoted by J.L.Linsky (see Giampapa 1986), MUSICOS promoted by B.H.Foing (1989), and ODIN promoted by B.R.Pettersen (1989), will certainly give sufficient momentum for a decisive step forward in our understanding of a relevant astrophysical phenomenon, whose cosmic significance has not yet received due recognition.

Acknowledgements. Solar and stellar activity research at Catania University and Observatory is supported by the "Ministero dell'Università e della Ricerca Scientifica", the National Research Council ("C.N.R.: Gruppo Nazionale di Astronomia") and the Italian Space Agency (A.S.I.), that are gratefully acknowledged.

REFERENCES

Andrews, A.D.: 1989, Astron. Astrophys., in press.

Agrawal, P.C., Rao, A.R., Sreekantan, B.V.: 1986, Montly Not. Roy. Astron. Soc. 219, 225.

Beskin, G.M., Mitronova, S.N., Neizvestnji, S.I., Gershberg, R.E.: 1989a, in Solar and Stellar Flares, Poster Papers, Proc. IAU Coll. 104, B.M. Haisch, M. Rodonò (eds.), Publ. Catania Astrophys. Obs., Special Volume, p. 95.

Beskin, G.M., Plakhotnichenko, V.L., Pustil'nik, L.A., Schwartsman, V.F., Gershberg, R.E.: 1989b, in Solar and Stellar Flares, Poster Papers, Proc. IAU Coll. 104, B.M. Haisch and M. Rodonò (eds.), Publ. Catania Astrophys. Obs., Special Volume, p. 99.

Beskin, G.M., Neizvestnyj, S.I., Plakhotnichenko, V.L., Pustil'nik, L.A., Schvartsman, V.F.: 1989, in Solar and Stellar Flares, Poster Papers, Proc. IAU Coll. 104, B.M. Haisch and M. Rodonò (eds.), Publ. Catania Astrophys. Obs., Special Volume, p. 103

Butler, C.J., Rodonò, M., Foing, B.H., Haisch, B.M.: 1986, Nature 321, 679.

Butler, C.J., Rodonò, M., Foing, B.H.: 1988, Astron. Astrophys. 206, L1.
Byrne, P.B.: 1989, in Solar and Stellar, Proc. IAU Coll. 104, B.M.Haisch and M.Rodonò (eds.), Solar Phys. 121, p. 61.
Byrne, P.B., Rodonò, M. (eds.): 1983, Activity in Red Dwarf Stars, Proc. IAU Coll. 71, Reidel Publ. Co., Dordrecht, The Netherlands.
Dennis, B.R., Schwartz, R.A.: 1989, Solar and Stellar Flares, Proc. IAU Coll. 104, B.M. Haisch and M. Rodonò (eds.), Solar Phys. 121, 75.
Doyle, J.G., Butler, C.J.: 1985, Nature 313, 378.
Doyle, J.G., Butler, C.J., Byrne, P.B., Van den Oord, G.H.J.: 1988, Astron. Astrophys. 193, 229.
Emslie, E.G.: 1989, in Solar and Stellar Flares, Proc. IAU Coll. 104, B.M. Haisch and M. Rodonò (eds.), Solar Phys. 121, 105.
Foing, B.H.: 1989, MUSICOS Circ., Inst.Astrophys.Spatiale, Paris.
Forbes, T.G.: 1988, in Activity in Cool Star Envelopes, O. Havnes, B.R. Pettersen, J.H.M.M. Schmiddt, and J.E. Solheim (eds.), Kluwer Acad. Publ., Dordrecht, The Netherlands, p. 115.
Genet, R.M., Boyd, L.J., Hayes, D.S., Baliunas, S.L., Crawford, D.L., Hall, D.S., Genet, D.R.: 1987, in Fifth Cambridge Work. on Cool Stars, Stellar Systems and the Sun, J.L.Linsky and R.E.Stencel (eds.), Springer-Verlag, Berlin, p. 473.
Gershberg, R.E., Shakhovskaya, N.I.: 1983, Astrophys. Space Sci. 95, 235.
Giampapa, M.S. (ed.): 1986, The SHIRSOG Workshop, Nat.Opt. Astron. Obs., Tucson AZ, USA.
Gibson, D.M.: 1983, in Activity in Red-Dwarf Stars, Proc. IAU Coll. 71, P.B. Byrne and M. Rodonò (eds.), Reidel Publ. Co., Dordrecht, The Netherlands, p. 273.
Gondalhekar, P.M. (ed.): 1986, Flares: Solar and Stellar, Astron. Astrophys. Workshop, Rutherford Appleton Lab., RAL-86-085.
Gurzadian, G.A.: 1977, Astrophys. Space Sci. 52, 51.
Haisch, B.M.: 1989, Astron. Astrophys. 219, 317.
Haisch, B.M., Rodono, M. (eds.): 1989a, Solar and Stellar Flares, Proc. IAU Coll. 104, Solar Phys. 121.
Haisch, B.M., Rodonò, M. (eds.): 1989b, Solar and Stellar Flares, Poster Papers, Proc. IAU Coll. 104, Publ. Catania Astrophys. Obs., Special Volume.
Houdebine, R.E.: 1989, in 6th Cabdridge Workshop on Cool Stars, Stellar Systems and the Sun, Publ. Astron. Soc. Pacific, Conf. Ser., in press.
Houdebine, R.E., Butler, C.J., Panagi, P.M., Rodonò, M., Foing, B.M.: 1989a, Astron. Astrophys. (submitted).
Houdebine, R.E., Foing, B.M., Rodonò, M.: 1989b, Astron.

Astrophys. (submitted).
Katsova, M.M., Kosovichev, A.G., Livshits, M.A.: 1981, Astrofizika 17, 285.
Katsova, M.M., Livshits, M.A.: 1986, in Flare Stars and Related Objects, L.V. Mirzoyan (ed.), Publ. Armenian Acad. Sci, USSR, p. 183.
Linsky, J.L., Neff, J.E., Brown, A., Gross, D.B., Simon, T., Andrews, A.D., Rodonò, M., Feldman, P.A.: 1989, Astron. Astrophys. 211, 173.
Mirzoyan, L.V. (ed.): 1986, in Flare Stars and Related Objects, Publ. Armenian Acad. Sci., Erevan, USSR.
Moffett, T.J.: 1972, Nature Phys. Sci. 240, 41.
Mullan, D.J.: 1976a, Astrophys. J. 204, 530.
Mullan, D.J.: 1976b, Astrophys. J. 206, 672.
Mullan, D.J.: 1989, in Solar and Stellar Flares, Proc. IAU Coll. 104, B.M. Haisch and M. Rodonò (eds.), Solar Phys. 121, 239.
Parker, E.N.: 1989, in Solar and Stellar Flares, Proc. IAU Coll. 104, B.M. Haisch and M. Rodonò (eds.), Solar Phys. 121, 271.
Pettersen, B.R.: 1989, ODIN Circ., Inst.Theor.Astrophys., Oslo, Norway.
Rodonò, M.: 1974, Astron. Astrophys. 32, 337.
Rodonò, M.: 1978, Astron. Astrophys. 66, 175.
Rodonò, M.: 1986, in Flare Stars and Related Objects, L.V. Mirzoyan (ed.), Publ. Armenian Acad. Sci., Erevan, USSR, p. 19.
Rodonò, M., Pucillo, M., Sedmak, G., De Biase, G.A.: 1979, Astron. Astrophys. 76, 242.
Rodonò, M., Cutispoto, G.: 1988, in Activity in Cool Star Envelopes, O.Havnes, B.R.Pettersen, J.H.M.M.Schmiddt, and J.E. Solheim (eds.) , Kluwer Acad. Publ., Dordrecht, The Netherlands, p. 163.
Rodonò, M., Houdebine, E.R., Catalano, S., Foing, B., Butler, C.J., Scaltriti, F., Cutispoto, G., Gary, D.E., Gibson, D.M., Haisch, B.M.: 1989a, in Solar and Stellar Flares, Poster Papers, IAU Coll. 104, B.M. Haisch and M. Rodonò (eds.), Publ. Catania Astrophys. Obs., Special Volume, p. 53.
Rodonò, M. et al.: 1989b, work in progress.
Shakhovskaya, N.I.: 1984, in Solar and Stellar Flares, Proc. IAU Coll. 104, B.M. Haisch and M. Rodonò (eds.), Solar Phys. 121, 375.
Simon, T., Linsky, J.L., Schiffer, F.H. III: 1980, Astrophys. J. 239, 911.
Uchida, Y.: 1986, in Highlights of Astronomy, J.P.Swings (ed.), Reidel Publ. Co., Dordrecht, The Netherlands, p. 461.
Uchida, Y., Sakurai, T.: 1983, in Activity in Red-Dwarf Stars, Proc. IAU Coll. 71, P.B.Byrne and M.Rodonò (eds.), Reidel Publ.Co., Dordrecht, The Netherlands, p. 629.

MIRZOYAN: You don't say anything about the evolutionary status of flare stars. Maybe the title of your paper do not permit you to do this. But I must express my disagreement with your opinion that the Sun is the only source to interpret stellar flares. I think that no less important for this problem is the study of flares in the star clusters and associations and of fuor-like changes. I hope that this study can contribute to the interpretation of stellar flares and may be even flares on the Sun.

RODONO: My talk was concerned with the physics of flare phenomena and the difficulties of interpreting them, and the evolutionary aspect was reviewed by you at the beginning of this Symposium. When I said with Parker's words that the Sun is the Rosetta Stone of flare studies, I did not intend to say that it is the only source of inspiration to study stellar physics. Actually, the study of flaring events on different stars will have a key role in this respect.

PALLAVICINI: Are there any observations of solar flares in the infrared showing a similar phenomenon as that reported by you for stellar flares?

RODONO: None as far as I know. White light flares on the Sun have been observed only recently in a systematic way and up to about 6000 A.

PALLAVICINI: Could the blue-shifted components you found in UV lines be due to evaporation of chromospheric material as sometimes observed for the Sun?

RODONO: The amount of mass implied by the observed blue-shift seems to be too large to be accounted for by chromospheric evaporation?

LANG: The solar analogy cannot be pushed too far when it comes to the radio emission of flare stars. When radio flares are seen at the same time as optical or X-ray flares in stars,the radio emission is very weak and sometimes absent. At other times there are strong radio flares with undetectable emission in other spectral domains. In contrast, the Sun radiates strong flares at the same time at optical,radio and X-ray wave-lengths. Also, the quiescent emission from YZ CMi is a few mJy at 6 cm, and it is nearly always present, so it is stretching the imagination to associate emission at this level with optical and X-ray flares.

Closing Remarks

Charles J. Lada
Steward Observatory, University of Arizona, Tucson, Arizona, USA 85721

A number of years ago, the nobel-prize winning Chilean poet Pablo Neruda paid a visit to the Byurakan and wrote these comments in his *Memoirs* (1978:Penguin Books, New York, pg. 243.): "I shall never forget my visit to the astronomical observatory of Byurakan, where I saw the writing of stars for the first time. The trembling light of the stars was picked up; very fine mechanisms were taking down the palpitation of the stars in space, like an electrocardiogram of the sky. In those graphics I observed that each star has its own distinct way of writing, tremulous and fascinating, but unintelligible to the eyes of an earthbound poet" Unintelligible to the eyes of a poet, but as we have seen from this symposium, to the eyes of the astronomer, the phenomenon of flare stars is becoming more and more comprehensible. Before I came here to Byurakan I must confess that I (like Neruda) knew very little about the flare star phenomenon. As a result of this symposium I have learned much about these objects and now have a keen appreciation for their importance for studies of early stellar evolution. In this regard, I find it a particular priviledge and extremely appropriate to have a meeting on the subject of flare stars here at the Byurakan Observatory, where so much seminal work on this topic has been done. I have been impressed by the dedicated efforts of the astronomers at the Byurakan Observatory in flare star research. Their efforts represent an important contribution to galactic astronomy.

Finally, on behalf of the all the participants and lecturers, I would like to congratulate the Scientific Organizing Committe for putting together a very exciting and interesting meeting on Flare Stars in Star Clusters, Associations and the Solar Vicinity. I would like to express particular gratitude to the local organizing committee who deserve special commendation for their heroic efforts to hold this meeting in light of the recent devastating earthquake and the current rail blockade by Azerbaijan. This committee has done an outstanding job, in the face of such trying times, to insure the smooth and effective exchange of ideas at this very stimulating international scientific gathering.

L. V. Mirzoyan et al. (eds.), Flare Stars in Star Clusters, Associations and the Solar Vicinity, 393.

Closing Remarks

[illegible]
[illegible] History, University [illegible] Thesis, [illegible]

A number of years ago [illegible] the [illegible] [illegible] the [illegible] [illegible] [illegible] the [illegible] [illegible] history [illegible] the [illegible] the [illegible] the [illegible] of the [illegible] of writing [illegible] [illegible] the eyes of a [illegible] the eyes of the [illegible]

[illegible] the [illegible] the [illegible] [illegible] and [illegible] the [illegible] [illegible] international scientific [illegible]

[illegible] Netherlands.

AUTHOR INDEX

A

Aanesen, T. 15
Ambartsumian, V.A. 163
Ambruster, C.W. 15
Aniol, R. 85
Antov, A. 27
Appenzeller, I. 209
Avgoloupis, S. 15

B

Bastian, T.S. 139
Bastien, P. 179
Benz, A.O. 139
Berdyugin, A.V. 37
Bondar, N.I. 55
Broutian, G.H. 63
Bruevich, E.A. 317
Butler, C.J. 153, 313, 325

C

Caillault, J.P. 159
Chavushian, H.S. 63
Chugainov, P.F. 189

D

Doyle, J.G. 325
Duerbeck, H.W. 85, 99

E

Epremian, R.A. 261

F

Fürst, E. 139

G

Gahm, G. 193
Garibjanian, A.T. 59, 95, 121
Gasparian, L.G. 253
Gershberg, R.E. 19
Getov, R.G. 19
Gosachinskij, I.V. 275
Grinin, V.P. 299, 343
Gudel, M. 139
Gyulbudaghian, A.L. 279
Götz, W. 215

H

Hambarian, V.V. 59, 95, 121
Harutyunian, H.A. 333
Havnes, O. 15
Hayrapetyan, V.S. 329, 333
Herbst, W. 169
Herouni, P.M. 145
Hojaev, A.S. 81
Houdebine, E.R. 313
Hovhannessian, R.Kh. 261

I

Ibrahimov, M.A. 257
Ilyin, I.V. 19, 41
Ishida, K. 43, 225
Ivanova, M.S. 15, 19, 27

K

Kandalian, R.A. 275
Karapetian, A.A. 33
Katsova, M.M. 321
Kelemen, J. 67
Kiang, T. 325
Krelowski, J. 293

L

Lada, C. 347, 393
Lang, K.R. 125
Leto, G. 19
Livshits, M.A. 317

M

Magakyan, T.Yu. 267
Matveyenko, L.I. 271
Mavridis, L.N. 15
Melikian, N.D. 31
Melkonian, A.S. 25, 253
Menard, F. 179
Mirzoyan, A.L. 59, 77, 113, 121
Mirzoyan, L.V. 1, 59, 95, 121
Mitskevich, A.S. 343
Mnatsakanian, M.A. 77, 113
Montmerle, T. 363
Movsessian, T.A. 267, 283

N

Natsvlishvili, R.Sh. 101
Nazaretian, F.S. 275
Nesterov, N.S. 19
Nikoghossian, A.G. 329

O

Ohanian, G.B. 109, 253
Olah, K. 15
Olsen, Ö. 15
Oskanian, V.S. 145

P

Pallavicini, R. 147
Panov, K.P. 15, 19, 27
Papaj, J. 293
Papas, C.H. 337
Parsamian, E.S. 109, 253
Petrov, P.P. 219
Pettersen, B.R. 15, 49
Pigatto, L. 117
Pointon, L. 139

R

Reipurth, B. 229
Rodonó, M. 371
Rodriguez, L.F. 279
Romanjuk, Ya.O. 35

S

Sanamian, V.A. 275
Schwartz, R.D. 221, 279
Seiradakis, J.H. 15
Seitter, W.C. 85
Shakhovskaya, N.I. 41, 53, 185
Shevchenko, V.S. 31, 257, 263
Simnett, G.M. 139
Solheim, J.E. 15
Stella, L. 147
Sundland, S.R. 15
Svyatogorov, O.A. 35
Szecsenyi-Nagy, G. 71

T

Tagliaferri, G. 147
Tovmassian, H.M. 33, 261
Tsikoudi, V. 287
Tsvetkov, M.K. 19, 85, 99, 105
Tsvetkova, K.P. 105

V

Valtaoja, E. 15
Valtaoja, L. 15
van den Oord, G.H.J. 325
Vardanian, R.A. 177

W

Wegner, W. 293

Y

Yakubov, S.D. 263
Yudaeva, N.A. 275

Z

Zajtseva, G.V. 173
Zalinian, V.P. 33
Zhilyaev, B.E. 35
Zoonematkermani, S. 159

SUBJECT INDEX

A

accretion 210, 219, 241, 364
age 117, 121
aggregates 85, 113
associations 1, 109, 226, 348
ASTRON 301

B

Balmer decrement 317, 321
Bok globules 226
broad-band radio observations 141
bursts 125, 139
BY Dra stars 189

C

catalog 101, 105
chromosphere 313, 317, 321
circular polarization 127, 179
circumstellar
-disk 181
-dust 181
-shells 215, 293, 366
corona 126

D

decrement 317, 321
density 130, 300
disks 179, 181, 209, 240
dMe stars 49, 134
dust 181
dynamic spectra 134

E

EINSTEIN 159, 302
electron density 130, 300
electron-cyclotron masers 125, 139
energy transport 337
EXOSAT 147, 153, 302
extinction 293

F

flare
  -impulsive 148, 306, 313
  -radio 125, 139, 145, 275, 302, 364, 382
  -slow 9, 82, 109, 257, 377
  -spike 27, 31, 33, 35, 132
  -stellar 15, 25, 31, 41, 43, 49, 193, 209, 321, 333, 371
  -thermal 155, 313
  -X-ray 147, 153, 159, 302, 330, 363, 375
flare colours 44
flare heating 304
flare models 299, 321, 325, 329, 333
flare stars 1, 71, 81, 85, 105, 117, 121, 125, 139, 163, 209, 263, 321, 371
  -new 67, 91, 93
flare statistics 17, 46, 50, 53, 59, 195, 322
frequency drifts 125
fuors 166, 229, 253, 257, 261

G

gas/dust complexes 271
globules 226
gyroresonant radiation 129
gyrosynchrotron radiation 125, 365

H

Hα emission 20, 82, 95, 153, 215, 300
heating 304
H_2O maser 271, 275
Herbig-Haro objects 165, 221, 225, 229, 279, 284

I

infrared excess 181, 287, 364
interferometry 126
IRAS 279, 283, 287
IR emission 283, 287
IUE 301, 314

J

jets 221, 245

K

K-band photometry 381
kinetics 317, 321

L

linear polarization 179, 185
line profiles 219, 231, 316, 343, 386
low mass stars 49, 347, 363
Lorentz force 327, 337

M

magnetic field 130, 179, 366, 374
mass-age-activity relation 85
mass loss 165, 215
microflaring 148

N

naked T Tau stars 185, 189
narrow-band emission 134
NLTE code 315
non-thermal radiation 129, 365

O

OSO III 153
oscillations 143, 325, 379
ouflow 166, 219, 267, 356, 375

P

photographic observations 71, 77
-chain method 81, 85
-trail method 63, 67
photometry 15, 19, 25, 31, 33, 35, 43, 49, 173, 177, 193, 253
-Hα, Hβ 25, 300
-UBVRI 15, 19, 25, 27, 31, 33, 35, 41, 185, 215, 300
-Walrawen 37
polarization 177, 179
-linear 179, 185
-circular 127, 179
post-T Tau stars 6, 287
precessing jet 221
prominences 219, 325
pulsation 125, 379

Q

quasi-periodic fluctuations 133
quiescent radio emission 128
quiescent X-ray emission 147

R

radio emission 125, 365
radio flares 125, 139, 145, 275, 302, 364, 382
radio observations 19, 125, 139
red dwarfs 41, 49, 55, 59, 121, 321, 347, 371
relaxation oscillations 143
rotation period 170, 175, 181, 185, 189

S

slow variations 55, 169, 173, 193
solitons 338
spectroscopy 19, 95, 99, 177, 219, 253, 293, 381
star clusters 1, 71, 109, 117, 350
star formation 347, 363
starspots 20, 55, 169, 174
stellar evolution 2, 117, 246

T

Taurus-Auriga complex 204, 283
thermal flare 155, 313
TiO feature 96, 100, 171
trail method 63, 67
T Tau stars 6, 81, 163, 169, 173, 179, 185, 189, 193, 209, 215, 219, 221, 226, 229, 257, 279, 284, 343, 363

U

U-filter monitoring 15, 25, 27, 31, 41, 49, 197, 300
UBV 33, 35, 43, 215
UBVRI 19, 185
unpolarized radio emission 128
uvby 194
UV Cet stars 1, 43, 59, 197, 299

V

variability 55, 148, 160, 169, 173

W

Walrawen photometry 37
wave solutions 337
winds 210, 243, 343

X

X-ray emitters 363
X-ray flares 147, 153, 159, 302, 330, 363, 375

Y

young stellar objects (YSO) 179, 354

Z

Z-pinch 304

OBJECT INDEX

AA Tau 189
AB Dor 189
AD Leo 15, 27, 43, 126, 139, 315, 382
AL Ori 93
AR Ori 93
AT Mic 126
AU Mic 126, 189
BBW 76 236
BD-16 6218 189
BD+26 730 56
Bernes 48 267
BF CVn 56, 189
BP Tau 189, 197
BY Dra 56, 189
B42 106
B45 106
B59 106
CC Eri 189
CI Tau 82, 195
CM Dra 41
Coalsack 86
CoKu Tau/1 267
Coma 86
CO Ori 170
Cygnus 86
DF Tau 258
DF UMa 189
DG Tau 199, 209, 285
DH Leo 189
DH Tau 189
DI Cep 199
DI Tau 189, 197
DK Leo 189
DN Tau 189
DO Cep 56
DR Tau 239, 258
DT Vir 56
DX Cnc 50
EI Cnc 50
EI Eri 189
EQ Peg 126
EQ Vir 56, 189
eta Gem 145
EV Lac 19, 25, 27, 33, 35, 43, 189
EX Lup 239, 258
FF And 189
FK Aqr 189
FL Vir 50
FS Tau 285
FU Ori 229, 258

FY Tau 82
FZ Tau 82
G 9–38 50
G 12–43 50
G 51–15 50
G 100–28 50
G 157–77 50
G 171–10 50
G 272–61 50
Gliese 171.2 A 41
GW Ori 197
HD 1835 189
HD 8358 189
HD 10516 295
HD 20336 294
HD 22403 189
HD 32343 294
HD 41335 295
HD 44458 294
HD 45910 295
HD 57150 294
HD 59878 294
HD 60606 294
HD 63462 294
HD 65875 294
HD 68980 295
HD 82558 189
HD 88661 294
HD 91816 189
HD 143313 189
HD 153261 295
HD 200775 295
HD 202904 294
HD 269665 261
HD 283447 189
HD 283572 185, 189
HH And 50
HL Tau 267
HP Tau 189
H II 5 65
H II 34 189
H II 152 189
H II 296 189
H II 324 189
H II 335 189
H II 625 189
H II 686 189
H II 727 189
H II 738 65
H II 739 189
H II 879 189

H II 882 189
H II 1124 189
H II 1332 65, 189
H II 1454 65
H II 1531 189
H II 1553 65
H II 1883 65, 189
H II 2034 65, 189
H II 2244 65, 189
H II 2381 65
H II 2411 75
H II 2870 65
H II 2927 189
H II 2984 65
H II 3096 65
H II 3163 189
H II 3197 65
II Peg 189
II Tau 75
IC 1318 86, 105
IC 1848 86
KO Ori 93
KZ And 189
K2 106
K3 106
LH 332-20 189
Lynds 1228 226
L1551 IRS 5 235
L723 267
Mon OB1/R1 226
M42/L1641 227
NGC 2264 3, 86, 118, 215, 264
NGC 7000 3, 86, 105, 232, 264
NGC 7023 86
NU Ori 177
omicron Vel field 86
Ori B/L1630 227
Orion 3, 85, 86, 101, 109, 114, 121, 159, 215, 226, 253, 264, 271, 275
OU Gem 189
Pleiades 3, 63, 67, 71, 77, 86, 95, 109, 114, 118, 121, 189, 215, 264
Praesepe 3, 86, 109, 118
PZ Mon 56
R CrA field 86
RNO 43N 267
ROX 21 189
ROX 29 189
RU Lup 194, 197
RW Aur 170, 199, 209, 258
RY Lup 189

RY Tau 170, 177, 179, 189, 197, 219, 258
SAO 76753 290
SAO 91750 290
SAO 98967 290
SAO 132301 290
SAO 157323 290
SAO 159682 290
SAO 174199 290
SAO 196352 290
SAO 207208 290
SAO 218788 290
SAO 240367 290
Sco-Oph 86
SU Aur 170, 197
SY Cha 189
Tau-Aur complex 204, 283
Taurus Dark Cloud 3, 81, 86, 215
The 12 197
T Tau 170, 177, 189, 221, 258
TW Hya 199
TZ CrB 189
UX Tau 189
UV Cet 31, 50, 126, 148, 300, 303
UZ Tau 258
VY Ari 189
VY Ori 258
VY Tau 82, 258
V1285 Aql 56
V350 Cep 258
V1057 Cyg 232, 258
V1381 Cyg 106
V1396 Cyg 56, 106
V1424 Cyg 106
V1494 Cyg 106
V1495 Cyg 106
V1496 Cyg 106
V1497 Cyg 106
V1498 Cyg 106
V1499 Cyg 106
V1513 Cyg 106
V1515 Cyg 233, 258
V1522 Cyg 106
V1526 Cyg 106
V1528 Cyg 106
V1529 Cyg 106
V1530 Cyg 106
V1534 Cyg 106
V1536 Cyg 106
V1537 Cyg 106
V1538 Cyg 106
V1539 Cyg 106

V1544 Cyg 106
V1545 Cyg 106
V1550 Cyg 106
V1581 Cyg 106
V1586 Cyg 106
V1587 Cyg 106
V1588 Cyg 106
V1589 Cyg 106
V1590 Cyg 106
V1591 Cyg 106
V1592 Cyg 106
V1593 Cyg 106
V1594 Cyg 106
V1595 Cyg 106
V1596 Cyg 106
V1587 Cyg 106
V1598 Cyg 106
V1599 Cyg 106
V1600 Cyg 106
V1601 Cyg 106
V1602 Cyg 106
V1603 Cyg 106
V1604 Cyg 106
V1605 Cyg 106
V1606 Cyg 106
V1607 Cyg 106
V1608 Cyg 106
V1609 Cyg 106
V1611 Cyg 106
V1612 Cyg 106
V1613 Cyg 106
V1695 Cyg 106
V1698 Cyg 106
V1699 Cyg 106
V1700 Cyg 106
V1709 Cyg 106
V1710 Cyg 106
V1712 Cyg 106
V1713 Cyg 106
V1714 Cyg 106
V1717 Cyg 106
V1735 Cyg 234, 258
V1750 Cyg 106
V1752 Cyg 106
V1753 Cyg 106
V1754 Cyg 106
V1755 Cyg 106
V1756 Cyg 106
V1757 Cyg 106
V1758 Cyg 106
V1772 Cyg 106

V1777 Cyg 106
V1778 Cyg 106
V1779 Cyg 106
V1780 Cyg 106
V1781 Cyg 106
V1785 Cyg 106
V1787 Cyg 106
V1790 Cyg 106
V1791 Cyg 106
V1793 Cyg 106
V1795 Cyg 106
V1796 Cyg 106
V1797 Cyg 106
V1798 Cyg 106
V1799 Cyg 106
V1924 Cyg 106
V1926 Cyg 106
V1927 Cyg 106
V1928 Cyg 106
V1929 Cyg 106
V1930 Cyg 106
V639 Her 56
V654 Her 56
V722 Her 189
V775 Her 189
V815 Her 189
V478 Lyr 189
V577 Mon 56
V346 Nor 234
V368 Ori 93
V714 Ori 93
V1005 Ori 189
V1118 Ori 239, 253
V1143 Ori 253
V1216 Sgr 56
V410 Tau 169, 189, 200
V819 Tau 189
V826 Tau 189
V827 Tau 189
V830 Tau 189
V833 Tau 42, 189
V836 Tau 189
V780 Tau 50
Wolf 424 302
Wolf 630 126
xi Boo A 189
XZ Tau 267
YY Gem 126, 189, 325
YZ CMi 43, 126, 189
Z CMa 235, 258